THIRD E

OUR SOCIETY

HUMAN DIVERSITY IN CANADA

Edited by Paul U. Angelini
*Sheridan College Institute of Technology
and Advanced Learning*

Our Society: Human Diversity in Canada, Third Edition
Edited by Paul U. Angelini

Associate Vice President, Editorial Director:
Evelyn Veitch

Editor-in-Chief, Higher Education:
Anne Williams

Executive Editor:
Cara Yarzab

Executive Marketing Manager:
Kelly Smyth

Senior Developmental Editor:
Rebecca Rea

Permissions Coordinator:
Lynn McIntyre

Content Production Manager:
Lara Caplan

Copy Editor:
Rodney Rawlings

Proofreader:
Gail Marsden

Indexer:
Belle Wong

Manufacturing Coordinator:
Loretta Lee

Design Director:
Ken Phipps

Interior Design:
Katherine Strain

Cover Design:
Sasha Moroz

Cover Images:
Shutterstock.com

Compositor:
Integra

Printer:
Thomson/West

COPYRIGHT © 2007 by Nelson, a division of Thomson Canada Limited.

Printed and bound in the United States
1 2 3 4 09 08 07 06

For more information contact Nelson, 1120 Birchmount Road, Toronto, Ontario, M1K 5G4. Or you can visit our Internet site at http://www.nelson.com

Statistics Canada information is used with the permission of Statistics Canada. Users are forbidden to copy this material and/or redisseminate the data, in an original or modified form, for commercial purposes, without the expressed permissions of Statistics Canada. Information on the availability of the wide range of data from Statistics Canada can be obtained from Statistics Canada's Regional Offices, its World Wide Web site at <http://www.statcan.ca>, and its toll-free access number 1-800-263-1136.

ALL RIGHTS RESERVED. No part of this work covered by the copyright herein may be reproduced, transcribed, or used in any form or by any means—graphic, electronic, or mechanical, including photocopying, recording, taping, Web distribution, or information storage and retrieval systems—without the written permission of the publisher.

For permission to use material from this text or product, submit a request online at www.thomsonrights.com

Every effort has been made to trace ownership of all copyrighted material and to secure permission from copyright holders. In the event of any question arising as to the use of any material, we will be pleased to make the necessary corrections in future printings.

Library and Archives Canada Cataloguing in Publication Data

Our society : human diversity in Canada / edited by Paul U. Angelini. —3rd ed.

Includes bibliographical references and index.
ISBN 0-17-640670-0

1. Pluralism (Social sciences)—Canada—Textbooks. 2. Canada—Social conditions—1991– —Textbooks. I. Angelini, Paul Ubaldo, 1962– II. Title.

HN103.5.O97 2006   306.0971
C2006-902990-3

CONTENTS

PREFACE v

PART I   AN OVERVIEW OF DIVERSITY IN CANADA

CHAPTER 1   Regionalism in Canada: The Forgotten Diversity 4
Paul U. Angelini

CHAPTER 2   Demographic Trends in Canada 38
Michelle Broderick

PART II   THE MANY FACES OF DIVERSITY

CHAPTER 3   Social Inequality and Stratification in Canada 66
Eddie Grattan

CHAPTER 4   Race and Ethnicity: The Obvious Diversity 94
Paul U. Angelini and Michelle Broderick

CHAPTER 5   Aboriginal Peoples 126
John Steckley

CHAPTER 6   Religion as Meaning and the Canadian Context 156
Mikal Austin Radford

CHAPTER 7   Disability as Difference 194
Nancy Nicholls

CHAPTER 8   Diversity and Conformity: The Role of Gender 216
Leslie Butler

CHAPTER 9   Sexuality: Emergent Understanding in Diversity 240
Brigitte Guetter

CHAPTER 10   Diversity in Canadian Families: Traditional Values and Beyond 272
Geoff Ondercin-Bourne

PART III   THE TREATMENT AND PERCEPTION OF DIVERSITY

CHAPTER 11   The Medium Diversifies the Message: How Media Portray Diversity 304
Grant Havers and Paul U. Angelini

CHAPTER 12   Perceptions of Diversity in Canadian Literature 330
Maureen Coleman

Contents

GLOSSARY 364
SELECTED BIBLIOGRAPHY 380
BIOGRAPHIES 401
INDEX 403

PREFACE

This third edition of *Our Society: Human Diversity in Canada*, like the first, was written for students who have had little or no exposure to the issues surrounding diversity in Canada. As an introductory, topic-oriented text, *Our Society* is designed to give readers a panoramic view of diversity in Canada framed within the dominant theoretical paradigms of the day. For readers who have already examined diversity, this edition provides up-to-date statistics, analyses of recent events, and a bibliography that covers a wide variety of issues in each subject area.

New in the third edition is a balanced, non-partisan chapter dealing with religion in Canada that provides a sound basis for rational discussion of recent developments in this area. New topics in other chapters include racial profiling, the Gomery Inquiry, media behaviour in Canada, and Edward Herman and Noam Chomsky's "Propaganda Model" for understanding media.

All postsecondary schools and high schools have at least one course that attempts to explain life in Canada. Courses range from the general, such as Introduction to Canadian Studies, to the specific, such as Race and Ethnic Relations in Canada. Increasingly, students have demanded more comprehensive analyses of life in Canada. For example, the course I teach has evolved from dealing exclusively with race and ethnicity to dealing with many issues in Canadian life, including regionalism, gender, and social stratification. This course, renamed Human Diversity and Interactions in Canada, better reflects the interests of my students.

This text focuses on human diversity in Canada. Diversity in simple terms refers to the differences that set people apart from each other. Therefore, in introducing students to the diversity of Canadian life, the text addresses the following questions: How do Canadians differ from each other? What are the scope and the range of each difference? How have differences evolved over time? How do we as Canadians view diversity? Rather than encouraging uniformity, the authors intend, through careful analysis, to promote tolerance and understanding and to show that differences can and do bring Canadians together. In the end, students will appreciate the differences that characterize Canadian life and will have a fuller understanding of what it means to be Canadian. The more we understand about ourselves and the more comfortable we become with this understanding, the more likely we are to accept others without prejudice. After all, difference should be celebrated, not lamented.

The reasons for preparing this third edition are the same as those for the original undertaking in 1997. First, students were vocal in their demands for more comprehensive courses dealing with the differences that characterize and shape life in every corner of Canada. The text, therefore, is student-driven.

Second, few texts dealing with this subject matter are written for the target audience of postsecondary and high-school students. Several features make this book uniquely suited for this target audience:

- The text is learning-centred, meaning that the concerns of students come before those of teachers. The writing style, organization, and level of analysis are introductory, with only the most important conceptual jargon included. Reading this text does not require specific prior knowledge.
- The book is student-friendly. Important terms are bolded and featured in the end-of-text Glossary for easy reference. In each chapter, numbered Boxes highlight important or complementary information. In addition, each chapter includes five to six numbered Critical Thinking Boxes that ask students to answer questions critically and constructively.
- The topics covered are truly inclusive. We have tried, as much as possible, to include both those topics demanded by students and those that instructors believed were necessary.
- The text is unabashedly Canadian! It is not an American text with Canadian information added; rather, it is a text written by Canadians and for Canadians, and it provides students with a balanced introduction to life in Canada.

All students—at college or university or in high school—are at a time in their lives when they begin to make definitive assessments about the world around them. This is especially important after the events of 11 September 2001 and the subsequent international fallout that continues well into 2006. We hope the treatment of the subjects covered in *Our Society: Human Diversity in Canada* will make their decisions informed and responsible ones.

ACKNOWLEDGMENTS

The work involved in putting together this edition was no less arduous than that involved in the previous ones. Again, without the guidance and support of many people, this text would still be just an idea. As a group we would like to thank the reviewers: Ted Dionne (Durham College), Slobodan Drakulic (Ryerson University), Lee Farnworth (Algonquin College), and Michael Whatling (McGill University).

We would also like to thank a number of people at Nelson for helping to make the third edition a reality. We owe a great debt to executive editor Cara Yarzab for her sound advice, expert counsel, and neverending support. A similar debt is owed to senior developmental editor Rebecca Rea, whose cogent editing, knowledge of copyright issues, and exceptional organizational skills kept this project on course and on time! We also appreciate the assistance of Lara Caplan, production editor; Rodney Rawlings, copyeditor; and Gail Marsden, proofreader.

I would like to thank Paul Saundercook for originally convincing me that a project of this nature was both viable and needed. His understanding of the postsecondary marketplace is second to none. I would also like to thank Hopie Palmer and Victor Montgomery for their research and graphic assistance.

Paul U. Angelini
Hamilton, Ontario
October 2006

THIRD EDITION
OUR SOCIETY
HUMAN DIVERSITY IN CANADA

PART I

AN OVERVIEW OF DIVERSITY IN CANADA

PART I

Part I takes a macro approach to diversity in Canada. Using the analogy of a house, these first two chapters are the building blocks, the foundation, the walls, and the roof. Together they provide a structure—they "frame" the discussions that take place in Part II.

Chapter 1 looks at regionalism. It begins with a definition of regionalism and then briefly outlines the different regions in Canada. It finishes by examining how the federal government in Ottawa has increased the tensions between regions and by assessing the actions taken by the same government to lessen regional inequality.

Chapter 2 introduces the reader to the study of demography in Canada. In addition to outlining the terms necessary for the study of demography itself, this chapter examines specific demographic trends critical to understanding developments in Canadian diversity.

CHAPTER 1

Regionalism in Canada: The Forgotten Diversity

Paul U. Angelini

Two things hold this country together. Everybody hates Air Canada coffee and everybody hates Ontario.
— Late New Brunswick Premier Richard Hatfield

Regions usually have some concrete, physical foundation. . . . But to some extent regions are also a state of min
— Political scientist Rand Dyck

Objectives

After reading this chapter, you should be able to

- define regionalism
- appreciate the role of regionalism in Canadian social life
- outline briefly some of the suspected causes of regionalism
- understand the sociopsychological component to regionalism
- understand the role of the federal government in creating and attempting to lessen regional differences, and appreciate the aspects of our political system that intensify regional differences

INTRODUCTION: HOW IS REGIONALISM A FORM OF DIVERSITY?

The purpose of this chapter is threefold: (1) to outline what is meant by **regionalism**, (2) to briefly explain some of the suspected causes of regionalism, and, most importantly, (3) to make the case that there is an important sociopsychological component to regionalism that is seldom acknowledged. Simply put, people have a profound effect on their region and on the world around them.

Let us stress that our purpose here is not to attempt to find the causes of regionalism. These attempts are usually bogged down in theoretical and ideological debates that, in the end, tell us little about the people living in these areas. We will, however, briefly outline what some other writers and researchers consider the causes of regionalism. You will have to assess for yourselves which make the most sense. When in doubt you always have the option of doing more research.

Canada is a country characterized by difference. From coast to coast, there are differences in physical terrain, climate, population, distribution of natural resources, percentage of people living in urban areas, ethnicity, religion, occupation, and income. These are real and identifiable differences—but what about "subjective" differences? How do people living in each region feel about themselves, their region, and their fellow inhabitants? What influences how people answer these questions? These are important considerations, because the answers help us more fully understand what is meant by regionalism, and they help bring the "human" aspect to this study. After all, regions are made up of people and people make regions.

WHAT IS REGIONALISM AND WHY STUDY IT?

Regionalism examines the people living in different areas in Canada and the different feelings they have regarding themselves, the people living in other areas, and the federal and provincial governments. Regionalism, therefore, is most of all an attitude. We can define regionalism as an attitude that reflects a long, deep, certain feeling held by the citizens of a particular geographical area that they have their own, unique identity. Often they feel that they have not been given adequate recognition for their hard work and sacrifices. Their dissatisfaction is focused in three specific areas:

1. They believe that people in the federal and provincial governments have not accurately recognized their contribution to the life of their region.
2. They believe they have not been given due recognition for their contribution to building this country called Canada.

3. They believe that their interests have not been adequately represented by the government in Ottawa and that this is one of the principal reasons why some regions receive far more money from the federal government than others; over the years this has meant that economic differences in regional development have been made worse.

In short, while Ontario and Quebec, the central region of Canada, receive almost everything, the Western and Atlantic regions receive very little. Ontario and Quebec get much of the recognition for building Canada; in comparison, the Western and Atlantic provinces get very little.

Before continuing, however, we must address three problems with regional analysis.

PROBLEMS WITH REGIONAL ANALYSIS

Three central problems exist in discussing regionalism:

1. How are geographical regions defined?
2. Are provinces necessarily regions?
3. Are regions the appropriate tool to study the people living in Canada?

How Are Geographical Regions Defined?

Some dispute exists concerning what physical characteristics should be used to designate a region. Historically, we in Canada have used geography to designate four regions (see Figure 1.1).

Perhaps the best example of dispute is over what is commonly referred to as the "West." Geographically, British Columbia is radically different from the three Prairie provinces because of the Cordilleran mountain system (the Canadian Rockies). As a result, many people in British Columbia do not believe that they should be included in the "Western" region. This belief is reinforced by the fact that the economies of these four provinces are also different: The B.C. economy is not based on farming, but the economies of Manitoba and Saskatchewan are. British Columbia has a huge forest industry and a significant fishing industry, whereas Alberta is the centre of Canada's oil and natural gas industries. Similar economic and physical differences exist within the other regions of Canada, too. This is true of Newfoundland and Labrador; the last province to join Confederation (1949) has always considered itself distinct from the rest of the Atlantic provinces.

Are Provinces Necessarily Regions?

Questions arise about the belief that a province is a region. Within provinces there tend to be different regions that share certain characteristics that make them distinct from

Figure 1.1 Regions of Canada

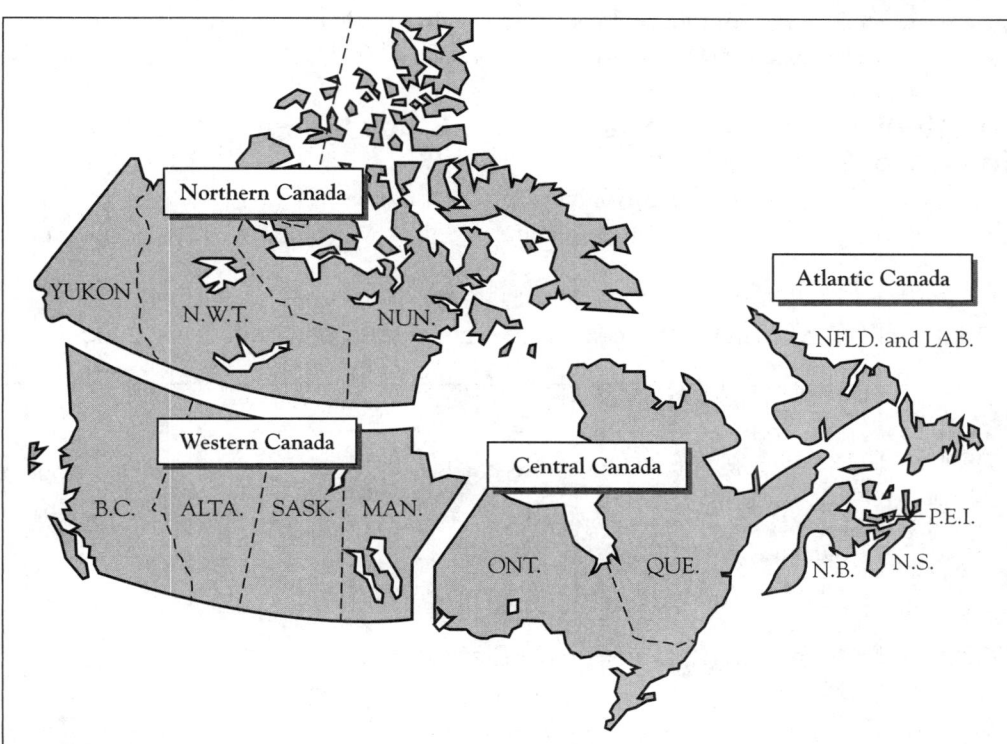

other parts of the province. The best examples of such differences are those between northern Ontario and southern Ontario and between northern Quebec and southern Quebec. In the case of Ontario, the northern part has always differed with respect to economic growth (much slower, if at all), unemployment, per capita income, and types of jobs. Unemployment is one area in which the differences are very visible. Northern Ontario has traditionally experienced higher unemployment rates than the rest of the province, in large part because of the resource focus of industry in northern Ontario, the seasonal nature of such work, and the existence of one-industry towns that shut down when the industry closes. It is not uncommon for unemployment in northern Ontario to be double the provincial average.

In short, the economic prosperity of Ontario is usually only experienced in Metropolitan Toronto, other parts of the **Golden Horseshoe**, which stretches along the coast of Lake Ontario from Niagara Falls to Oshawa, and some parts of south-central Ontario. A survey by Statistics Canada shows that eight of the top ten cities according to median employment income in Canada (Thunder Bay and London are tied for tenth place) are located in Ontario; including Ottawa and excluding Thunder Bay, seven are located south of Ottawa (see Table 1.1).

Are Regions the Appropriate Tool to Study the People Living in Canada?

Because so many questions surround what constitutes regions, issues have arisen regarding the use of regions as an explanatory tool. A wide variety of other tools could

Table 1.1 Median Employment Income of Selected Canadian Cities, 2003

	Income ($)
Oshawa, Ontario	32 900
Ottawa-Gatineau, Ontario	32 500
Kitchener, Ontario	29 800
Windsor, Ontario	29 400
Hamilton, Ontario	29 400
Calgary, Alberta	28 900
Toronto, Ontario	28 500
Edmonton, Alberta	27 800
Regina, Saskatchewan	27 600
London, Ontario	27 000
Thunder Bay, Ontario	27 000
Quebec City, Quebec	26 600
Halifax, Nova Scotia	26 000
Victoria, British Columbia	26 000
Kingston, Ontario	25 700
Vancouver, British Columbia	25 400
Montreal, Quebec	25 200
Winnipeg, Manitoba	25 000
Canada	**24 800**

Source: Statistics Canada, "Median Total Income and Median Employment Income by Census Metropolitan Area, 2003," *The Daily*, Thursday, 5 May 2005 <www.statcan.ca/Daily/English/050505/d050505b.htm>, accessed 1 June 2005.

be used to explain social life in Canada, including social class, occupation, elites, ethnicity, economic development, and individual choice. Marxists and neo-Marxists stress that analyzing social class and power will best explain the reality of living in Canada. Other writers, such as Richard Simeon, dismiss the use of regions as an analytical tool altogether.[1]

PROVINCES AS REGIONS

With questions surrounding the use of regions as a tool for understanding the Canadian experience, and if provinces are not necessarily regions, can we realistically treat physically and economically different provinces as regions? Our belief is that we can. Specifically, it is important to realize that since 1945, provincial governments have consistently increased their own control over their populations and have continued to challenge the authority of the federal government with respect to the economic and political leadership of their citizens. It is possible that provinces possess the political tools and the will to create a shared regional identity.[2] Our approach does not deny that different regions exist within provinces, and it does recognize provinces as a vital part in the life of the people living in them. As we will see in the section dealing with government actions, Canadian history is full of attempts on the part of the federal government to address provincial demands and to try to bring economic development and prosperity to different parts of the country, with varying degrees of success.

THEORIES ABOUT THE CAUSES OF REGIONALISM

Many explanations have been offered for the causes of regionalism, for why some regions have prospered and others have not, and for the effect this has had on the people who live there. The following is a brief sketch of some of the more prominent of these explanations.[3]

The Natural Resources Approach

The natural resources approach is more often referred to as "staples theory."[4] Generally, this theory asserts that the key to economic prosperity is the availability of natural resources. A short list of such resources includes oil, natural gas, fish, lumber, fur (beavers), coal, and various minerals.

This approach comprises two critical beliefs. The first asserts that some areas in Canada prosper while others do not because of the availability and marketability of their natural resources. In short, how much of a particular resource does your region possess, and does your region have companies or other countries prepared to buy that resource? The second critical belief concerns the external limits on developing your resource. Specifically, how is your resource affected by the fluctuations in price caused by changes in the international marketplace? Some examples will help make this point. How do the oil-producing countries located primarily in the Middle East (e.g., Saudi Arabia, Kuwait, Qatar, and Bahrain) affect the price of oil extracted from oil fields in Alberta? How do changes in world grain prices affect grain growers in Saskatchewan? In large part the prices are beyond the control of the people who produce the product. This helps explain why Western grain farmers and Atlantic fishermen demand compensation and assistance from the federal government in Ottawa to offset international price changes.

To conclude, the staples approach asserts that economic prosperity is largely determined by the availability of natural resources. The problem with this approach is that there is a tendency to exaggerate the likelihood that a region will prosper economically if natural resources are located there in extremely large quantities. Some regions in Canada have an abundance of natural resources yet have not experienced the economic prosperity that the theory would predict. As Ralph Mathews has written, Southern Ontario, for example, has no strong natural resource base, yet it is wealthy. Meanwhile, although the Atlantic provinces have a rich base of iron ore, coal, gold, forests, fish, and hydroelectric power, they have remained poor throughout most of the period since Confederation.[5]

Market Approaches

Market approaches assert that some regions are more prosperous than others because of interference, usually by governments, in the local market that results in market failure. Types of government interference include subsidizing companies that need help to survive and subsidizing the wages of employees.

Market approaches emphasize that for economic development to take place, governments must not interfere with the way the free market functions. Wages should be allowed to fall, taxes collected from companies should be minimal, labour (workers) should move to where jobs are, and money (capital) should not be restricted or penalized if and when it decides to relocate—when companies decide to stop operations, lay off workers, and set up shop elsewhere.

In the 1980s and 1990s market approaches were quite popular as solutions to the problem of regional differences in economic development and as solutions to the economic problems of countries as a whole. The United States and the United Kingdom have vigorously pursued these policies. In Canada, the Progressive Conservative premier of

Alberta, Ralph Klein, and the Progressive Conservative former premier of Ontario, Mike Harris, were both fiercely committed to allowing the free market to operate unmolested by their respective governments.

Interventionist Approaches

Generally, interventionist approaches are based on the belief that some regions have prospered while others have not because of the many political (politicians and government) and economic (the development strategies of companies) forces that have historically favoured some regions at the expense of others.[6] As a general solution, these interventionist approaches regard government involvement as absolutely essential to overcoming regional problems.

Interventionist approaches differ from the staples and market theories in two specific respects. First, staples theory regards regional differences in economic prosperity as natural; interventionist approaches do not. Second, market theories see problems as avoidable if, in large part, governments simply allow the free market to operate. Interventionist approaches do not see the problems as avoidable.

In the post–World War II era, the Canadian federal government has practised interventionism quite extensively. The most visible forms of this intervention are federal government **transfer payments** (worth billions of dollars each year) to the provinces to help the poorer regions of the country (see Table 1.2).

Marxist Approaches

From a Marxist perspective, regional economic inequality is a natural outcome of the class and power differences in a society dominated by capitalism, because the driving force of capitalism is the accumulation of profit. Accumulating profit includes maximizing the value of your company for the shareholders who buy and sell shares in your company and exploiting the people, places, and things used during the production of the goods or services.

Marxism regards regional economic differences as "functional" (i.e., beneficial) to the operation of capitalism for three reasons. First, capitalism uses workers in poorer regions as reserve or surplus labour to hire and lay off whenever the capitalist sees fit. In other words, when times are good, workers from poor regions travel to more prosperous ones to work; when times are bad, these same workers are simply let go, and they return to the region from which they came. Second, underdeveloped regions provide raw materials that are processed in more developed regions. In this way money (capital) is drained from the poor regions to the richer ones. Third, poorer regions provide a market for the goods and services produced in the more developed regions.[7]

Capitalists exploit poorer regions in another way that is usually implicit in Marxist analysis and needs to be reinforced here: capitalists use the existence of poorer regions as a threatening tool to get concessions from workers (e.g., to accept less money and fewer

Table 1.2 Major Transfers to Provinces and Territories (millions of dollars)

Province/Territory	2004–2005	2005–2006
Prince Edward Island	439	456
Yukon	508	534
Northwest Territories	736	778
Nunavut	796	844
Saskatchewan	1 859	1 481
Newfoundland	1 367	1 531
New Brunswick	2 220	2 339
Nova Scotia	2 466	2 620
Manitoba	3 045	3 201
Alberta	4 325	4 818
British Columbia	6 136	6 653
Quebec	13 865	15 572
Ontario	16 745	18 609
Total	**54 509**	**59 435**

Note: All figures comprise Equalization, Territorial Formula Financing and Health and Social Transfers.

Sources: "Federal Transfers to Provinces and Territories (March 2005)," Department of Finance Canada, available <www.fin.gc.ca/fedprov/mtpe.html>, accessed 26 March 2006.

benefits) and governments (e.g., to collect less tax money) in wealthier regions. Just as companies threaten to move their operations (including jobs) to the developing world (e.g., Mexico) if unions, workers, and governments do not provide concessions to them, so too do companies threaten to do the same within Canada. In attempts to lure business to their provinces (and out of other provinces) provincial premiers have been known to actively encourage this behaviour.

To conclude, in the above four ways Marxists regard regional imbalance as a normal outcome of capitalism.

FACTORS THAT INTENSIFY REGIONALISM

The Canadian political system operates in ways that intensify regional feelings and economic differences. Two of the most important are federalism and our electoral system.

Federalism

When the Dominion of Canada was created in 1867, a time we refer to as Confederation (see Box 1.1), the politicians who wrote the *British North America Act* (since 1982 the *Constitution Act of 1867*) decided that there should be two levels of government to govern the Canadian people. The first level is federal and refers to the national government located in Ottawa. The second level is provincial and refers to the provinces and territories and their governments.

These same politicians outlined what level of government would be responsible for what area of social life. Section 91 of the Constitution outlines federal responsibilities and

---

BOX 1.1

Confederation in Chronological Order

Date and Province or Territory	Seat of Government
1 July 1867	
Ontario	Toronto
Quebec	Quebec City
Nova Scotia	Halifax
New Brunswick	Fredericton
15 July 1870	
Manitoba	Winnipeg
Northwest Territories	Yellowknife (originally Winnipeg)
1 July 1871	
British Columbia	Victoria
1 July 1873	
Prince Edward Island	Charlottetown
13 June 1898	
Yukon Territory	Whitehorse
1 September 1905	
Saskatchewan	Regina
Alberta	Edmonton
1 April 1949	
Newfoundland	St. John's
1 April 1999	
Nunavut	Iqaluit

> **CRITICAL THINKING BOX 1.1**
>
> Which theory do you believe best explains regionalism in Canada? Would Canadians in other provinces share your belief? If so, why? If not, why?

section 92 outlines provincial ones. Historically, when disputes arose over a new policy area, such as atomic energy or aviation, the Supreme Court decided which level of government assumed the responsibility for that particular policy area. The federal government was given control of the armed forces, and provincial governments were given control over health care and education. **Federalism** can be defined as a system of government that divides responsibilities between two levels of government, with each level being unable to abolish the other. The two levels of government must cooperate with each other. For roughly the past twenty years, this cooperation has taken the form of meetings between the provincial premiers, their key Cabinet ministers and advisors, and the Prime Minister along with his or her key Cabinet ministers and advisors. The making of political decisions and policies jointly by federal and provincial Cabinet ministers, senior bureaucrats, premiers, and prime ministers is called **executive federalism**.

Executive federalism creates problems. Senior members of government and the bureaucracy make important decisions, and many of those members have the interests of specific regions or other interests at heart, not the interests of Canada as a whole. This is of greater concern when decisions are made behind closed doors without properly informing Canadians of what is going on and without providing Canadians with opportunities for criticism, revision, and input. The **Meech Lake Accord** (1987) was an example of this type of decision making. Then Prime Minister Brian Mulroney thought it would all be a matter of simply "rolling the dice." Canadians were so upset at having been shut out of the negotiations that few tears were shed when three years later (1990) the Accord did not receive the necessary provincial approval in Manitoba, Newfoundland, and New Brunswick to be adopted by the government in Ottawa.

Executive federalism compounds the problems associated with regionalism when some regions feel they are not being represented in Ottawa and when some provinces feel other provinces have too much influence there. This has certainly been the case regarding perceptions of Ontario and Quebec. Both the Western and the Atlantic provinces have always claimed that the two central provinces have far too much representation in Ottawa. Historically, members of Parliament from Ontario and Quebec have dominated important key Cabinet positions, such as foreign affairs, finance, justice, and international trade, as well as senior positions in the bureaucracy and government agencies.

The Western provinces have always believed that the citizens of Ontario and Quebec have not adequately recognized their sacrifices and contributions to building Canada. They also believe that their interests have not been vigorously represented in Ottawa. This was especially true during Pierre Trudeau's years as Prime Minister (1968–79 and 1980–84). For the better part of 14 years, Trudeau's Liberal governments did not have one member elected to Parliament west of the city of Winnipeg, leaving Saskatchewan, Alberta, and British Columbia with virtually no federal political representation in government! Consider how Ontario and Quebec would react if they ever experienced the same lack of representation.

Before the First Ministers' Conference in Victoria, British Columbia, on 1 August 2001, Newfoundland Premier Roger Grimes complained loudly about the return of a per capita funding formula to federal transfers to the provinces. In 1999, at the urging of then Ontario Premier Mike Harris, Ottawa's Canadian Health and Social Transfer (CHST) was calculated using a per capita formula; such a formula benefits the more populous provinces. Premier Grimes put forward the idea that the Atlantic provinces were being asked to put more money into the CHST so that Ontario could expand the services it offered its citizens, which were already beyond what was being delivered in the Atlantic provinces. In effect, the poorer provinces ended up subsidizing the richer ones.

Our Electoral System

The electoral system in Canada contributes to making regionalism worse. Canada's electoral system is a single-member plurality system, or what some call "first past the post." In this system a party receives one seat for every riding it wins. The number of seats in each province depends on its population—the provinces with the most people receive the most seats. We call this **representation by population**. Therefore, the party that wins the most seats wins the election. However, the candidate who wins the riding does not necessarily need more than 50 percent of the votes. To win, a candidate simply needs more than anyone else running against him or her. An example will better illustrate this point. Candidate X receives 46 percent of the votes, and Candidate Y receives 30 percent, while Candidate Z receives 24 percent. Candidate X wins, with only 46 percent of the votes

---

CRITICAL THINKING BOX 1.2

How would you feel if your province were not represented in the federal government for almost 15 years? What might be the sociopsychological consequences of such an experience? How did the citizens of Saskatchewan, Alberta, and British Columbia feel and act during Pierre Trudeau's years as Prime Minister?

(which is not a majority) and in spite of the fact that 54 percent of the people (the total votes of Candidates Y and Z) voted against Candidate X. Now, if this scenario is repeated in riding after riding, the party that Candidate X belongs to will win the election, even though in the election more people voted for other parties. The January 2006 and June 2004 federal elections are excellent examples of this (see Box 1.2).

Our electoral system contributes to regionalism when the party that wins the election wins because it has won more seats than any other party in Ontario and Quebec. Other

---

BOX 1.2

Federal Election, 18 January 2006*

Party	Elected Members	Percentage of Popular Vote
Conservative	124	36.3
Liberal	103	30.2
Bloc Québécois	51	10.5
New Democratic Party	29	17.5
Green Party	0	4.5
Other	1	1.0
Total	**308**	**100.0**

Federal Election, 24 June 2004**

Party	Elected Members	Percentage of Popular Vote
Liberal	135	36.7
Conservative	99	26.9
Bloc Québécois	54	12.4
New Democratic Party	19	15.7
Other	1	5.6
Total	**308**	**100.0**

Sources: *Adapted from "2006 Canadian Election Results," Andrew Heard home page, Political Science Department, Simon Fraser University site <www.sfu.ca/~aheard/elections/results.html>, accessed 26 March 2006; **Adapted from "Electoral Results by Party," Parliament of Canada Web Site, February 16, 2006, <www.parl.gc.ca/information/about/process/house/asp/PartyElect.asp?lang=E&Hist=N>, accessed 26 March 2006.

Note: Totals may not add up to 100% because of rounding.

regions of the country lose when they do not vote for the same party that wins the most seats in Ontario and Quebec. In the 2006 federal election, the Bloc received only 10.5 percent of the popular vote, but because these votes were regionally concentrated in the province of Quebec, the Bloc were allotted a whopping 51 seats. The NDP received 17.5 percent of the popular vote and only 29 seats, while the Green Party, with a healthy and respectable 4.5 percent of the popular vote, won no seats at all! Of the 308 seats contested in the federal election held 18 January 2006, 181 belong to Ontario (106) and Quebec (75). The winning political party, therefore, will spend a lot of time, effort, and money pleasing the people and monied interests that elected them. This is inevitable, because the majority of Canada's financial, banking, and manufacturing interests are overwhelmingly concentrated in these two provinces, especially in Ontario. A leisurely drive on the major highways that lead to and from Metropolitan Toronto (Highways 401, 403, 407, and the Queen Elizabeth Way) will certainly confirm this fact.

THE REALITY OF REGIONALISM AND ITS OUTCOMES

Perhaps the most visible signs of regionalism and, therefore, of the discontent felt by people living in some regions, are the political parties that were created to represent the interests of a particular region. These parties are a vivid representation of the concerns, expectations, attitudes, and fears of an entire region. Some examples include the rise of the Bloc Québécois in Quebec, the Reform Party (formerly the Canadian Alliance and now known nationally as the Conservative Party) in western Canada, and the Confederation of Regions Party in New Brunswick. These parties were created for at least two reasons. The first concerns the basic similarity between the Liberal and Progressive Conservative parties in terms of ideology, policies, and organization. If the two main parties look the same and sound the same, the concerns of some people are not being addressed. Starting your own party overcomes these problems. Second, belonging to these new parties represents a rejection of the two other parties—it is a protest against the traditional way of doing things. Joining a new party is a way of saying "If you don't listen to us, we will find other ways to voice our discontent and push forward our interests." If you are electorally successful, the other political parties must take you seriously. The Canadian Alliance/Reform Party is an excellent example of this.

The Case of Quebec: Winning Hearts?

Without question the greatest threat to Canadian federalism is Quebec separatism. While other issues may occupy the minds of Canadians on a daily basis, such as racial profiling by police in Toronto, Asian immigration in Vancouver, farming costs in Saskatchewan, the size of fish stocks in the Atlantic provinces, they do not threaten to break up the country. Canadian unity has been at the top of every federal government agenda especially since the Parti Québécois victory in 1977. The federal government has spent billions of dollars trying to win the hearts and minds of all Quebecers. Two of the most visible, divisive, emotional, and symbolic attempts were the 1987 Meech Lake Accord and the 1995–2002 Sponsorship Program.

The Meech Lake Accord (1987) was Prime Minister Brian Mulroney's attempt to have Quebec formally sign the Constitution, something it did not do when the Constitution was repatriated by the Liberal government of Pierre Trudeau in 1982. Quebec City agreed to accept the accord if the Prime Minister and provincial premiers agreed to five key demands (see Box 1.3). The most contentious of these was recognition of Quebec as a "distinct society." Accepting the Accord with such a clause was seen by many, including former Prime Minister Pierre Trudeau, as having dire consequences for the survival of Canadian federalism.[8] Trudeau believed a Parti Québécois government intent on separating from Canada could potentially use this clause as "proof" that Quebec was "special" and needed independence from Canada to maintain its "distinctiveness." In spite of the fact that Prime Minister Mulroney and all 10 premiers accepted the demands, the Meech Lake Accord died in June 1990 when ratification was delayed in Manitoba by Aboriginal MLA Elijah Harper and when newly elected governments in New Brunswick and Newfoundland and Labrador failed to ratify it.

BOX 1.3

Quebec's Five Demands

- Constitutional recognition of Quebec as a "distinct society"
- a role in appointments to the Supreme Court of Canada
- a veto for Quebec on constitutional amendments
- a more influential role in immigration
- limits on federal power in new federal–provincial shared-cost programs—the right to opt out, with full financial compensation, from future shared-cost programs in areas of exclusive provincial jurisdiction

Source: Marjorie Montgomery Bowker, *The Meech Lake Accord: What It Will Mean to You and to Canada* (Hull, QC: Voyageur Publishing), 1990. For his reaction to the Charlottetown Accord, see Pierre E. Trudeau, *A Mess That Deserves a Big NO* (Toronto: Robertson Davies Publisher, 1992).

The most recent attempt to win the hearts of Quebecers was the $250 million Sponsorship Program and Advertising Activities initiated in 1995. The Program was put together in the wake of Quebec's 1995 sovereignty referendum. The strength of the separatist vote convinced the federal government that something had to be done to weaken separatist feelings in Quebec.

The official purpose of the Sponsorship Program was to raise the profile of the federal government in Quebec by "sponsoring" a number of events and activities across the province. In return for cash, the Canadian flag would be prominently displayed at every event. Such events included but were not limited to the Montreal Grand Prix, hunting and fishing shows, cultural festivals, and television programming. Senior members of the federal Liberal party believed the program was a rational response to the strength of separatist sentiment in Quebec. The public was made aware of potential pecuniary mismanagement when Allan Cutler, a career civil servant in Public Works and Government Services, "blew the whistle" on his bosses. The program became a scandal when in 2002, federal Auditor General, Sheila Fraser, recommended the RCMP investigate how advertising money was handed out to different advertising agencies in Quebec (see Box 1.4). Slowly, a picture

---

BOX 1.4

Sponsorship Program and Advertising Activities, 1995–2002

- Put in place after the 1995 Quebec referendum.
- The size of the Program is estimated at $250 million.
- The fund was organized by the Public Works department headed by Alfonso Gagliano, at the time Prime Minister Jean Chrétien's Quebec Lieutenant.
- Senior government officials in Quebec mishandled millions of dollars since 1995.
- Five Crown corporations are involved: the RCMP, VIA Rail, Canada Post, the Business Development Bank of Canada, and the Old Port of Montreal.
- More than $100 million in fees and commissions were paid to different communications agencies.
- In the majority of cases, agencies did little more than hand over cheques.

Source: Adapted from "Auditor General's Report 2004," *CBC News Online*, 11 February 2005, <www.cbc.ca/news/background/auditorgeneral/report2004.html>, accessed 27 March 2006.

began to emerge of phoney invoices, double billing, and exorbitant fees. Under intense public pressure, Prime Minister Paul Martin established a public inquiry in February 2005 headed by Justice John Gomery.

The Gomery Report was made public on 1 November 2005. Judge Gomery minced no words and put the blame for the scandal squarely on the shoulders of former Prime Minister Jean Chrétien and his senior advisors, members of Cabinet, and the bureaucracy. The major findings included:

1. Prime Minister Paul Martin was cleared of any personal blame.
2. Former Prime Minister Jean Chrétien was found partly responsible for the Program, while his chief of staff, Jean Pelletier, was blamed for mismanaging it.
3. The Quebec wing of the Liberal party had benefited from financial kickbacks.
4. The Public Works minister at the time, Alfonso Gagliano, was blamed for the fraudulent behaviour of his staff, since he was directly involved in partisan decision making.
5. Public officials had been fearful of going public because of the close relationships between senior civil servants and senior political officials.[9]

In response, Jean Chrétien immediately accused the judge of biased and unfair reporting, and said he would take his claims to the Federal Court and ask for a judicial review of the judge's findings.

Meech Lake and the Sponsorship Program are two items on a long list of attempts by federal governments of all political colours to win the hearts and votes of Quebecers. The Canadair debacle, discussed below, is another.

The Case of Western Canada

Western Canada has always held the belief that its interests are continually undervalued and sometimes simply ignored by the government in Ottawa. The growth of the Reform Party (later the Canadian Alliance, and now the Conservative Party of Canada) is directly linked to this belief. Some of the more prominent examples will demonstrate this point.

The first example goes back to the implementation of the **National Policy** in 1879. The National Policy of Prime Minister John A. Macdonald was an attempt to build a country out of many different geographical regions and to change the very nature of the Canadian economy from one based on extracting natural resources to one based on manufacturing and other nonresource activities.[10] The tool used to begin this change was the *tariff* on imported goods, which caused them to become more expensive than similar goods produced in Canada.

The purpose of the tariff was to protect Canadian manufacturing companies, which were located primarily in central Canada. But the practical consequence of the National Policy was that, for example, Western farmers had to buy their manufactured goods, such as tractors and combines, from more expensive producers in central Canada because the cheaper ones produced in the United States were now even more expensive with the tariff. This cost Western farmers huge amounts of money, and they have never forgotten it. Generally, Westerners believe that they have contributed enormously to Ontario's economic development and prosperity.

The second example concerns the financial institutions, primarily banks, which are located overwhelmingly in Ontario. By jacking up interest rates to control inflation, the actions of these financial institutions have penalized those who live and work outside Ontario. In the 1980s the Bank of Canada attempted to control inflation in Ontario's Golden Horseshoe (its most intensely developed region) by hiking interest rates regardless of how these higher interest rates would negatively affect economic growth in other parts of Canada. This meant that Western grain farmers, Atlantic fishermen, and people outside the Golden Horseshoe would have to pay more, because of the higher interest rates, to work and live (see Figure 1.2). In the early part of the 1980s, interest rates climbed to more than 20 percent! Would you want to borrow money to buy a new tractor or fishing boat at that rate? (Compare with interest rates in 2006.)

The third example concerns the multimillion-dollar maintenance contract for Canada's high-technology CF-18 fighter aircraft. In 1987 three companies submitted bids: Bristol Aerospace of Winnipeg, IMP of Halifax, and Canadair of Montreal. Originally the contract was awarded to Bristol Aerospace of Winnipeg, because it was more technologically capable of handling the sophisticated aircraft and because the bid was between 8 and 12 percent cheaper. When Canadair began to publicly complain and when members of Parliament from Quebec began to put huge pressure on Prime Minister Brian Mulroney by reminding him that Quebec had voted overwhelmingly for his Progressive Conservative party in the federal election of 1984, Mulroney reversed the decision and awarded the contract to Canadair. The pressure of Quebec MPs was especially intense because another election was just around the corner in 1988. The Prime Minister believed it was more important to keep Quebec happy than Manitoba because Quebec has 75 seats in Parliament whereas Manitoba has only 14. The people of Manitoba were punished once again when the federal Liberals announced in 1995 that Air Force Headquarters would be moved from Winnipeg to Ottawa in 1996. In fact, during World War II, Industry Minister C.D. Howe established almost 50 Crown corporations in Ontario and Quebec. These industries continued to benefit both economies long after the war ended.

Perhaps no issue in the new millennium will have a greater impact on Western Canada than what can only be called "the farm crisis." It is virtually unknown outside

Figure 1.2   Western Perceptions of Their Contribution to Canadian Development

farming communities and has been consistently ignored by federal and provincial governments. This crisis has hit Western farming communities particularly hard. Put simply, international price fluctuations leading to low commodity prices, corporate agribusiness mergers, high production costs, and poor weather conditions have contributed to a steady decline in the number of farm families and has returned net farm income to Depression-era levels.[11] For example, annual realized net farm income has fallen to those levels for grain and hog producers in Alberta, Ontario, and across Canada.[12] Canadian farmers face the prospect of operating 21st-century farms with Depression-era net incomes.[13] Farm families have resorted to increasing their off-farm income in order to raise overall household income.

Another example, similar to those above, concerns the Atlantic provinces. The opening of the St. Lawrence Seaway in 1957 was greeted with great fanfare in Ontario and

Quebec—but not in New Brunswick and Nova Scotia, to whom the Seaway was simply another way economic interests in central Canada took business away from the entire Atlantic region, especially the port cities of Saint John and Halifax.

Our Northern Experience

No region of Canada has been more universally ignored and misunderstood than the land above the 60th parallel, collectively known as Canada's North. Our northern reaches are so far removed from our daily consciousness that we lack a basic understanding of the land and the people who live there. Here is a quick quiz. (The answers are in Box 1.5.) What is the name of the territory that came into existence on 1 April 1999? What is its capital city? How big is the newest territory? What is the name of the dominant ethnic group that lives there? What language do they speak? What does the name of the territory mean in their language? Why is the new territory important to its inhabitants?

Since two-thirds of Canadians live within 320 kilometres of the Canada–U.S. border, any location beyond 320 kilometres is considered "up north." For inhabitants of Winnipeg, Edmonton, and Regina, any journey north is considered to be heading "up north," yet it would take days (depending on the mode of transportation) to reach the 60th parallel, let alone the borders of Nunavut, Yukon, or the Northwest Territories. Perhaps the best example of this southern-centrism is found in southern Ontario. For the inhabitants of the

---

BOX 1.5

Nunavut

- Nunavut is the territory that came into existence on 1 April 1999.
- Nunavut's capital city is Iqaluit.
- It covers approximately 1 994 000 square kilometres.
- Approximately 80 percent of the population is Inuit.
- Inuktitut is the Inuit language.
- Nunavut means "our land" in Inuktitut.
- The Inuit believe this land is their ancestral home and have always referred to it as Nunavut. A territorial government (a form of self-government) now speaks for the rights, needs, interests, and desires of the people and their ancestral home.

Source: "Basic Facts: The New Territory," Nunavut.com site <www.nunavut.com/basicfacts/english/basicfacts_1territory.html>, accessed 26 March 2006.

> **BOX 1.6**
>
> *One-Industry Towns: Shefferville, Quebec*
>
> "One-industry town" is the term for Canadian communities whose economic activity is dominated by one particular industry, such as logging, mining, or fishing. The prosperity of each town is directly linked to the prosperity of its major employer. When these companies experience economic troubles or close their doors, the surrounding community suffers accordingly. An example is Shefferville, Quebec. When iron ore mining was operating at capacity, it was the vibrant home of about 4500 people. When the iron ore owners based in Cleveland, Ohio shut down operations, it was reduced to a ghost town of less than 500 people. Can you name any other one-industry towns?

Golden Horseshoe, especially Toronto, driving to cottages located in the Muskoka and Kawartha Lakes is considered going "up north." In fact, some consider Barrie (located in central Ontario) to be "up north." Few Ontarians have any understanding of cities anywhere north of the Horseshoe.

Here is another quick quiz. These answers you must find for yourself. What college is located in Barrie? What are the names of colleges located in the cities of Thunder Bay, Sault Ste. Marie, Sudbury, Timmins, and North Bay? Are there universities located in any of these cities? If so, what are their names? What industries dominate the lives of people living in the northern reaches of Ontario, Quebec, Manitoba, Saskatchewan, Alberta, and British Columbia? What is the meaning and significance of a "one-industry town"? (See Box 1.6.)

As Canadians we must make a concerted effort to have an awareness and appreciation of those living in our country's northern reaches.

> **CRITICAL THINKING BOX 1.3**
>
> What are the logical consequences of the farm crisis for Canadians? Why is this issue not at the top of all government agendas? Why have the media essentially ignored this issue?

THE ACTIONS OF THE FEDERAL GOVERNMENT

The federal government has always recognized that regionalism does exist. Since Confederation it has spent much time and billions of dollars attempting to reduce the gap between the economically prosperous regions of Canada and those that are not so well off. The federal government has historically attempted to accomplish this in three distinct ways: (1) by reducing physical distances, (2) by instituting programs (spending money), and, perhaps most significantly, (3) by concentrating on sociopsychological phenomena (i.e., attitudes and people's perceptions of each other and of other regions).

Reducing Physical Distances

The early attempts to reduce regional differences and isolation were physical in nature. The first was the construction of the Canadian Pacific Railway (CPR) by Canada's first Prime Minister, John A. Macdonald. The purpose of the railway was to unite the different and far-off regions from the Atlantic Ocean to the Pacific Ocean. Building the CPR was also a precondition for British Columbia joining Confederation; without the railway, there is much doubt that British Columbia would have joined.

The next noteworthy attempt came with the creation of Trans-Canada Airlines (TCA) in 1937, later renamed Air Canada. The creation of TCA was based on the realization by the government that air travel would be the quickest and most efficient way to service the large, outlying, and sparsely populated regions of Canada, in addition to linking the larger metropolitan areas. Private airline companies, it was believed, would not fly to these remote areas because it would not be profitable. This remained the rationale for Air Canada until it was privatized (sold to private investors—the government no longer owns it) in 1989. However, many would argue that Air Canada ceased to operate according to its original mandate long before it was privatized.

Other examples of attempting to physically link the people of Canada were the construction of the Trans-Canada Highway and the creation of Via Rail, a Crown corporation to run railway passenger service, in 1977–78. The vicious budget cuts to Via Rail carried out by the Progressive Conservatives during the mid-to-late 1980s disproportionately penalized Atlantic Canada relative to any other region. Protests of these cuts were widespread, and many believed the cuts to be based on poor research and intentional attempts by the federal government, which continually portrayed Via Rail in a negative fashion, to justify the cuts.[14]

Spending Money

Economically, the federal government has provided money, in many different ways, to the 10 provincial governments to help minimize economic differences.[15] In large part the federal government accomplishes this through the use of transfer payments. Transfer payments take place when the federal government collects money through taxation, such as personal income tax, and then hands over or "transfers" a certain percentage of the money collected, as agreed to by the provinces, to the provincial governments. The provinces use this money to help pay for programs such as health care and education. As already mentioned, federal transfer payments are worth billions of dollars every year. It can be argued that transfer payments are more important to the poorer provinces than to the richer ones. A brief look at Table 1.3 shows that on a per capita basis, the poorer provinces receive a greater proportion of transfer payments than the more prosperous ones do. In 2005–06 Prince Edward Island received total transfer payments of $456 million. In 2004–2005 federal transfers accounted for about 42 percent of the province's revenues. In 2005–2006 federal transfers worked out to be $3291 per person, the highest of any

Table 1.3 Federal Government Transfer Payments to the Provinces and Territories, 2005–2006

Province	Amount	Per Capita
Prince Edward Island	$456 m	$3 291
New Brunswick	$2.330 b	$3 111
Newfoundland	$1.531 b	$2 966
Nova Scotia	$2.620 b	$2 794
Manitoba	$3.201 b	$2 717
Quebec	$15.572 b	$2 052
British Columbia	$6.653 b	$1 570
Saskatchewan	$1.481 b	$1 487
Ontario	$18.609 b	$1 487
Alberta	$4.418 b	$1 486
Territory		
Nunavut	$844 m	$28 061
Northwest Territories	$778	$17 951
Yukon	$534	$16 818

Source: "Federal Transfers to Provinces and Territories," Department of Finance Canada site, 31 March 2005 <www.fin.gc.ca/fedprov/mtpe.html>, accessed 27 March 2006. Reproduced with permission.

Note: Equalization payments are especially important to the life of social programs throughout Canada. See Errol Black and Jim Silver, "Equalization: Financing Canadians' Commitment to Sharing and Social Solidarity," Canadian Centre for Policy Alternatives site, March 2004 <www.policyalternatives.ca/documents/Nova_Scotia_Pubs/NSequalization.pdf>, accessed 27 March 2006.

province. Ontario received a total transfer payment of $18.6 billion, which works out to only $1487 dollars per person. Ontario's transfers in 2004–2005 accounted for only 21 percent of Ontario's estimated revenues. On a per capita basis, Alberta's $4.418 billion transfer payment translates to $1486 dollars per person, the lowest in Canada. In spite of federal government attempts to reduce spending by transferring less money to the provinces, transfer payments are still worth billions of dollars. Without transfer payments, Canada's three territorial governments would be inoperable. In 2004–2005, major federal transfers accounted for about 91 percent of Nunavut's revenues!

The federal government has also formed and used government departments and agencies, as well as legislation, to study and to help stimulate economic growth in poorer regions. Some of the more notable examples include the *Agricultural and Rural Development Act* (1965), the Fund for Rural Economic Development (1966), the Department of Regional Economic Expansion (1969), the Department of Regional Industrial Expansion (1982), the Department of Industry, Science, and Technology (1987), the Atlantic Canada Opportunities Agency (1987), Western Diversification (1987), Enterprise Cape Breton (1987), the Canadian Polar Commission (1991), the Canadian Rural Partnership (1998), and the First Nations and Inuit Health Branch of Health Canada (2000).

In 1995 regional representation can be seen in the names given to three federal government departments: Public Works and Atlantic Canada Opportunities Agency, Indian Affairs and Northern Development, and Finance (with responsibility for Quebec regional development). Other departments and agencies with regional responsibilities in 2006 are: the Prairie Farm Rehabilitation Administration, Atlantic Pilotage Authority Canada, the Cape Breton Growth Fund (CBGF), Indian and Northern Affairs Canada, Broadband for Rural and Northern Development, FedNor (Federal Economic Development Initiative in Northern Ontario), Marine Atlantic, the Northern Pipeline Agency Canada, and Pacific Pilotage Authority Canada.

Promoting Understanding Among Canadians

The final way the federal government has tried to minimize the differences between regions is by promoting understanding between Canadians. Because of Canada's large physical size (9 970 610 square kilometres) and small population (approximately 31.7 million; see Box 1.7) the federal government has taken responsibility for connecting distant regions to each other, emotionally and attitudinally—specifically, by emphasizing what it meant, or means, to be "Canadian" and by educating Canadians about Canada itself, our history, people, places, and attitudes.

The federal government in Ottawa has attempted to educate and inform Canadians about each other by using new technology as it became available (radio and TV) and by the use of **royal commissions**.

BOX 1.7

Canada in a World Perspective

Country	Area (km²)	Population (2001 estimated)	Population (per km²)
Russia	17 075 400	145 000 000	8.5
Canada	**9 970 610**	**31 700 000**	**3.1**
China	9 556 100	1 328 000 000	136.0
United States	9 529 100	250 000 000	29.9
Brazil	8 511 965	173 900 000	20.3
Australia	7 628 300	19 600 000	2.5

In response to new radio technology, the federal government formed the Canadian Broadcasting Corporation (CBC) in 1932. CBC Radio provided an opportunity for Canadians to talk to each other and learn about each other. It also provided Canadian musicians, social commentators, sports broadcasters, newsreaders, and talk show hosts an opportunity to develop their creative talents. Hockey was first broadcast on CBC Radio. In fact, hockey play-by-play commentator Foster Hewitt coined one of the most well-known phrases in all of sport when he described a goal being scored as simply, "He shoots, he scores!" The CBC later did the same when television technology evolved, and in 1952 it formed CBC Television. The purpose of CBC Television was or is to emphasize things that are "Canadian" and to connect people to their community and region. Over the years this has resulted in Canadian programs such as *Hockey Night in Canada*, *Road to Avonlea*, *Degrassi High*, *The Beachcombers*, and *Rita and Friends*. CBC television also gave birth to the regionally oriented newscasts that usually follow the national news.

Before TV, the federal government was already involved in the making of films and documentaries. In 1939 it created the National Film Board (NFB). For decades now NFB documentaries and short films have been seen across Canada, especially in schools. Canadians everywhere are familiar with the 30-second NFB vignettes broadcast between programs on the CBC. The work of the NFB has been awarded many international honours. Overall, the purpose of the CBC and NFB has been to educate and inform Canadians, and the federal government has spent hundreds of millions of dollars over the years to promote this education. There is no better, and perhaps no more ambitious, example of this than the 2001 CBC documentary *Canada: A People's History*. It was an immediate hit with Canadians, averaging 2.5 million viewers for the first six episodes.[16]

But CBC-TV producer Mark Starowicz could only entice one company to buy advertising time.[17] CBC proceeded to make the $25 million series in spite of corporate Canada's indifference. It seems Canadians do want to know about each other!

The federal government's heavy involvement in radio, TV, and film is based in part on the belief that what Canadians think, feel, and believe about each other differs from region to region and that this has an important impact on Canadian unity. In fact, in 2002–2003 the federal government spent $3.4 billion on culture (that includes but is not limited to broadcasting, film, video, sound recording, and book and periodical publishing).[18] This belief has prompted the federal government to investigate specific problems in Canada or the likely effects some government policies might have on some regions. The tool used to do this is the royal commission. Royal commissions, headed by a person appointed by the federal government, utilize the expertise of people who work in the public sector or the private sector, academics, and, if necessary, experts outside Canada. They do not implement policy, but simply suggest directions for policy.

To do this, they study, investigate, and accumulate information on important issues or matters of government policy. In the past, there have been two royal commissions concerned with certain aspects of regionalism as they travelled across Canada. The first was the 1937 Royal Commission on Dominion–Provincial Relations, which produced an in-depth study of federal–provincial financial relations—how the federal government transferred money to the provincial governments. The second was the 1981–85 Royal Commission on the Economic Union and Development Prospects for Canada. This Commission was important because it came out in favour of free trade with the United States, in spite of the fact that some experts who presented before the Commission concluded that free trade would be harmful to the poorer regions, especially the Atlantic provinces.

It is not uncommon for the federal government to ignore the recommendations of royal commissions. In 1997, the *Royal Commission Report on Aboriginal Peoples* tabled almost 500 recommendations. The principal recommendation was to increase spending immediately and commit new resources to Aboriginal life in Canada. The Commission suggested spending an average of almost $2 billion per year for the next 20 years. This extra spending would be in addition to the $5 billion to $7 billion already spent annually. Ottawa has yet to implement this principal recommendation. In January 1998, Ottawa announced a $350 million program to deal with the mental and physical damage caused by the residential school program established and partly administered by the federal government between 1867 and 1945.

Royal commissions are popular with the federal government, and it is common for one to last several years. Since 1867 there have been approximately 400 of them[19]—an average of three per year!

THE SOCIOPSYCHOLOGICAL DIMENSION TO REGIONALISM: THE REALITY

Clearly profound economic and social differences exist among the regions in Canada. It is equally clear that there are sociopsychological differences as well. That is, people in different regions do think and feel differently about each other and about the federal government in Ottawa.

What does the **sociopsychological dimension to regionalism** mean? It means more than economic differences between regions with respect to money, companies, investment, income, government policy, elections, and federalism. The sociopsychological dimension is concerned with how individuals living in different regions feel about themselves, their community, other regions, and the federal government. Regional differences in these areas are commonly thought to exist, say, only between the French-speaking majority of Quebec and the English-speaking majority of Quebec and the rest of Canada. Differences, however, go much further than just language and culture. Every year, public opinion polls, quality-of-life surveys, government surveys, and other forms of research indicate that Canadians have different beliefs, opinions, and attitudes about living in Canada and about who benefits the most from government policy. Generally, Ontario and Quebec are considered to be the big winners and Atlantic Canada the big loser. These beliefs have generated much envy. The most recent manifestation of this envy (some would say anti-Ontario attitude) appeared in national headlines before the First Ministers Conference held in Victoria, British Columbia, on 1 August 2001. Nova Scotia Premier John Hamm suggested that the equalization formula be redesigned to put more money in Atlantic coffers. Former Ontario Premier Mike Harris responded by comparing the Nova Scotia Premier to a welfare cheat! University of Moncton Professor Donald Savoie, one of Canada's leading authorities on regional issues, believes the Atlantic provinces got a raw deal with Confederation and that "the region is not doing well because of federal government policy."[20] In fact, he believes that equalization payments are designed to ensure Atlantic Canadians have enough money to buy goods made in Ontario and Quebec—in short, the Atlantic provinces have been kept poor by central Canada.

It is not surprising that people living in different parts of Canada have different attitudes concerning the Canadian experience. These differences are largely because the region you live in, the job you do, and the language you speak all affect the way you think and feel. Our socialization and life experiences have a profound influence on the way we come to understand each other, the federal government, and ourselves. Three examples will illustrate this point: the 2005 flag-lowering protest in Newfoundland and Labrador; the 1995 fishing dispute between Canada and Spain; and the cross-Canada contempt for central Canada and Metropolitan Toronto.

The 2004–2005 Flag Removal in Newfoundland and Labrador

Throughout 2004 and in early 2005 federal–provincial negotiations over oil and natural gas revenues between Ottawa and the Liberal government of Newfoundland and Labrador led by Danny Williams were progressing slowly. The talks centred on sharing offshore oil and natural gas revenues. Sharing revenues has always been a contentious issue in Newfoundland and Labrador. Many in that province believe keeping all or almost all revenues would give the province the base to significantly reduce its reliance on government transfer payments for fiscal survival. In short, Newfoundland and Labrador, a "have not" province, would become more like Alberta, a "have" province.

When talks stalled, Premier Williams ordered all Canadian flags to be removed from government buildings. This contentious move was widely supported across the province. Even Memorial University, traditionally neutral in such disputes, decided to follow the government's lead.[21] Removing the flags was widely discussed in the Atlantic provinces, but it went largely unnoticed outside Atlantic Canada. The premier's actions did lead to a resumption of talks that concluded with a significant victory: the passage of Bill C-43, which allows Newfoundland and Labrador to keep an additional $2.6 billion per year in revenues from offshore oil and natural gas.

The 1995 Fishing Dispute

The 1995 fishing dispute centred on Canadian claims that Spanish fishing vessels waited just beyond Canada's 200-mile (322-kilometre) boundary and fished to excess using illegal nets and taking even the smallest turbot (a species of fish). Earlier, in an attempt to allow fish stocks to replenish themselves, the federal government had banned cod fishing (before the 1992 election), and later it drastically reduced the turbot quotas for Canadian fishermen. Ottawa believed Spanish fishing would eventually lead to the complete collapse of turbot stocks. When Canadian Coast Guard vessels arrested two Spanish fishing vessels, the citizens of the Atlantic provinces, particularly Nova Scotia, were celebrating everywhere, organizing support rallies, carrying signs, and just plain being happy. Finally, the federal government was seen by them as acting on their behalf. The CBC, CTV, *Maclean's*, the *Toronto Star*, and the *Globe and Mail* all reported extensively on the dispute. This attention was heartening for many living in the Atlantic region during these tough times. Barbara Nees from the sociology department of Memorial University in St. John's remarked that it gave people a positive sense of community and a feeling that they were not suffering alone.[22] Nevertheless, the same happiness and pleasure was not shared to the same degree by people living in Toronto, Calgary, Saskatoon, Dryden, Laval, or other cities that do not depend on fishing or fish processing for their livelihood.

The issue for this century for the Atlantic provinces may turn out to be the bulk sale of fresh water to markets in the United States. Atlantic-province governments look upon water as another resource to be exploited—another commodity to be sold in the international marketplace where demand is high, especially in the American southwest. Newfoundland Premier Roger Grimes is at the forefront of this push. The federal government, however, is under increasing pressure from the general population and many environmental groups to do the opposite. With economic uncertainty still plaguing the Atlantic fisheries, the bulk sale of fresh water has taken on new importance.

The Contempt for Central Canada, Especially Metropolitan Toronto

Perhaps nowhere is the sociopsychological dimension to regionalism more evident than with respect to the contempt felt by most Canadians for central Canada and Metropolitan Toronto. Rivalries certainly exist between the big cities in Canada, but although there may be intraprovincial rivalries between cities like Calgary and Edmonton or Regina and Saskatoon, and there may be rivalries across provinces between cities like Halifax and Saint John, what all these cities have in common—along with all other cities across Canada—is that they resent the privileged position of Toronto. This resentment is primarily based on central Canada's virtual dominance of economics, politics, and social life.

BOX 1.8

Even the Supreme Court of Canada?

Canada's highest court is not immune to charges of regional favouritism. Sauvageau, Schneiderman, and Taras report that the notion of the Supreme Court as "objective" and "impartial" is not shared nationally. Those in Quebec regard it as a "Leaning Tower" that always leans in the same direction—a centralist one (thus favouring English Canada). In fact, Guy Laforest saw the introduction of the 1982 Charter of Rights and Freedoms as the "updating of the [English] conquest [of Quebec]." Similar sentiments are found in the Western provinces where Canada's highest court remains a potent "symbol of the entrenched power of the East."

Sources: Florian Sauvageau, David Schneiderman, and David Taras, *The Last Word: Media Coverage of the Supreme Court of Canada* (Vancouver: UBC Press, 2005), pp. 24–25; Guy Laforest, *Trudeau and the End of the Canadian Dream* (Montreal and Kingston: McGill-Queen's University Press, 1995), pp. 180–181, quoted in Sauvageau et al., op. cit., p. 25.

Economic Dominance

The economic dominance of central Canada, especially Metro Toronto, is evident in the overwhelming presence of the head offices of the most dominant corporations in Canada. According to the *Globe and Mail*[23], in 2005 28 of the top 50 private corporations in terms of revenue were located in Ontario and another 10 in Quebec. A total of 38 of the top 50 private companies were located in central Canada. Moreover, 8 of the top 10 were located in the Greater Toronto Area (GTA). In fact a total of 58 of the top 110 companies were located in Ontario and all but 8 in the GTA! If you include those located in Quebec, a whopping 81 of 110 were located in central Canada. This leaves only 29 in the rest of Canada.

Politically, this is reflected in the fact that the first non-Ontarian to occupy the office of Minister of Finance was Jean Chrétien, a Quebec native, in 1979. Can you name another finance minister who did not originate from Ontario or Quebec?

A simple mention of "Bay Street" brings to mind huge office towers, money, and economic power. Again, the economic dominance of the Golden Horseshoe of Ontario is clearly visible with a simple drive along the highways that run through it: the Queen Elizabeth Way and the 401, 403, 407, and 427.

Political Dominance

Ontario and Quebec virtually dominate Canada in terms of representation in Parliament, with 178 out of 301 seats and appointments to key cabinet posts such as the departments of finance, foreign affairs, justice, and international trade. The trend has been for the federal government to choose a finance minister who is a lawyer from Bay Street (who must first be elected, of course). Usually, this lawyer returns to a job on Bay Street at the conclusion of his or her political career.

Cultural and Social Dominance

Culturally, the media outlets of the CBC, Baton Broadcasting, Global TV, and City TV are all located in Metro Toronto. Canada's "national" newspapers, the *Globe and Mail* and the *National Post*, and Canada's largest-circulation daily the *Toronto Star*, are located in Metro Toronto. Metropolitan Toronto is also the centre of publishing and, along with Montreal, the centre of fashion, entertainment, and the arts. In particular, Metro Toronto is the home of the Air Canada Centre, SkyDome (now called the Rogers Centre), the CN Tower, Ontario Place, the Canadian National Exhibition, the Hummingbird Centre, Canada's Wonderland, Caribana, the Royal Ontario Museum (ROM), the Royal Alexandra Theatre, Second City, the Pantages Theatre, Roy Thomson Hall, and countless other sites and exhibits. In 2005, the Canadian Soccer Association (CSA) announced it was building its new "national" soccer stadium, fit for hosting the most

> **CRITICAL THINKING BOX 1.4**
>
> Have the Western provinces received their share of recognition for helping to build this country? What about the Atlantic provinces? Have Ontario and Quebec received too much?

prestigious and largest international soccer games, in Toronto on the York University campus. For all the reasons mentioned, it is little wonder that other regions of Canada envy and resent central Canada and Metro Toronto in particular.

CONCLUSION

Regionalism is an important diversity in Canada, so important in fact that examining the nature of regionalism is critical if we are to have a complete understanding of social life in Canada. Canadians from different regions are diverse in a great many ways. From a regional perspective, for example, Canadians find themselves different in the areas of geography, climate, income, ethnicity, social class, and attitudes toward each other and toward the federal government. The last two differences confirm for us that regionalism does have an important sociopsychological component.

What does the future hold for these differences? Probably much of the same. It is extremely unlikely that the nature of the Canadian political system will change any time soon, if at all. This is especially true of federalism and our electoral system, two key elements that intensify regional feelings and differences. As a consequence, it is highly unlikely that the behaviour of the federal government will change. It is equally unlikely

> **CRITICAL THINKING BOX 1.5**
>
> To what extent does regionalism make Canadians different from each other? Are regional differences more or less important than other differences such as language, social class, gender, sexual orientation, family structure, race, and ethnicity? List what you consider the five most important differences. Why did you choose the differences you did? Would a student in another province choose the same ones? Why or why not?

> **CRITICAL THINKING BOX 1.6**
>
> How do you explain and reconcile the paradox that the federal government created vast regional disparities (economically and sociopsychologically) and later committed itself to reducing the disparities it had created?

that the nature of capitalism will change. As the saying goes, "Money will go where it will get more money." That means money (and jobs) will travel indiscriminately between regions, within regions, and, when necessary, even outside Canada.

These two developments, or lack thereof, do not bode well for the future of Canada. As Canadians, we must be vigilant and insist that all levels of government treat all Canadians living in all regions fairly and equally. If any level of government fails to behave in this manner, it should be reminded that the ballot box is never more than a few years away.

CHAPTER SUMMARY

Understanding regionalism is important to understanding life in Canada. There are economic, political, and sociopsychological components to regionalism. To understand regionalism in its entirety, we must pay particular attention to the sociopsychological component, because it is usually not addressed in the traditional literature. There are many different theories dealing with the causes of regionalism, and it is up to you to decide which of the four dominant theories presented here are best. The federal government has a role in both creating and attempting to minimize regional differences. Finally, we must be aware of and appreciate that certain aspects of our political system—federalism and our electoral system—can and do intensify regional differences.

KEY TERMS

executive federalism, p. 14
federalism, p. 14
Golden Horseshoe, p. 7
Meech Lake Accord, p. 14
National Policy, p. 20
regionalism, p. 5

representation by population, p. 15
royal commission, p. 27
sociopsychological dimension to regionalism, p. 30
transfer payments, p. 11

DISCUSSION QUESTIONS

1. What is *regionalism*? Why is regionalism important to understanding the Canadian experience?
2. Briefly outline four theoretical explanations for regionalism. Which do you think makes the most sense? Would Canadians from different provinces choose different explanations? If so, why? If not, why not?
3. How do federalism and our electoral system compound and intensify regional differences?
4. What do we mean by the *sociopsychological dimension* to regionalism?
5. What role(s) has the federal government played in both creating and helping to reduce regional differences? Can you think of any current examples?
6. Is the idea of regionalism "real" to you? If so, why? If not, why?

NOTES

1. See Richard Simeon, "Regionalism and Canadian Political Institutions," in O. Kruhlak, R. Schultz and S. Pobihushchy, eds., *The Canadian Political Process* (Toronto: Holt, Rinehart and Winston, 1979).
2. See Harry H. Hiller, *Canadian Society: A Macro Analysis* (Scarborough, ON: Prentice-Hall, 1991), p. 11.
3. The explanations of these approaches draw heavily from the following: Ralph Mathews, "Understanding Regionalism as Effect and Cause," in *Social Issues: Sociological Views of Canada*, 4th ed. (Scarborough: Prentice-Hall, 1988), pp. 60–72; Ralph Mathews, *The Creation of Regional Dependency* (Toronto: University of Toronto Press, 1983), pp. 37–55; Janine Brodie, *The Political Economy of Canadian Regionalism* (Toronto: Harcourt Brace Jovanovich, 1990), pp. 21–36; and Donald J. Savoie, *The Canadian Economy: A Regional Perspective* (Toronto: Methuen, 1986), pp. 9–24. The names of some approaches have been altered.
4. This approach is drawn from Mathews, "Understanding Regionalism," pp. 64–65, and Mathews, *The Creation of Regional Dependency*, pp. 45–46.
5. Mathews, "Understanding Regionalism," p. 65.
6. This explanation is based on the one provided by Brodie, *The Political Economy of Canadian Regionalism*, pp. 27–34. The explanation includes Keynesianism, regional science, and developmental approaches.
7. Mathews, "Understanding Regionalism," p. 69.
8. Trudeau went national with his objections, which were published in a short treatise edited by former Cabinet minister Donald Johnston. See Donald Johnston, ed., *With a Bang, Not a Whimper: Pierre Trudeau Speaks Out* (Toronto: Stoddart Publishing, 1988).
9. *Toronto Star*, 2 November 2005.

10. For an excellent explanation of the National Policy, see Desmond Morton, *A Short History of Canada* (Edmonton: Hurtig, 1983), pp. 92–105. This brief history of Canada is both readable and enjoyable. We highly recommend it to anyone interested in the development of Canada.
11. Darrin Qualman, *The Farm Crisis and Corporate Power*, Canadian Centre for Policy Alternatives site, p. 13, April 2001 <www.policyalternatives.ca/documents/National_Office_Pubs/farm_crisis.pdf>, accessed 27 March 2006.
12. Statistics Canada, *The Daily*, 4 May 2005.
13. Qualman, op cit., p. 6.
14. For a disturbing account of the decimation of Via Rail, see Jo Davis, ed., *Not a Sentimental Journey: What's Behind the Via Rail Cuts, What You Can Do About It* (Toronto: Gunbyfield Publishing, 1990).
15. For an explanation of the evolution and different types of transfer payments, see Garth Stevenson, *Unfulfilled Union: Canadian Federalism and National Unity*, 3rd ed. (Toronto: Gage, 1988), pp. 124–50.
16. Linda McQuaig, "Just One Sponsor, but Canadians Love CBC *People's History*," StraightGoods.com, 21 December 2000 <http://goods.perfectvision.ca/ViewFeature.cfm?REF=23>, accessed 8 February 2002.
17. Ibid.
18. Statistics Canada, *The Daily*, 27 January 2005.
19. Keith Archer, Roger Gibbins, Rainer Knopff, and Lesie A. Pal, *Parameters of Power: Canada's Political Institutions* (Toronto: Nelson, 1995), p. 287.
20. "Maritimes Kept Poor by Ontario," *The Hamilton Spectator*, 30 July 2001.
21. Robert Adamec, *Memorial Gazette* 37(9) (27 January 2005).
22. "Conflicting Emotions," *Maclean's*, 27 March 1995.
23. "Top 300 Private Companies," Globeinvestor.com <www.globeinvestor.com/series/top1000/tables/private/2005>, accessed 27 March 2006.

CHAPTER 2

Demographic Trends in Canada

Michelle Broderick

Except in the case of vaccination against small pox . . . it is unlikely that immunization or therapy had a significant effect on mortality from infectious diseases before the twentieth century.
— T. McKeown, *The Modern Rise of Population*

In technologically advanced human societies, virtually every female now survives to reproductive age. This is a biologically novel situation.
— F. Fenner, "Foreword," in *The Structure of Human Populations*

Objectives

After reading this chapter, you should be able to

- understand what the field of demography entails

- identify what kinds of information are used in the demographic study of populations

- understand the methods used by demographers

- appreciate that demographic variation can often account for social, cultural, and economic diversity within a population

- recognize that the study of the demographic history of Canadian populations touches on a wide variety of subjects, from illegitimacy to epidemics

DEMOGRAPHY: AN INTRODUCTION

Demography is the scientific study of human populations. It focuses on the size and composition of a population, which depend on such factors as fertility, nuptiality, mortality, and migration. Demography demonstrates how social and economic factors affect these demographic parameters and, hence, human behaviour. Variations in demographic characteristics can be observed at many different levels, for example, township, city, province, or nation, and they are often linked to social and economic features of these population units. Unique demographic histories experienced by populations in different regions are often the main cause of social, cultural, and economic diversity observed today. By studying the history of populations, we not only learn how they have diversified over time, but also, because of the greater time depth (several generations), find it easier to link social and economic factors to demographic patterns. This chapter will examine and define the variables used in demography, followed by some examples of contemporary and historical studies on Canadian populations.

Some Important Definitions: Demographic Variables

A **variable** is a characteristic that differs or varies among groups, such as age and religious affiliation. In demographic research, the variables examined are those that affect the growth of a population: fertility, nuptiality, mortality, and migration.

Fertility

Fertility refers to the number of **live births** occurring in a population within a specific period (either one year or aggregates of years). Fertility differs from **fecundity**, which is the maximum number of children that a woman can produce during her lifetime, that is, the potential of childbearing. The figure for fecundity is based on the length of the female reproductive period, usually defined as between the ages of 15 and 44, during which time a woman can produce a maximum of 15 to 20 children; however, few women ever reach this potential because of the biological and social factors that reduce fertility. For instance, the length of the reproductive period can be affected by the age at which a woman marries—if reproduction does not occur outside of marriage. If a woman marries and has her first child at the age of 25, her reproductive period will be approximately 19 years. During this time she could produce on average 13 children. If that same woman married and had her first child at the age of 35 instead of 25, her reproductive period would be shorter, and the number of children she could produce would be considerably lower. Other factors affecting female fertility are the use of contraceptives and abortifacients; by preventing or terminating unwanted births, fewer children are born. Social attitudes toward reproduction can also affect fertility. For instance, if a woman's role in society is that of homemaker, where childbearing is seen as a valuable contribution, women will tend to have more children. Illness and disease can also limit fertility, by either preventing fertilization or by inducing spontaneous abortions (miscarriages).

Several methods are used to calculate fertility. The easiest method is the **crude birth rate**, which is the number of live births in a given population (see Figure 2.1). This method is "crude" in the sense that it is based on the entire population, not just the women who are capable of reproducing; therefore, this figure can be misleading because the number of women between 15 and 44 years of age can vary among populations. A more refined and accurate method is the use of the **fertility rate**, which is based on the number of women of reproductive age (15 to 44 years).

Nuptiality

Nuptiality, or marriage, is an important variable for two reasons: (1) marriage is related to fertility, in that the age at which marriage occurs can define the length of the female reproductive period and (2) the timing of marriage is affected by social and economic factors. For example, patterns of economic activity, such as farming, coupled with religious restrictions, such as Lent, have in the past resulted in a seasonal pattern of marriage, with most marriages occurring between October and December (i.e., after the harvest but

Figure 2.1 Crude Birth Rate in Canada, 2001 (live births per 1000 population)

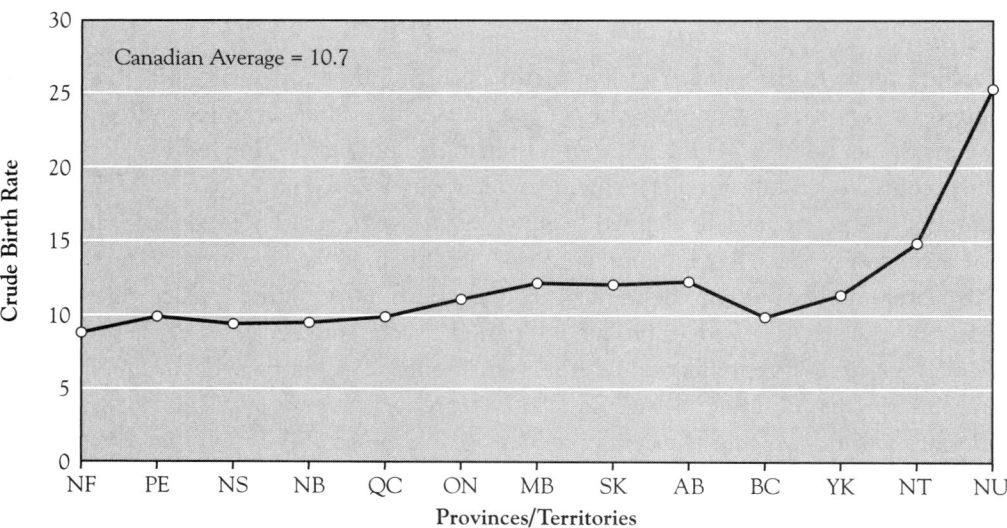

The pattern of fertility is not uniform across Canada. The lowest birth rate was recorded in Newfoundland and the highest in Nunavut. (Nunavut was declared a separate Territory from the Northwest Territories on 1 April 1999.)

Source: Adapted from Statistics Canada, *Births*, Catalogue No. 84F0210XPB, 2001, 11 August 2003, p. 4, Table 1.4.

before religious festivals). Today, this pattern is no longer observed in Canada. It is also common for people to delay marriage until they have achieved economic independence from their parents. The examination of nuptiality not only includes looking at those individuals who marry, but also at those who do not. For instance, if the majority of men and women in a population marry, the potential for population growth through births increases; however, if a large portion of the adult population do not marry, and, hence, may not reproduce, population growth will be slowed. The incidence of **celibacy**, the proportion of individuals in a population who never marry (and presumably, never reproduce), is often related to economics, especially to inheritance practices. For example, in Ireland farmers traditionally passed on their wealth (land) to one son only. The other children (both sons and daughters) either remained on the homestead and did not marry, or they left to seek their fortunes elsewhere.[1] Such a practice would also influence emigration and, hence, the size of the population.

Typical methods used to measure marriage include the **crude marriage rate**, which is the number of marriages recorded in a population (see Figure 2.2), and the **age-specific marriage rate**, based instead on the number of single individuals, who never married, in a given age interval. **Age at first marriage** can also be calculated; it provides, as

Figure 2.2   Crude Marriage Rate in Canada, 2001 (marriages per 1000 population)

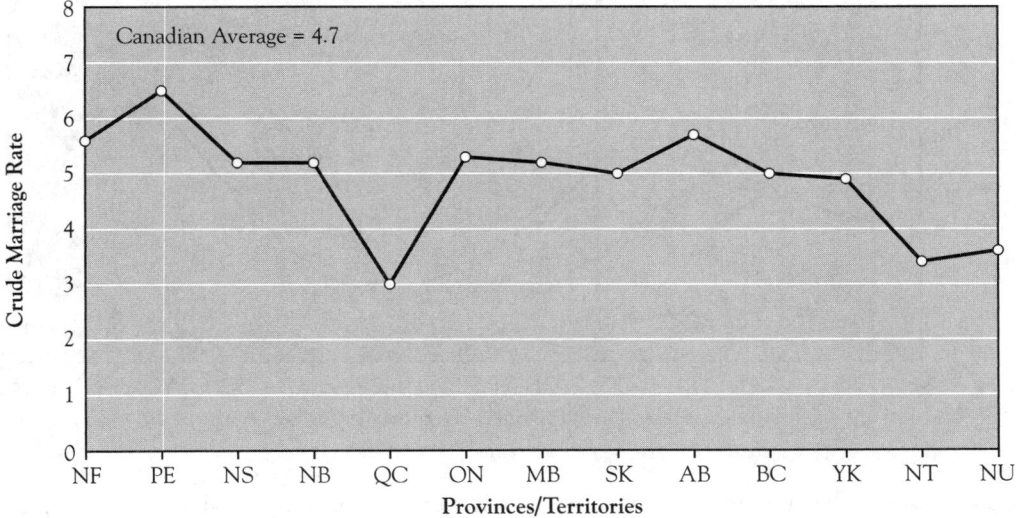

Prince Edward Island, Quebec, the Northwest Territories, and Nunavut deviate the most from the Canadian marriage pattern.

Source: Adapted from Statistics Canada, *Marriages*, Catalogue No. 84F0212XPB, 2001, 20 November 2003, p. 4, Table 4.

already mentioned, information on the length of the female reproductive period (see Figure 2.3). This particular measure has been found to be highly sensitive to economic factors.

Mortality

Mortality refers to the number of people in a population dying in a given period. Two important aspects of mortality are life span and life expectancy. **Life span** refers to the maximum age that a human has ever lived. This figure is currently 122 years: Jeanne Louise Calment was born in Arles, France, on 21 February 1875 and died on 4 August 1997.

As with fecundity, very few humans reach this potential. The **life expectancy** is the age to which most humans can expect to live and is based on the average age at death. In Canada, this figure is 75 years for males and 81 years for females.[2] A variety of biological and social factors can affect mortality. For instance, exposure to disease and toxic substances (such as pollution), nutrition, physical labour or exercise, stress, and access to health care can all affect mortality in a population.

Figure 2.3   Average Age at First Marriage in Canada, 2001

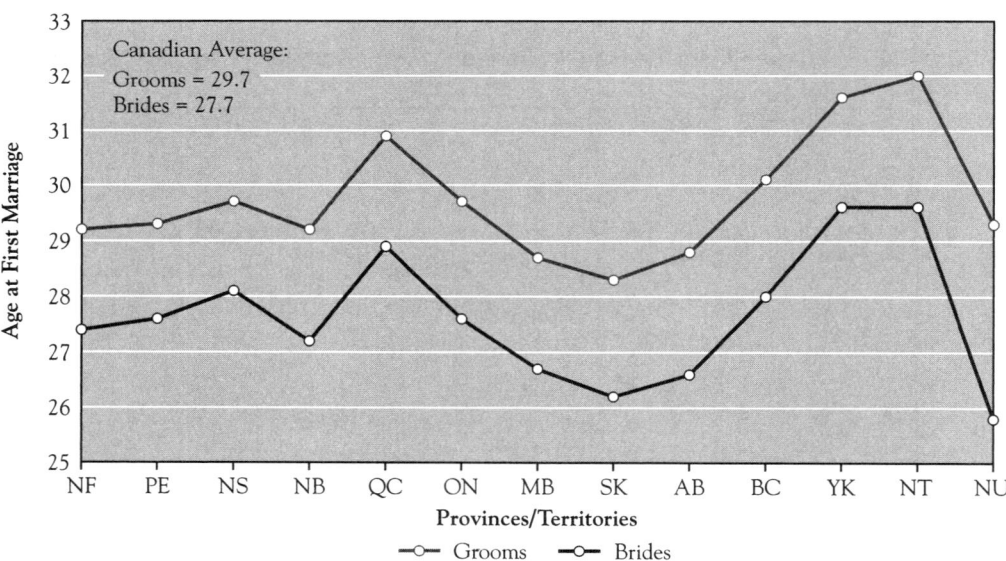

The Canadian pattern of age at first marriage is not unique. It is common to find that men marry at slightly older ages than women. Variation is observed across Canada, with people marrying at younger ages in Manitoba, Saskatchewan, and Alberta, and at older ages in Quebec, British Columbia, and the Yukon and Northwest Territories.

Source: Adapted from Statistics Canada, CANSIM Marriage Database, Canadian Vital Statistics, Table 101-1002 <http://cansim2.statcan.ca>, accessed 15 October 2005.

Mortality is measured in a variety of ways, such as the **crude death rate**, which is the number of deaths in a given population (see Figures 2.4 and 2.5). Again, this is a crude method that does not reveal details of the mortality experience. A more detailed picture is achieved by examining mortality in different age groups, that is, **age-specific mortality rates**, and it helps to identify major risk factors affecting different segments of the population. One of the most useful age groups in which to study mortality is of those aged less than one year, that is, the **infant mortality rate** (see Figure 2.6). This measure is extremely useful as an indicator of the general health status of the population, because infants are more susceptible to environmental factors, such as food consumption, medical care, and public sanitation. In other words, the infant mortality rate reflects the standard of living in a population. Infant mortality is also associated with life expectancy. When infant mortality is low, life expectancy is high, and vice versa. For instance, in developed countries life expectancy is much higher than in developing countries, and this reflects the standard of living. Analysis of infant mortality is sometimes divided into two age groups: under six months of age and between six and

Figure 2.4   Crude Death Rate in Canada, 2000 (deaths per 1000 population)

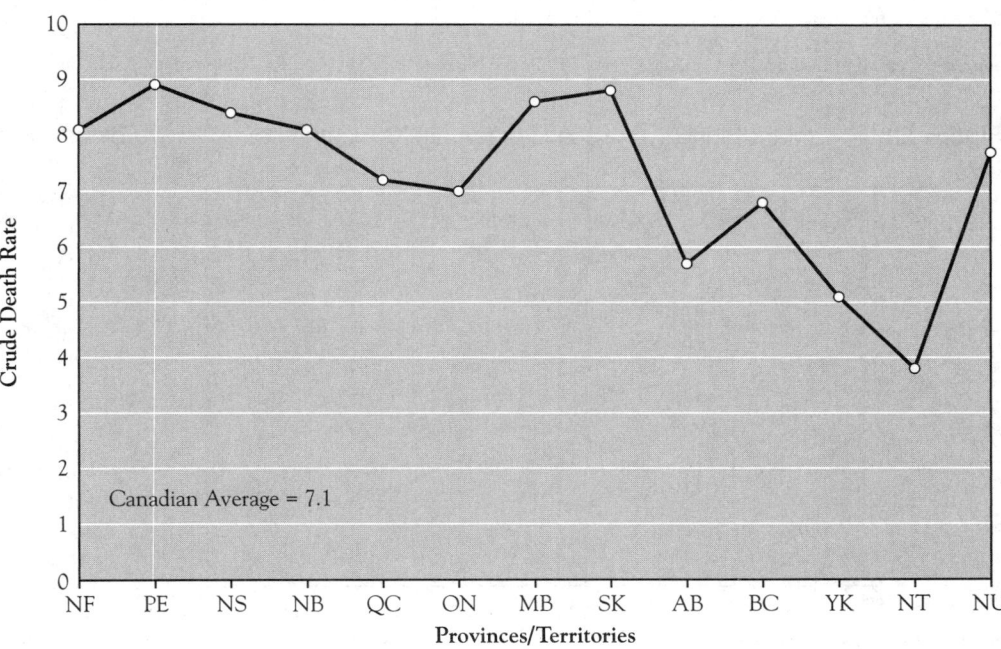

The crude death rate varies across Canada. The highest death rate occurs in Prince Edward Island, while the lowest occurs in the Northwest Territories.

Source: Adapted from Statistics Canada, *Deaths*, Catalogue No. 84F0211XPB, 2000, 25 September 2003, Table 4.

An Overview of Diversity in Canada

Figure 2.5 Sex-Specific Mortality Rates in Canada, 2000 (deaths per 1000 males and females in the population)

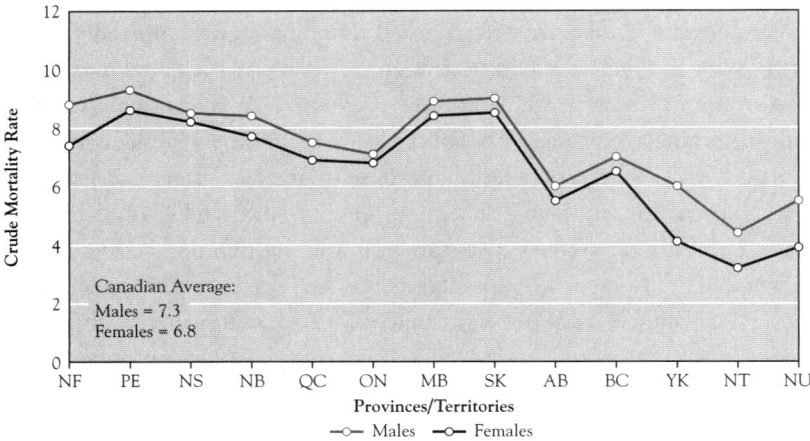

By comparing the crude mortality rates between the sexes, you can see that males are dying at a higher rate than females. This pattern is reflected in their life expectancies, which are 75 years for males and 81 years for females.

Source: Adapted from Statistics Canada, *Deaths*, Catalogue No. 84F0211XPB, 2000, 25 September 2003, Table 4.

Figure 2.6 Infant Mortality Rate in Canada, 2001 (deaths per 1000 live births)

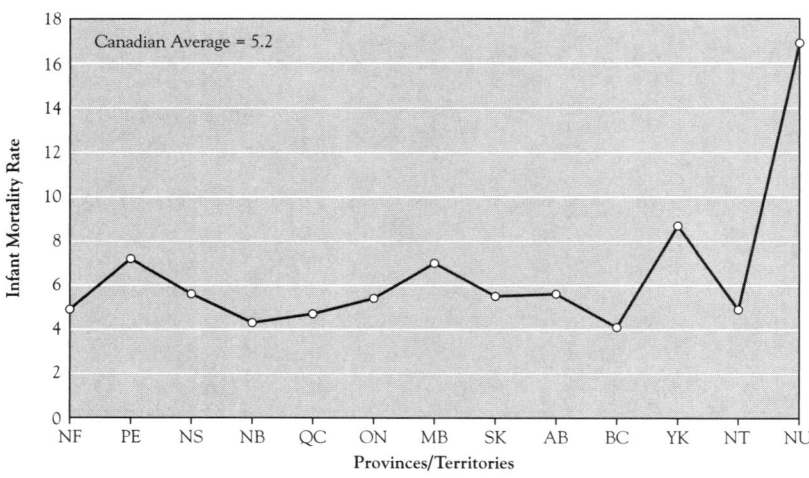

The infant mortality rate also displays a certain amount of variation across Canada. It is at its lowest in British Columbia and at its highest in Nunavut.

Source: Adapted from Statistics Canada, "Infant Mortality Rate in Canada, 2001 (Deaths per 1000 Live Births)" <http://www40.statcan.ca/l01/cst01/health21a.htm>, accessed 15 October 2005.

twelve months. Death under six months tends to be associated with problems that a child was born with, whereas death between six to twelve months is usually related to external or environmental factors, such as disease or nutrition.

Morbidity, the incidence of particular diseases in a population, is also of interest, because this is related to both the social and physical environments. For instance, in Canada we enjoy a temperate climate with warm summers and cold winters. As a result many microorganisms and disease-causing parasites cannot survive year-round, and we do not experience high rates of certain diseases seen elsewhere (e.g., malaria). The age profile of a population will also influence morbidity. In Canada the proportion of the population over the age of 64 almost tripled between 1921 and 1996, increasing from 4.8 percent to 13.0 percent[3] (see Figure 2.7 and Critical Thinking Box 2.1). This increase is in part the result of lower fertility (fewer children are being born) and of higher life expectancy or survivorship (more people are living longer). Because we have a larger portion of older individuals, we also have an increase in diseases associated with aging, such as cancer and degenerative bone diseases.

Figure 2.7 Proportion of the Population over the Age of 64 Years, 1951 and 2001

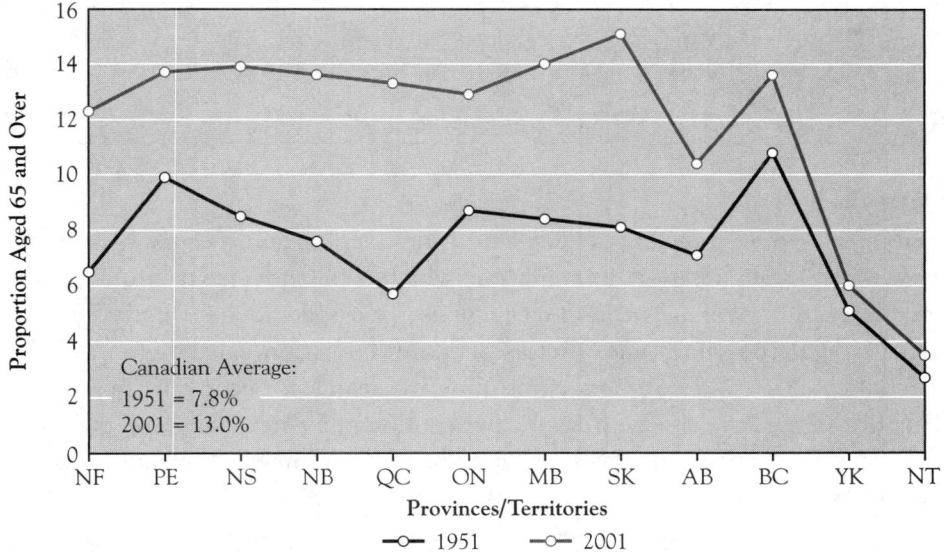

Regional variation in the proportion of the population over the age of 64 is marked. An increase in the number of seniors is also noted over time. This variation across time and space is related to changes in fertility, mortality, and migration.(Data on Nunavut, separate from the Northwest Territories, was not available prior to 1999; therefore, it was combined with the data.)

Source: Adapted from Statistics Canada, *Age, Sex, and Marital Status*, Data Products: Nation series: 1991 Census Population, Catalogue No. 93-310, Table 1, 6 July 1992 (1951 data), and from the Statistics Canada site <www12.statcan.ca/english/census01/products/highlights/AgeSex/HighlightsTables.cfm?Lang=E>, accessed 15 October 2005.

> **CRITICAL THINKING BOX 2.1**
>
> *The Aging of Canada's Population*
>
> The **dependency ratio** reflects the proportion of the population who are under the age of 15 and above the age of 64. This group is viewed as being dependent on society, directly or indirectly. The aged are directly supported by our society through the government pension plan and discounted services, whereas children are only indirectly supported, because it is the responsibility of their parents to care for them.
>
> Whereas the dependency ratio has decreased over time, from 64.43 in 1921 to 47.14 in 2001, the contribution of those over the age of 64 has increased more than threefold, from 12.2 percent to 40.4 percent. This increase reflects the fact that our Canadian population is aging. This trend is likely to continue, and we run the risk of increased poverty and illness among the aged, since our society may not have the resources to support those services needed by the elderly.
>
> Can we begin to plan for this contingency today?
>
> *Source:* Adapted from Statistics Canada, "Dependency Ratio 2001," available <www12.statcan.ca/english/profil01/CP01/Index.cfm>, accessed 15 October 2005.

Migration

Migration refers to the movement of people into and out of specific geographical areas. There are two types of migration, **immigration**, which is the movement into a specific area, and **emigration**, which is the movement out of a specific area. Both forms of migration occur at the same time. Factors affecting migration can be grouped into three general categories: (1) economic, meaning the search for economic opportunities; (2) political, meaning escape from the persecution of a particular group of people and from social institutions such as slavery; and (3) environmental, meaning escape from the effects of earthquakes, floods, and famines. In examining the factors affecting migration, you must remember that push and pull factors occur at the same time, that is, some factors pull people toward an area while others push them away. For instance, an urban setting generally offers a higher frequency and wider variety of jobs; therefore, it attracts people. However, the cost of living in an urban setting tends to be much higher than in a rural area and acts as a deterrent. Migration can also be viewed as either voluntary or involuntary. Voluntary migration occurs when the decision to move or not is up to the individual. For example, if you were offered a well-paying job in Calgary, the decision to move would be entirely yours. Migration is involuntary when the individual is forced to move or stay (see Critical Thinking Box 2.2), as during World War II when many were forced to flee Europe while others were prevented from doing so by the Nazi regime.

Several measurements of migration are commonly used. The in-migration or **immigration rate** consists of the number of people entering an area. The out-migration or **emigration rate** consists of the number of people leaving an area. The **gross migration rate** reflects the total number of people who both enter and leave an area. The **net migration rate** is the annual increase or decrease in the size of a population, based on the number of people entering an area minus the number who leave. Migration is often the single most important factor affecting the size—growth or decline—of a

---

CRITICAL THINKING BOX 2.2

Slavery and the Arrival of Black People in Nova Scotia

The first black people in Canada arrived here as slaves. The institution of slavery was formally acknowledged through numerous royal proclamations, beginning in Quebec in 1689, and continued well into the early 19th century. Black people were imported in large numbers into Nova Scotia primarily as a source of labour in the construction of the city of Halifax. The next major wave of black immigrants into Nova Scotia came after the American Revolution and the War of 1812.

Although many slaves fled from the United States to Canada via the *underground railroad*, between 1787 and 1800 they also fled from Canada into New England, where slavery had already been abolished. It was not until 1833 that slavery was abolished in Canada and the rest of the British Empire, and many now believe that this was more in the nature of an economic decision than a moral one, that is, related to the high monetary cost of maintaining slavery.

Black people, like many other ethnic groups who either voluntarily or involuntarily migrated to Canada, have made important contributions to our country. Why is their early presence in Canada downplayed at best or ignored at worst?

Source: T. Johnson, "The Canadian Black Population and Immigration," *Anthropos* 73(1978): 588–92; S.E. Williams, "Two Hundred Years in the Development of the Afro-Canadians in Nova Scotia, 1782–1982," in J.L. Elliott, ed., *Two Nations, Many Cultures: Ethnic Groups in Canada*, 2nd ed. (Scarborough: Prentice-Hall, 1983).

An Overview of Diversity in Canada

population, and it can play an important role in the spread of disease. For example, during the 19th century Canada suffered repeated epidemics of cholera, which were introduced by infected migrants.[4]

Migration can involve internal and external migrants. **Internal migrants** are those who move within a specific area. **External migrants** are those who move from outside a specific area. These areas can be defined at many different levels, such as a neighbourhood, province, or nation. For instance, in a sample from the 2001 Canadian census,[5] it was found that during the previous year 86.2 percent of migrants moved within the same province, 6.8 percent moved into a different province, and 7.0 percent immigrated from outside Canada (see Figure 2.8). These figures are different from those several decades ago. In 1976, 68.1 percent of migrants moved within the same province, 17.8 percent moved into a different province, and 14.0 percent immigrated from outside Canada.[6] It is important to note, however, that the rate of migration from outside Canada is under strict management with yearly quotas, whereas no such formal constraints exist on

Figure 2.8 Mobility in Canada, 2001

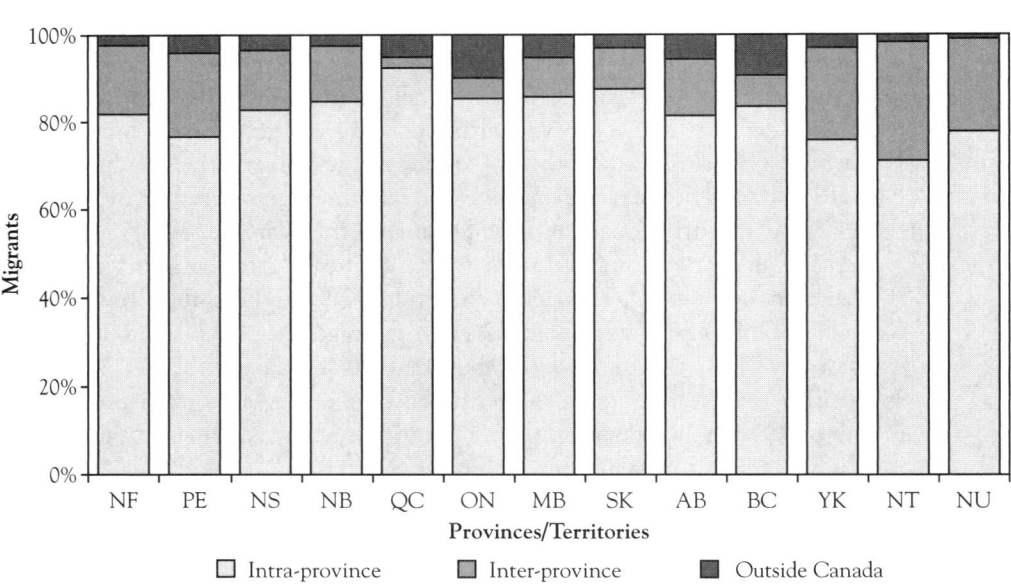

In most cases, the majority of migrants (individuals in this case who have migrated in the previous year) are moving within the same province. Migrants from outside Canada tend to congregate in provinces that have a more diverse economy, such as Quebec, Ontario, Manitoba, Alberta, and British Columbia.

Source: Adapted from Statistics Canada, "Mobility in Canada, 2001," <www12.statcan.ca/english/census01/products/highlights/Mobility/Index.cfm?Lang=E>, accessed 15 October 2005.

internal migration. Management of external migration can dictate the occupational, ethnic, age, gender, and political profiles of migrants. It can also severely limit their choices of an ultimate destination (see Critical Thinking Box 2.3). The majority of external migrants originate from Europe (42 percent), Asia (37 percent), Central and South America (6 percent), the Caribbean and Bermuda (5 percent), Africa (5 percent), and the United States (4 percent).[7]

Sources of Demographic Information

Numerous sources of information (data) are used in demography. All births, deaths, and marriages that occur in a population are usually recorded in vital registries. Two sources of registries are available: (1) **civil registries**, which are compiled by government, and (2) **ecclesiastical registries**, which are compiled by individual religious groups. The amount or detail of information recorded in these sources varies. For instance, a birth

---

CRITICAL THINKING BOX 2.3

Chinese Immigration to Western Canada

Chinese immigration to Western Canada began with the British Columbia gold rush of 1858. Between 1881 and 1885 large numbers of Chinese men were recruited to help build the Canadian Pacific Railway. Immigration of Chinese was limited demographically because the Canadian government was primarily interested in attracting migrant labourers, not settlers. As such, restrictive immigration policies created an unusual demographic profile among Chinese communities: they were predominantly male. With the growth of organized labour in Western Canada, numerous policies were introduced that were designed at first to discourage new migrants (e.g., British Columbia levied a $50 tax on each new Chinese immigrant), then to end the arrival of Chinese migrants. (The *Chinese Immigration Act* of 1923 was only repealed in 1947; between those years, no Chinese migrants could legally enter Canada.) These policies were in response to a perceived threat to organized labour of cheap Chinese labour.

Can we trust government policymakers to have enough insight into the effects of their policies concerning immigration, or will they create irrevocable damage to individuals, cultures, and societies?

Source: J.L. Elliott, "Canadian Immigration: A Historical Assessment," in J.L. Elliott, ed., *Two Nations, Many Cultures: Ethnic Groups in Canada* (Scarborough: Prentice-Hall, 1983); P.S. Li, "Chinese Immigrants in the Canadian Prairie, 1910–47," *Canadian Review of Sociology and Anthropology* 19 (1982): 527–40.

record usually includes the name of the child, the date of birth or baptism (depending on the source), the names of the parents, and the place of birth or baptism. Other information that may be included is the occupations of the parents, the place of birth, and religious affiliation. Death records typically record the name of the deceased, place of death, cause of death, and age at death. Other information sometimes available includes occupation and place of residence. Marriage records generally include the names of the bride and groom, the date of their marriage, their marital status, the place where they were married, their occupation, age, and religious affiliation. On occasion, their place of birth and the names of their parents are also recorded.

Censuses, compiled by government, are also an important source of information. A **census** is a list of all people who reside in a particular geographical area during a specific period. The first Canadian census after Confederation was taken in 1871, although censuses were recorded before Confederation (the earliest of these dates to 1666). The **decennial census**, that is, a census taken every 10th year, was first established while Canada was still under British rule.[8] Information usually recorded in a census varies considerably, both over time and across space, but it usually includes household or dwelling, name, age, sex, relationship to the head of the household (e.g., wife, son, niece, boarder), religious affiliation, and occupation. From census records we can examine household composition and formation. A **household** is a group of individuals who live together; this need not be a family, and it can include hired servants and boarders. When combined with vital registries, census records can be used to reconstruct individual families, that is, to create **genealogies**.

Other sources that can complement those already mentioned are cemetery data, voters' lists, military records, wills, and personal journals, to name but a few. The type of sources used will often depend on the topic of research and the availability of records. For instance, if you want to study fertility, you would need access to birth registries, either civil or ecclesiastical; if possible the use of both sources is recommended, as this provides a check on the accuracy of the data. However, both the government and ecclesiastical institutions can restrict access to **nominative records**, which list names. Therefore, you might not be able to reconstruct the fertility experience of specific women; instead, you would generalize about fertility on the basis of the number of women aged 15 to 44 in a given population and the number of births registered within a specific period of time.

Problems with Sources of Demographic Information

Numerous problems exist with using the mentioned sources. First, these sources were collected for reasons other than demographic research, and so they may contain hidden biases. **Crosschecking** the data—that is, using as many sources as possible—can minimize biases. For example, linking individuals from death records to both birth and census records will allow you to check the accuracy of the initial information, such as age, spelling of name, and so on. When linking various sources, there are often problems, many of which are related to the fact that the spelling of people's names and place names changes over time. Second, the quality of the data often varies because of errors in copying the initial information, level of literacy, method of initial data collection, and falsification of data (e.g., age is often misreported in censuses). Third, problems can also be related to changes in geographical boundaries, both ecclesiastical and civil. Therefore, a change in fertility might only be an artifact; that is, it might have been caused by the loss or gain in the number of communities included in a given parish or district.

Because of the variety of limitations associated with sources of demographic information, several techniques have been developed to assess them. In small populations with, say, fewer than one hundred **vital events**—births, marriages, and deaths—each year, a comparison of baptisms to marriages can be used. If this ratio is seven or eight to one, then it is possible that marriages were underregistered, because this ratio should be between four and five to one. The sex ratio of baptisms can also be used. If this ratio exceeds 110 over several years, then female baptisms are underregistered, because the ratio of males to females should be around 105 at birth. When examining large populations, those with hundreds of vital events (or more) recorded each year, you can use one of three methods to evaluate the quality of the data. The **documentary method** involves using the expertise of other researchers who have worked with those same sources—in other words, others may be able to attest to the accuracy and completeness of the records. The **expectation method** involves calculating expected proportions of vital events based on such factors as economic conditions, marriages, and migration. The **comparative method** entails calculating trends of vital events based on data from nearby populations. Overall, the more intimately researchers come to know their populations, the better they will be able to identify and assess any discrepancies.

Analyzing Demographic Information

The population is the focus of demographic research. Each **population** can be defined based on four aspects: (1) biological, involving shared genes; (2) ecological, involving shared environment; (3) social, involving a common cultural heritage; and (4) demographic, meaning shared time and space.[9] For instance, ethnicity is often used as a social factor defining a population. In 2001 the bulk of the Canadian population (57.1 percent) consisted of those of British, French, or European origin. The remainder identified their ethnic origin as Asian (7.2 percent), Aboriginal (3.2 percent), Caribbean (1.2 percent), Arab (0.8 percent), African (0.7 percent); Latin, Central, and South American (0.6 percent), and Other (29.2 percent—this includes those who identified themselves as North American)[10] (see Figure 2.9). Once the population has been identified, the data are

An Overview of Diversity in Canada

Figure 2.9 Ethnic Origins of Canada's Population, 2001

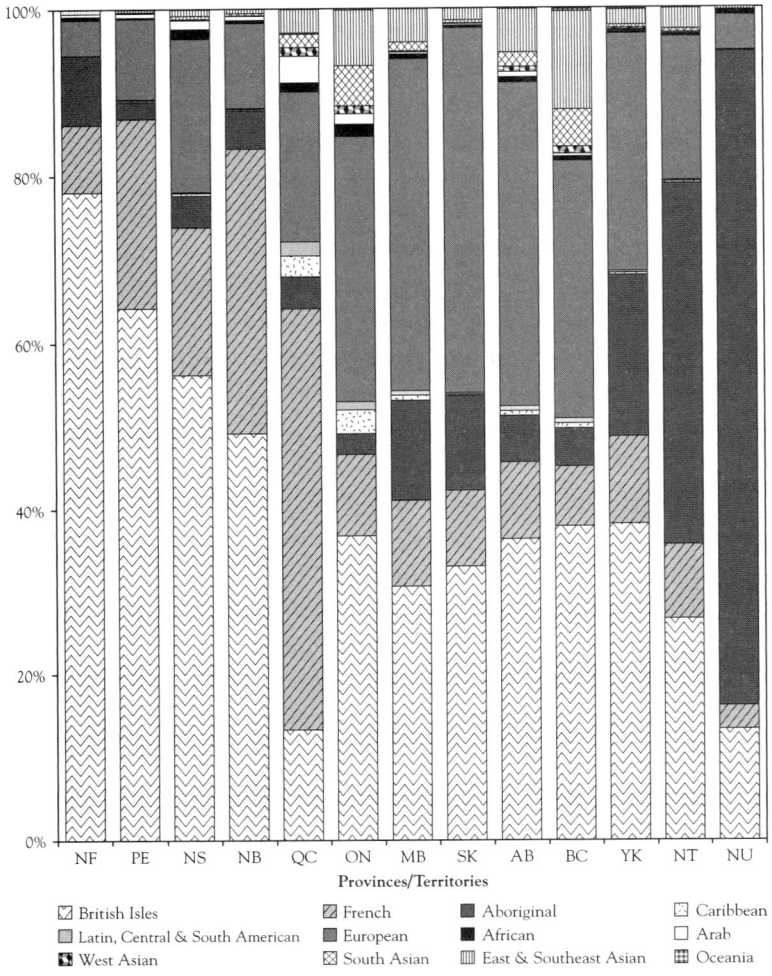

A great deal of variation exists across Canada regarding ethnic origin. A particular ethnic group dominates in Newfoundland, Prince Edward Island, Nova Scotia, New Brunswick, Quebec, and Nunavut. Those of British descent dominate in the east, ranging from 78 percent in Newfoundland to 49 percent in New Brunswick. In Quebec, individuals of French descent make up 51 percent of the population, while in Nunavut, 79 percent of the population are of Aboriginal descent. Due to the variation in its ethnic makeup, each province has a unique cultural mosaic. (Excluded are those who identified their ethnic origins as North American, which includes American, Canadian, Newfoundlander, Quebecois, and other provincial or regional groups.)

Sources: Adapted from Statistics Canada, "Ethnic Origin (232), Sex (3) and Single and Multiple Responses (3) for Population, for Canada, Provinces, Territories, Census Metropolitan Areas and Census Agglomerations, 2001 Census—20% Sample Data," Catalogue No. 97F0010XCB2001001, 21 January 2003, available <www.statcan.ca/bsolc/english/bsolc?catno=97F0010X&CHROPG=1>, accessed 15 October 2005; also <www12.statcan.ca/english/census01/home/Index.cfm>.

collected, which often involves thousands of entries (births, marriages, deaths, and census entries). Therefore, before analysis can proceed, the data must be reduced to make them more manageable. This usually involves **aggregating** the data, that is, summarizing observations either over specific geographical areas, time periods, or both. For instance, obtaining an average age at marriage for each decade over the study period reduces the data and makes comparisons easier.

In analyzing the data, two approaches are available: longitudinal and cross-sectional. **Longitudinal analysis** involves following a birth or marriage cohort through time. A *cohort* is a group of people who share the timing of a vital event—for example, all individuals in a population who were born in 1890. **Cross-sectional analysis** involves dividing data into nonoverlapping age categories to make demographic inferences about segments of the population who are at different stages of life (e.g., child versus adult). For instance, to examine fertility patterns using the longitudinal approach, you would take all of the population born in a specific period and follow them from birth to death; using the cross-sectional approach, you would group your population into different age groups, using these different groups to represent the patterns experienced over a lifetime. Of the two, the cross-sectional approach is the quickest and easiest, because it is not necessary to collect demographic information (e.g., births) over a long period; however, the longitudinal approach is potentially the most accurate, because you can examine the cumulative effects of social and environmental factors on the same group of people over time.

Biological Inferences

One of the goals of demographic research is to make inferences about the genetic structure of a population. This is achieved by either looking at marriages or looking at births and deaths.

Choosing a Marriage Partner

Marriage, which infers reproduction and, hence, the passing on of genes, can be either random or nonrandom. In most populations, humans included, the **selection of a mate** is **nonrandom**—that is, a distinct preference in the choice of a mate is evident. This preference can be related to physical or socioeconomic characteristics, such as physical attractiveness, age, religious affiliation, socioeconomic status (social class), and ethnicity. Individuals tend to choose mates who share the same sociocultural background as they do. This preference may be related to a conscious decision not to marry outside of one's group (e.g., Jews and Catholics), or it may simply be the byproduct of geographical and social segregation of subgroups in a population. In other words, individuals belonging to a particular economic and ethnic group often live in the same general area and interact socially more within their group than outside it; therefore, the potential mates each is exposed to are in a sense "preselected" and have the same social background.

The task of the demographer is to identify these preferences in a population. Several methods are used to assess the frequency of nonrandom mating: genealogical analysis, isonomy, and kinship coefficients. **Genealogical analysis** requires years of effort, depending on the size of the population in question, because it involves reconstructing individual families. **Isonomy** is a faster method that involves looking at the frequency of marriages occurring between individuals who share a surname (e.g., John Martin marrying Susan Martin). Of course, this method involves several assumptions, not the least of which is that similar surnames have the same origin (e.g., all Martins ultimately descended from one family of Martins). **Kinship coefficients** are similar to isonomy in that surnames are also used; they differ in that they do not focus on marriages but simply look at the proportion of surnames in a population, calculating the probability that the individuals are related. Each method attempts to identify the existence of preferences in the exchange of mates between families; the demographer working with kinship coefficients is trying to determine whether individuals in certain families choose their marriage partners from specific families or whether they marry someone irrespective of that individual's family.

Evolution by Natural Selection

The disappearance of the dinosaurs and the appearance of anatomically modern humans are generally what come to mind when people think of evolutionary change. But **evolution** can also occur on a micro level that involves small changes in the gene frequency of a population. **Natural selection** refers to the preferential survival and reproduction of individuals in a population by virtue of possessing a genetic characteristic that gives them an advantage. For instance, if a certain blood type protected an individual from a deadly disease, those who shared that blood type would be more likely to survive than others and have a higher probability of successfully reproducing. The result would be that future generations would increasingly possess the advantageous blood type. The net result, when comparing generations, would be a change in gene frequency (e.g., an increase in individuals with blood type A and a decrease of those with blood type O—which has been tentatively associated with exposure to the bubonic plague).[11] Such a change can be referred to as "evolution." When this change results from the genetic adaptation to an environmental stress, we term it *natural selection*.

Natural selection is not easily detected in human populations.[12] The main reason for this is the length of the human generation, approximately 25 years. Therefore, we would need to study many generations in detail before we could detect natural selection. This is not possible, because researchers would not live long enough to complete their research, and it is unlikely that funding could be obtained for a project of such a long duration. The use of historical populations is also limited in this regard, because we are not able to obtain genetic samples (e.g., blood samples). However, a method has been developed that measures the potential for natural selection in human populations. This method is called **Crow's Index of Selection**. Crow's Index takes into account both deaths and births—mortality and fertility—and focuses on whether differences occur in subgroups of the population; subgroups can be defined using a variety of social factors, such as ethnicity, religious affiliation, and socioeconomic status. If differences do occur, the potential for natural selection exists, that is, a change in the frequency of a gene or genes that results because a segment of the population became genetically adapted to an environmental stress.[13]

HISTORICAL DEMOGRAPHIC ANALYSIS OF CANADIAN POPULATIONS

Common sources of demographic information include written documents, such as birth, death, and marriage registries. When focusing on historical sources in Canada, the time depth is limited to the arrival of people from Europe, Asia, and Africa. At the very best, some information is available as early as the 15th century. Early sources are geographically limited to what was referred to as New France and to several British colonies, such as New England and Rupert's Land; some of these areas were later known as Upper and Lower Canada. What follows is a survey of the types of demographic research on Canadian populations. Although it is excluded from this chapter because of space constraints, please note that a subfield of demography, **paleodemography**, focuses on nonwritten sources, such as skeletal remains, to study patterns of birth, death, and migration in prehistoric populations.

Regional Survey

Most demographic research in Canada has focused on Quebec and Ontario, with relatively less work done on other provinces and territories. This focus on central Canada is partly because of the availability of records; the majority of immigrants arrived and lived in central Canada, and thus this area has the longest time frame of demographic sources.

An exhaustive and complete survey of historical demographic studies of Canadian populations is not possible here. Instead, the intention is to introduce the topic of demography. It is hoped that interested readers will pursue specific areas in more depth (see the Selected Bibliography at the end of this book for names of journals that regularly feature a broad range of articles on Canadian historical demography). What follows is a brief examination of historical demographic studies of Canadian populations.

The study of migration in historical Canada is one of the most common types of research, which is easily explained by the fact that, except for North American Aboriginals (see Critical Thinking Box 2.4), Canada was settled by migrants. Such studies often focus on the contributions of particular ethnic groups, such as the Arabs,[14] Irish,[15] Ukrainians,[16] Italians,[17] and Armenians,[18] to name but a few. Others take a more general approach, focusing on trends over a wider area and including several ethnic groups.[19] Considerably less attention has been given to emigration from

---

CRITICAL THINKING BOX 2.4

The First People in North America

The first people to arrive in North America were the ancestors of today's Aboriginal population, who crossed from the Old World via a land bridge that linked northeastern Siberia to Alaska. This land bridge, called **Beringia**, was last in existence between 27 000 and 11 000 years ago. The Aboriginals are thought to have arrived in North America sometime between 20 000 and 15 000 years ago; we have evidence from archaeology that humans had reached central Mexico about 11 000 years ago. It seems likely that people crossed the land bridge in pursuit of migrating herds of animals, such as mammoths, caribou, and bison.

The archaeological record indicates that the North American Aboriginals had diversified into a variety of cultural groups, some of which attained complex levels of sociopolitical organization, that is, tribes and chiefdoms. However, the arrival of Europeans disrupted Aboriginal cultures, in some instances irrevocably. Today, many descendants of these early migrants are in the process of resurrecting their rich cultural heritage.

Because we are all immigrants to this country, and if length of residence has any meaning, then why are the land claims of North American Aboriginals not given precedence over our own?

Source: B.M. Fagan, *People of the Earth*, 7th ed. (New York: Harper Collins, 1992).

Canada.[20] Research on the growth of urbanism and urban centres in Canada is also of interest, and it often relates directly to patterns of immigration.[21]

Research on fertility and nuptiality examines such issues as birth control and abortion,[22] illegitimacy,[23] and changes in fertility patterns over time and across space.[24] In many cases, fertility is linked to a variety of economic factors. In some instances, marriage and fertility patterns are described for specific ethnic groups, which is useful in identifying sociocultural differences affecting these and other demographic variables.[25] A related focus of Canadian research is family and household composition, both of which are strongly influenced by local economic patterns.[26]

Studies of historical patterns of mortality often focus on epidemics, such as cholera,[27] smallpox,[28] typhoid,[29] and influenza.[30] This type of research not only examines the mortality associated with these diseases, but also looks at the sociocultural effects they caused. Other methods examine the more general patterns of mortality.[31] Again the importance lies in the comparisons with other populations to identify the role of human behaviour in the spread of disease, for example, the role of food preparation techniques, method of water storage, or crowding.

These studies illustrate the diverse nature of the Canadian population. Each geographical region has experienced a unique demographic history, which, in turn, has affected local, social, and economic conditions; it is important to note that this interaction continues to this day. We are dealing with an inherently dynamic system, and we can expect that the legacy created by past demographic behaviour will continue to affect future populations. By understanding this phenomenon we will be better equipped to identify future trends and to plan for potential problems, such as the stress on local economies when 50 percent of our population consists of people over the age of 64.

CONCLUSION

Demographic research can be useful in demonstrating the importance of social and economic factors on human behaviour. By looking at fertility, nuptiality, mortality, and migration, the effects of numerous variables, such as resource exploitation, patterns of occupation, and social class, can be uncovered. This information can then be combined with other sources, not only to build a profile of life in the past, but also to help us understand the dynamic nature of human populations today. Understanding the interrelationship between the biological and social aspects of human groups can have important implications for a variety of policies directed at living populations, ranging from the effective delivery of community health care to the development and introduction of population control measures. Such information can also contribute to our understanding of the interrelationship between human populations and their physical environments. Knowledge of complex systems may enable us to more effectively deal with, and perhaps avoid, a variety of crises—most of which are created by humans—such as the emergence and spread of new and virulent diseases.

CHAPTER SUMMARY

This chapter introduces the major concepts used in demography. Sources of information on populations vary and include records compiled by government, such as birth registries, ecclesiastical institutions, such as church marriage registries, and individual people, such as wills, journals, or diaries. However, because no source of demographic information is perfect, care should be taken to identify biases in the information. The methods used by demographers are sometimes dictated by the nature of the data. For instance, if researchers do not have access to data over a long period, they are necessarily limited to performing cross-sectional analyses.

Some sources and methods used in demographic analysis are examined by focusing on four key areas: fertility, nuptiality, mortality, and migration. Fertility and mortality affect the natural growth or decline in a population's size; the ratio between the two determines whether a population increases, decreases, or remains stable. Nuptiality, or marriage, influences fertility in several ways. For instance, the timing of marriage can define the length of the female reproductive period, and variation in the proportion of the population who marry, or do not marry, can affect fertility by limiting or increasing the number of couples who are in a position to produce children—assuming that reproduction outside of marriage is low. Migration is one of the most important factors affecting the growth or decline of a population; if more individuals enter an area than leave it, the size of the population will increase. The reverse is also true; if more individuals leave an area than enter it, the size of the population will decline. Each of these four variables (fertility, nuptiality, mortality, and migration) varies within and among populations. Over time they define the unique demographic nature of a population, and they are ultimately responsible for generating the degree of diversity currently observed, regardless of how that diversity is defined (e.g., culturally, politically, or economically).

This chapter briefly surveys the types of demographic research on historical Canadian populations. These include studies of immigration and emigration, birth control, illegitimacy, households, and epidemics. In most cases the aim of such research is not only to describe events in the past, but also to link them to social, economic, and (when possible) biological factors. The ultimate aim of such research is to understand the factors that can affect a population's demographic profile. This information may then be applied to current issues and problems, such as overpopulation and the spread of deadly viruses.

KEY TERMS

age at first marriage, p. 41
age-specific marriage rate, p. 41
age-specific mortality rates, p. 43
aggregating, p. 53
Beringia, p. 56
celibacy, p. 41
census, p. 50
civil registries, p. 49
comparative method, p. 51
crosschecking, p. 50
cross-sectional analysis, p. 53
Crow's Index of Selection, p. 55
crude birth rate, p. 40
crude death rate, p. 43
crude marriage rate, p. 41
decennial census, p. 50
demography, p. 39
dependency ratio, p. 46
documentary method, p. 51
ecclesiastical registries, p. 49
emigration, p. 46
emigration rate, p. 47
evolution, p. 54
expectation method, p. 51
external migrants, p. 48
fecundity, p. 39

fertility, p. 39
fertility rate, p. 40
genealogical analysis, p. 54
genealogies, p. 50
gross migration rate, p. 47
household, p. 50
immigration, p. 46
immigration rate, p. 47
infant mortality rate, p. 43
internal migrants, p. 48
isonomy, p. 54
kinship coefficients, p. 54
life expectancy, p. 42
life span, p. 42
live births, p. 39
longitudinal analysis, p. 53
migration, p. 46
morbidity, p. 45
mortality, p. 42
natural selection, p. 54
net migration rate, p. 47
nominative records, p. 50
nonrandom mate selection, p. 53
nuptiality, p. 40
paleodemography, p. 55
population, p. 51
variable, p. 39
vital events, p. 51

DISCUSSION QUESTIONS

1. If you were going to reconstruct your own family history (genealogy), what types of information would you need? Which sources would provide you with this information? What potential problems might arise the further back in time you proceeded?
2. What factors might affect the timing of marriage today (age at marriage)? Would these factors be similar to or different from those affecting marriage in the past? What effect could these factors have on fertility?
3. To meet the needs of a changing economy (new technology), our population needs to become more educated. What effects might increased educational attainment, that is, spending more time in school, have on age at marriage and fertility?
4. Until the 20th century, the growth of the Canadian population was not the result of natural growth (births exceeding deaths) but primarily the result of immigration. Which factors do you think might have attracted these early migrants? Which factors might be attracting immigrants today?
5. Compare and contrast the mortality experience of the Canadian population over a 100-year period. Obtain the information you need from Statistics Canada. Look at the different categories of causes of death (e.g., infectious disease, degenerative disease). Are there any differences between the sexes regarding the mortality experience? Are there any differences in the mortality experience across the provinces? If so, what might account for these differences? Has the pattern of mortality changed over time? If so, explain how (and perhaps why) it has changed.

NOTES

1. K.H. Connell, *The Population of Ireland 1750–1845* (Oxford: Clarendon Press, 1950), p. 47.
2. Statistics Canada, CANSIM, Tables 102-0018 and 102-0019, available <http://cansim2.statcan.ca/cgi-win/cnsmcgi.exe?CANSIMFile=CII/CII_1_E.HTM&RootDir=CII/&LANG=E>.
3. Statistics Canada, *2001 Community Profiles*, Catalogue No. 93F0053XIE, available <www12.statcan.ca/english/profil01/cp01/index.cfm>.
4. G. Bilson, *A Darkened House: Cholera in Nineteenth-Century Canada* (Toronto: University of Toronto Press, 1980), p. 23.

5. Statistics Canada, "2001 Census of Canada," available <www12.statcan.ca/english/census01/home/Index.cfm>.
6. Statistics Canada, *Population: Demographic Characteristics—Mobility Status* (Ottawa: Supply and Services Canada, 1978), p. 1.
7. Adapted from Statistics Canada, "Selected Places of Birth (85) for the Immigrant Population, for Canada, Provinces, Territories, Census Metropolitan Areas and Census Agglomerations, 1996 and 2001 Censuses—20% Sample Data," Catalogue No. 97F0009XCB2001003, 21 January 2003, available <www.statcan.ca/bsolc/english/bsolc?catno=97F0009X&CHROPG=1>.
8. Statistics Canada, *2001 Census Handbook* (Ottawa: Minister of Industry, 2003).
9. G.A. Harrison and A.J. Boyce, "Introduction: The Framework of Population Studies," in G.A. Harrison and A.J. Boyce, eds., *The Structure of Human Populations* (Oxford: Clarendon Press, 1972), pp. 3–4.
10. Statistics Canada, "2001 Census of Canada," available <www12.statcan.ca/english/census01/home/Index.cfm>. These figures are based on those who listed single and multiple responses.
11. A.K. Roychoudhury and M. Nei, *Human Polymorphic Genes: World Distribution* (Oxford: Oxford University Press, 1988).
12. T. Dobzhansky, "Natural Selection in Mankind," in G.A. Harrison and A.J. Boyce, eds., *The Structure of Human Populations* (Oxford: Clarendon Press, 1972), p. 232.
13. J.F. Crow, "Some Possibilities for Measuring Selection Intensities in Man," *Human Biology* 30 (1958): 1.
14. B. Abu-Laban, "Arab Immigration to Canada," in J.L. Elliott, ed., *Two Nations, Many Cultures: Ethnic Groups in Canada* (Scarborough: Prentice-Hall, 1979), p. 372.
15. A.G. Brunger, "Geographical Propinquity Among Pre-famine Catholic Irish Settlers in Upper Canada," *Journal of Historical Geography* 8(1982): 265.
16. V.J. Kaye and C.W. Hobart, "Origins and Characteristics of the Ukrainian Migration to Canada," in C.W. Hobart W.E. Kalbach, J.T. Borhek, and A.P. Jacoby, eds., *Persistence and Change: A Study of Ukrainians in Alberta* (Toronto: Ukrainian Canadian Research Foundation, 1978), p. 25.
17. R.F. Harney, "Men Without Women: Italian Migrants in Canada 1885–1930," *Canadian Ethnic Studies* 11 (1979): 29.
18. I. Kaprielian, "Immigration and Settlement of Armenians in Southern Ontario: The First Wave," *Polyphony* 4 (1982): 14.
19. See for example J.L. Elliott, "Canadian Immigration: A Historical Assessment," in J.L. Elliott, ed., *Two Nations, Many Cultures: Ethnic Groups in Canada* (Scarborough: Prentice-Hall, 1979), p. 289; W. Parker, "The Canadas," in A. Lemon and N. Pollock, eds., *Studies in Overseas Settlement and Population* (New York: Longman, 1980), p. 267; J.C. Weaver, "Hamilton and the Immigration Tide," *Families* 20 (1981): 197; G. Wynn, "Ethnic Migrations and Atlantic Canada: Geographical Perspectives," *Canadian Ethnic Studies* 18 (1986): 1.

20. See for example A.A. Brookes, "The Golden Age and the Exodus: The Case of Canning, Kings County," *Acadiensis* 11 (1981): 57; R. Crawley, "Off to Sydney: Newfoundlanders Emigrate to Industrial Cape Breton 1890–1914," *Acadiensis* 17 (1988): 27; Y. Lavoie, *L'Émigration des Québécois aux États-Unis de 1840 à 1930* (Quebec: Editeur officiel du Québec, 1979).
21. See for example C.M. Gaffield, "Boom and Bust: The Demography and Economy of the Lower Ottawa Valley in the Nineteenth Century," *Canadian Historical Association. Historical Papers* (1982): 172; M.B. Katz, M.J. Doucet, and M.J. Stern, "Population Persistence and Early Industrialization in a Canadian City: Hamilton, Ontario, 1851–1971," *Social Science History* 2 (1978): 208; P. Matwijiw, "Ethnicity and Urban Residence: Winnipeg, 1941–1971," *Canadian Geographer* 23 (1979): 45; Weaver, "Hamilton and the Immigration Tide," p. 197.
22. P. Gossage, "Absorbing Junior: The Use of Patent Medicines as Abortifacients in Nineteenth-Century Montreal," *The Register* 3 (1978): 1; A. McLaren, "Birth Control and Abortion in Canada, 1870–1920," *Canadian Historical Review* 59 (1978): 319–40.
23. R.D. Sharna, "Premarital and Ex-nuptial Fertility (Illegitimacy) in Canada 1921–1972," *Canadian Studies in Population* 9 (1982): 1.
24. H. Charbonneau, "Jeunes femmes et vieux maris: la fécondité des mariages précoces," *Population* 35 (1980): 1101; E.M.T. Gee, "Early Canadian Fertility Transition: A Components Analysis of Census Data," *Canadian Studies in Population* 6 (1979): 23; E. Roth, "Historic Fertility Differentials in a Northern Athapaskan Community," *Culture* 2 (1982): 63; J.E. Veevers, "Age Discrepant Marriages: Cross-National Comparisons of Canadian–American Trends," *Social Biology* 31 (1984): 118.
25. See for example J. Keyes, "Marriage Patterns Among Early Quakers," *Nova Scotia Historical Quarterly* 8 (1978): 299.
26. A.G. Darroch and M.D. Ornstein, "Family and Household in Nineteenth-Century Canada: Regional Patterns and Regional Economies," *Journal of Family History* 9 (1984): 158; F.K. Donnelly, "Occupational and Household Structures of a New Brunswick Fishing Settlement: Campobello Island, 1851," in R. Chanteloup, ed., *Labour in Atlantic Canada* (Saint John: University of New Brunswick, 1981), p. 55; P.S. Li, "Immigration Laws and Family Patterns: Some Demographic Changes Among Chinese Families in Canada, 1885–1971," *Canadian Ethnic Studies* 13 (1980): 58; S. Medjuck, "The Social Consequences of Economic Cycles on Nineteenth-Century Households and Family Life," *Social Indicators Research* 18 (1986): 233; D.A. Norris, "Household and Transiency in a Loyalist Township: The People of Adolphustown, 1784–1822," *Social History* 13 (1980): 399.
27. Bilson, *A Darkened House*, p. 23.
28. J.R. Gibson, "Smallpox on the Northwest Coast, 1835–1838," *BC Studies* 56 (1982): 61.
29. S. Lloyd, "The Ottawa Typhoid Epidemics of 1911 and 1912: A Case Study of Disease as a Catalyst for Urban Reform," *Urban History Review* 8 (1979): 66.
30. J.D.P. McGinnis, "The Impact of Epidemic Influenza: Canada, 1918–1919," *The Canadian Historical Association, Historical Papers* (1977): 121.

31. R. Bourbeau and J. Légaré, *Évolution de la mortalité au Canada et au Québec, 1831–1931: Essai de mésure par génération* (Montreal: Les Presses de l'Université de Montréal, 1982); H. Charbonneau and A. LaRose, eds., *The Great Mortalities: Methodological Studies of Demographic Crises in the Past* (Liège: Ordina Éditions, 1979); Robert W. Fogel, Stanley L. Engerman, James Trussell, Roderick Floud, and Clayne L. Pope, "The Economics of Mortality in North America, 1650–1910: A Description of a Research Project," *Historical Methods* 11 (1979): 75; Y. Landry, "Mortalité, nuptialité et canadianisation des troupes française de la guerre de Sept Ans," *Social History* 12 (1979): 298; B. Osborne, "The Cemeteries of the Midland District of Upper Canada: A Note on Mortality in a Frontier Society," *Pioneer America* 6 (1974): 46.

PART II
THE MANY FACES OF DIVERSITY

PART II

Part II takes a micro approach to diversity in Canada. Chapters 3 to 9 each examine a particular kind of diversity. The topics range from "traditional" diversities such as race, ethnicity, social inequality, and gender, to less traditional ones, including sexual orientation, the family, and disabilities.

Continuing with our house analogy, if Part I is the structure of the house, then Part II is the details of the interior. The text moves from the general, in Part I, to the specific, in Part II, as reflected in the title of Part II, "The Many Faces of Diversity."

Chapter 3 looks at social inequality in Canada. It introduces the reader to the inequalities of income and wealth and to the social structure in Canada. It reviews the major theories of inequality and stratification and examines their strengths and weaknesses.

Chapter 4 explores race and ethnicity in Canada. It outlines the major determinants of personal and group identity, the factors that influence societal interaction, and the problems of prejudice, discrimination, and racism. It also briefly outlines the history of immigration to Canada and the meaning of "race."

Chapter 5 looks at Aboriginal peoples. It vividly illustrates the heterogeneous nature of Aboriginal peoples, cultures, and languages. It also examines the deep historical presence of Aboriginal peoples in Canada, allowing you to better understand contemporary native issues including the legal circumstances of Aboriginal peoples today.

Chapter 6 deals with the religious experience in Canada. It is divided into two principal sections. The first discusses issues involved in forming a working definition of religion; the second deals with several aspects of the history of religion in Canada.

Chapter 7 deals with disability. After briefly looking at the history of disability in Canada, it explores the roles of several institutions in the lives of disabled people, including the government, religious organizations, schools, and the family. The chapter ends with a look at important social policy issues for disabled Canadians.

Chapter 8 examines the issue of gender. It defines the reality of gender identity and makes it clear that men and women occupy different spheres in society. Finally, it addresses the question of whether gender divisions are positive or negative, both for society and for the individual.

Chapter 9 looks at sexuality. It focuses primarily on the evolution of and variation in human sexuality. In dealing with variation, it points out that sexual diversity involves more than homosexuality. The chapter also provides some historical background, which allows you to assess recent studies.

Chapter 10 analyzes diversity in Canadian families. It challenges the notion of the traditional family, redefining the word *family* to help explain the different types of families that exist in Canada today.

CHAPTER 3

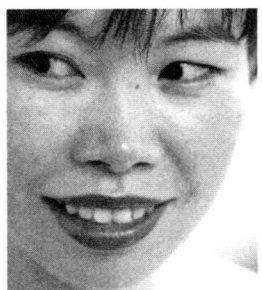

Social Inequality and Stratification in Canada

Eddie Grattan

Men are by nature unequal. It is vain, therefore, to treat them as if they were equal.
— James Anthony Froude, "Party Politics"

We hold these truths to be self-evident, that all men are created equal.
— Thomas Jefferson

Objectives

After reading this chapter, you should be able to

- describe the degree of social inequality in Canada
- present a summary of the major theories of social inequality
- detail the impact of social inequality on our daily lives
- list the variety of forms of social inequality
- describe the social class system of Canada

INTRODUCTION

All societies are characterized by social inequality and **stratification**. **Social inequality** is the varying degree to which different people have access to and control over valued resources, such as money, wealth, status, and power. In Canada, access and control are severely restricted by a person's social background, sex, and race or ethnicity. The differing degrees of access to and control over valued resources serve to divide Canadian society into recognizably distinct and unequal groups, or strata. Canada is a stratified society.

Let me provide an example from a personal experience. During the 2002 National Hockey League playoffs, I attended a hockey game at the Air Canada Centre. Although the game was quite exciting, having to sit in the so-called cheap seats (at $60 each!) often made the game difficult to follow. Spectators in the most expensive seats (the "platinums"), however, appeared to experience the event more intensely and personally: they could hear the players shouting to one another and see more clearly the players' expressions, as well as the hits, saves, and goals.

The difference between my experience and that of the people in the more expensive seats may appear to be an issue of little significance. Nevertheless, if I have a relatively low income, it is unlikely I will be able to afford a "gold" seat; if, however, my income is relatively high, it is likely I will be able to purchase a more expensive ticket and thus increase my enjoyment of the game.

It is evident that these differing (or unequal) experiences reflect wider patterns of inequality (in this case, income differences) in society.

BOX 3.1

Who Can Afford Tickets for Sports Events?

Over the past several years, ticket prices for major sporting events have increased dramatically. Tickets for the 2005 Grey Cup in Vancouver started at $80, increasing to as much as $240. Regular season tickets for the Toronto Blue Jays range from $9 to $59, and the Toronto Argonauts of the CFL offer tickets ranging from $15 to $60. The availability of many of these tickets, however, is limited, as the majority of seats in most arenas and stadiums belong to season-ticket holders, many of whom renew their seats annually.

The Many Faces of Diversity

CRITICAL THINKING BOX 3.1

How does the amount of income affect the types of sport a person will play?

The ability to pay is, of course, dependent largely on income level. The amount of income a family earns determines many aspects of its existence, including the types of food it consumes, where (or, indeed, if) to travel on vacation, where, and for how long, the children will attend an educational institution, as well as many of the family's values and beliefs.

In some instances, the ability (or inability) to pay literally can determine whether one lives or dies. This reality is illustrated by the sinking of the ocean liner *Titanic* in 1912, when about 1500 of the 2300 passengers lost their lives. Passengers in the more expensive, upper-deck cabins had a greater chance of survival than those in the cheaper, lower-deck cabins:

> Of those holding first-class tickets, more than 60 percent were saved, primarily because they were on the upper decks, where warnings were sounded first and lifeboats were accessible. Only 36 percent of the second-class passengers survived, and of the third-class passengers on the lower-decks, only 24 percent escaped drowning. On board the *Titanic* class turned out to mean much more than the quality of accommodations: it was truly a matter of life or death.[1]

In Canada, social inequality is strongly associated with a person's social background, race, ethnicity, and sex. Canadians born into families with low incomes will generally receive less education and, thus, find it difficult to obtain high-income positions. Immigrants to Canada, particularly those with little education, generally face greater obstacles to finding employment. Women, although sharing similar levels of education as men, on average earn significantly less than men do.

INEQUALITIES OF INCOME AND WEALTH IN CANADA

Inequality of Income

In everyday conversation the terms "income" and "wealth" are used interchangeably; however, although closely related, an important distinction exists between them. **Income** is the *flow* of money received over a specified period, usually a year. Ms. Smith, for example, when asked her income, will most likely reply X dollars per year. The largest part of income for most Canadians is received in the form of wages and salary. A smaller part derives from government financial assistance, such as unemployment and welfare benefits—although for some people, these constitute a significant portion of total income.

Other, less identifiable, forms of income are also important. These include gifts, money received from cashing in an insurance policy, and capital gains (e.g., money received from selling shares of stock at a price higher than their initial cost). It is often difficult to calculate precisely how much these contribute to the average income of Canadians, as the necessary information is difficult to obtain.

Canada is a wealthy country. In terms of per capita income, it ranks among the top in the world. In 2002, the average total income of Canadian families was $73 200 (see Table 3.1).

Average income, however, disguises significant income differences among families. An informative way of looking at these differences is to split all Canadian families into five equal groups, each referred to as a *quintile*. The groups are assembled in the following way. Imagine that all families are placed in a line, a family's place being determined by its income level. The poorest family is placed at the front of the line, followed by the next poorest, and so on, until the last family, with the highest income, is placed at the end. Next, the line is split into five equal groups. The first group, or quintile, is composed of the first 20 percent of the line. Obviously, this group will consist of the poorest people.

Table 3.1 Average Total and After-Tax Income of Canadian Families, by Income Quintile, 1997– 2002 (2002 dollars)

	1997	1998	1999	2000	2001	2002
Average Total Income						
	65 600	68 300	69 100	71 600	73 400	73 200
Average After-Tax Income						
Lowest quintile	19 100	19 800	20 600	20 800	22 400	22 300
Second quintile	33 900	35 000	36 300	37 000	38 600	39 000
Middle quintile	47 000	48 400	49 600	50 800	53 100	53 600
Fourth quintile	62 200	64 300	66 000	67 700	70 300	71 200
Highest quintile	101 000	105 500	106 400	111 500	117 300	116 400

Source: Adapted from Statistics Canada, "Social Indicators," *Canadian Social Trends* 76 (Spring 2005): 35, Catalogue 11-008.

The next group consists of the next 20 percent. A similar process occurs in selecting the third, fourth, and fifth groups. The fifth group, of course, comprises those families with the highest incomes.

Table 3.1 reveals a significant difference between the average after-tax income of the lowest and highest quintile. In 2002, the average after-tax income of the lowest quintile was $22 300, less than one-fifth that of the highest quintile, at $116 400. This difference has changed little over the period 1997 to 2002.

The amount of income an individual or family receives does not indicate the source of that income. Table 3.2 shows that people in the first, or lowest, quintile, on average, receive around a half of their income from government transfer payments, such as welfare benefits, children's benefits, and employment insurance. Single-parent families, as well as recently arrived immigrants, make up a large proportion of these families.

As income levels increase from the lowest to the highest quintiles, the proportion of income received from transfer payments decreases significantly. The highest family quintile, for example, consisting of those families with the highest incomes in Canada, received only 2.9 percent of its income from transfer payments in 2003.

A more revealing look at income inequality is provided by Table 3.3, which indicates that income inequality has not changed significantly over the past two decades. Of the total income received by households in Canada in 2000, the lowest quintile received only 5.2 percent. If Canada were an equal society, this quintile, composed of 20 percent of the population, would receive 20 percent of the total income. So, with only 5 percent of total income, it obtains only a small percentage of what it would receive in a perfectly equal

Table 3.2 Government Transfers and Income Tax, 2002 and 2003, as Percentage of Income by Income Quintile for Families and Unattached Individuals (dollar constant 2003)

	Families		Unattached Individuals	
	2002	2003	2002	2003
Lowest quintile	50.7	49.0	58.8	53.6
Second quintile	22.9	23.8	61.0	59.5
Third quintile	12.6	12.6	31.2	32.1
Fourth quintile	6.9	6.7	12.8	11.5
Highest quintile	2.8	2.9	4.0	3.6

Source: Adapted from Statistics Canada, CANSIM, Table 202-0301 and Catalogue No. 75-202-XIE.

Table 3.3 Distribution of Total Income in Canada by Quintile for All Households (in percentage*)

Sources	Lowest Quintile	Second Quintile	Middle Quintile	Fourth Quintile	Highest Income Quintile
1987	5.1	10.9	17.3	24.5	42.2
1992	5.0	10.8	17.1	24.7	42.4
1996	5.0	10.4	16.7	24.6	43.4
2000	5.2	11.3	16.7	23.3	43.6

*Percentages do not all total 100 percent because of rounding.

Source: Adapted from Statistics Canada, "Household Facilities by Income and Other Characteristics," Catalogue No. 13-218, March 1998; and Statistics Canada, "Average Income and Share of Census Families' Income by Income Deciles, Canada, 2000," *Income of Canadian Families*, Catalogue No. 96F0030XIE2001014.

society. The fourth and fifth quintiles, consisting collectively of the top 40 percent of income-earners in Canada, received a combined 66.9 percent of total income. This contrasts with 16.5 percent for the bottom 40 percent of income-earners.

The past 15 years have seen a widening of the disparity in income levels. Most markedly, the percentage of low-income-earners increased from 1989 to 1997, rising from 10 to 13.3 percent.[2] A major reason for this shift is the reduction in government transfers through the 1990s. The marked drop in social assistance recipients in the 1990s[3] stemmed from the implementation of stricter eligibility rules and reductions in benefit levels adopted by a number of conservative governments. In Ontario, for example, the social assistance rate fell from 12.1 percent in 1993 to 5.5 percent in 2003.[4] Such reductions obviously affect those at the lower end of the income scale much more severely than those at the top. This contrasts with the period of the early 1980s and mid-1990s, when income levels in Canada remained virtually stationary. Before this, from about 1920 to 1980, incomes in Canada quadrupled, with Canadians in 1980 being able to purchase four times as many goods and services as in 1920. The change since 1980 has forced many Canadians, particularly those entering the labour market, to adjust their expectations with the awareness that their level of income will be substantially below that of their parents and grandparents.

It is also unclear what the long-term impact will be of the terrorist attacks on the United States in September 2001. These attacks led to a significant downturn in the North American economy. In response, interest rates in both the United States and Canada were lowered swiftly in an attempt to boost economic performance. The North

American economy appears to be starting to rebound, if somewhat unevenly, from this downturn. However, concerns over oil supplies—and the consequent steep jump in fuel prices—as well as the increased participation of many Asian countries in the global economy are making it difficult to determine what kind of effect these developments will have on Canadian society—and social inequality—in the near future.

Inequality of Wealth

Wealth is the accumulation of assets, such as a house, car, savings, cottage, land, jewellery, and art objects. As already mentioned, there is a close relationship between wealth and income, since only those with a substantial income can accumulate wealth. Those with high incomes accumulate wealth because of excess **disposable income**—income above that required for necessities, such as food, clothing, and accommodation. For many others, especially those living near or below the poverty line, it is often difficult, if not impossible, to accumulate wealth because little disposable income remains after purchasing necessities.

Sociologist Alf Hunter provided a more detailed illustration of this in his classic work on social inequality in Canada.[5] Writing in 1981, Hunter noted that workers earning about $25 000 per year possess average wealth of half ($12 500) that amount; those earning more than $50 000 annually have wealth double ($100 000) that amount; and those earning more than $100 000 possess wealth more than ten times ($1 000 000) that amount. So, if a person's income increases by $75 000 ($25 000 to $100 000), his or her wealth might increase by almost $1 million! These large increases in wealth, resulting from relatively small increases in income, emerge from the increasing amounts of disposable income available as income increases. A relatively small increase in income can have extraordinary effects on wealth accumulation.

Table 3.4 shows that inequalities of net worth are more extreme than inequalities of income. This is not surprising, as data on income do not reveal financial assets such as investments in stocks and RRSPs and property ownership. The lowest 10 percent of families in 1999 had a median net worth of only $2000! The top 10 percent, by contrast, had a median net worth of $703 500!

These differences in net worth change over time. As Table 3.5 reveals, in recent years there has been a widening disparity in net worth. Between 1984 and 1999, the top 20 percent of all family units experienced a 39 percent increase in net worth and the second quintile a 2 percent increase. The inability of lower-income Canadians to accumulate wealth is evident in data on retirement savings. In 1999, of those reporting an income of less than $20 000, only 15 percent contributed to an RRSP or employer-sponsored registered pension plan (RPP). This contrasts with 6 percent for those earning between $20 000 and $39 999, and 92 percent of those reporting incomes of more than $60 000.[6]

Table 3.4 Distribution of Net Worth in Canada by Decile, 1999

Deciles*	Total Net Worth (%)	Median Net Worth** ($)
All family units	100	81 000
Lowest 10%	0	2 000
Second 10%	0	3 100
Third 10%	1	14 300
Fourth 10%	2	35 500
Fifth 10%	3	64 700
Sixth 10%	5	101 500
Seventh 10%	8	152 600
Eighth 10%	11	220 800
Ninth 10%	17	338 100
Tenth 10%	53	703 500

*Family units ranked by net worth.
***Median net worth* means, for example, that one-half of "All family units" have a net worth of more than $81 000 and one-half have a net worth of less than $81 000.

Source: Adapted from Statistics Canada, "The Assets and Debts of Canadians: An Overview of the Results of the Survey of Financial Security," 1999, Catalogue No. 13-595, 15 March 2001, p. 9.

Table 3.5 Change in Median Net Worth from 1984 to 1999, by Net Worth Quintile

	Median Net Worth (constant 1999 dollars)		Change from 1984 to 1999 (constant 1999 dollars)	
	1984 ($)	1999 ($)	($)	(%)
All family units	58 400	64 600	6 200	11
Lowest 20%	0	600	600	0
Second 20%	12 200	12 500	300	2
Third 20%	58 400	64 600	6 200	11
Fourth 20%	124 400	157 500	33 100	27
Fifth 20%	291 200	403 500	112 300	39

Source: Adapted from Statistics Canada, "The Assets and Debts of Canadians: An Overview of the Results of the Survey of Financial Security," 1999, Catalogue No. 13-595, 15 March 2001, p. 30.

Of the top 1 percent (about 76 000 families) of Canadian income-earners in 1993, average income was $295 300. Much of this income, an average of $68 000 per family, was received in the form of investment income. Also, the majority of high-income-earners worked in similar areas of the economy, mainly in managerial, administrative, medical, and health-related occupations. More specifically, of all physicians, surgeons, and dentists in Canada in 1993, more than 25 percent belonged to the top 1 percent of income-earners.

ASCRIPTION AND INEQUALITY

The inequalities of income and wealth discussed here reflect the stratified nature of Canadian society. Important in this respect are racial and ethnic relations, male and female relations, and a person's social background. Our ethnicity and race, sex, and social background have fundamental effects on our lives, often restricting our ability to achieve desired educational, occupational, and financial goals. We possess no control over our racial or ethnic identity, over whether we are male or female, or over our social background. For this reason, these factors are called **ascribed statuses**.

Increasingly, attention is being given to the role of age and physical or mental disability in affecting social inequality. Although these factors have traditionally been viewed as having less of an impact on inequality than ethnicity and race, sex, and social background, recently they have been shown in several instances to contribute significantly to social inequality.

Ascribed statuses affect our chances of success in society. Because education is essential to occupational and financial success, those ascribed statuses that hinder educational success are important in understanding inequalities of income and wealth. A close link exists between number of years of schooling and a person's occupation and income. Of the many factors involved in determining a person's occupation (especially her or his first), the most important is education. However, the type and amount of education a person receives is itself strongly determined by ascribed statuses.

Ethnicity and Race

Canadians of Asian, black, British, and Jewish ancestry have the highest average years of schooling; Indians, Inuit, and Italians tend to have the lowest.[7] This would, on the surface at least, suggest that ethnicity corresponds positively with income level. That is, British, Asians, blacks, and Jews will earn the most, Indians, Inuit, and Italians the least. Evidence suggests this is partially correct.[8] In 1985 men of Jewish ancestry earned the highest average income ($47 000). Men of British ancestry, surprisingly, made average incomes of between only $30 000 and $36 000, ranking about halfway. Blacks earned approximately $25 000, and Asians slightly more than $20 000. Those with the least education—Inuit, Métis, and Natives—also ranked near the bottom, between $23 000 and $25 000.

More recent data suggests that the situation may be changing slightly. Although there continue to be marked differences in the income earned by nonvisible and visible minorities, some nonvisible ethnic groups, such as Italians, Portuguese, and Poles, received average incomes in 1991 about equal to that of the British.[9] According to Gee and Prus, there exists a "racial divide" in Canada between Whites and non-Whites, with coloured Canadians "less likely to be full-time workers and more likely to be unemployed or out of the paid labour force."[10]

Blacks and Asians, despite their high average levels of postsecondary education, do not earn incomes equivalent to their educational credentials. Of Asian and black immigrants arriving in Canada between 1981 and 1991, 22 to 28 percent held university degrees, almost double the rate of the Canadian population. However, compared with the Canadian population, a greater proportion of Asians and blacks have education levels less than Grade 9.

Even if the high number of Asians and blacks with less than a Grade 9 education is taken into account, the low average earnings of the groups remain largely unexplained. This may be the result of several related developments.[11] First, more than other ethnic groups, blacks and Asians are the targets of racist and discriminatory employment and hiring practices. Second, having only recently arrived in Canada, immigrants are employed, like most new workers, at the lower end of business organizations. It may take several years for many of them to reach the upper levels of business organizations.

Furthermore, although these ethnic and racial groups may eventually succeed in achieving a significant degree of occupational, and hence financial success, it is unlikely that even a few will ever secure a position within the economic elite in Canada. Since the 19th century, the Canadian economic elite—the owners and controllers of the major corporations and banks—has been overwhelmingly of British origin.

By establishing control early in Canadian development, this elite has been able to consolidate its dominant position and prevent other ethnic groups from gaining access to top positions. In 1972, 86.2 percent of the economic elite was of British origin, compared with a British-origin population in Canada of 44.7 percent. People of French origin make up 28.6 percent of the population, and in 1971 they constituted only 8.4 percent of this elite. Canada's economy is in the hands of a relatively small group of men of British origin, who continue to pass on their privileges and positions of power to their children, in the process excluding others from elite positions. As the sociologist Wallace Clement observed,

> Top decision making positions in the economy in Canada are dominated by a small upper class. This provides them with a life style much different than that experienced by the vast majority of Canadians and the privileges that accrue to them are passed on to their children. Canada has not fulfilled its promise as a society with equal opportunity. As long as corporate power is allowed to remain in its present concentrated state, there is no hope for equality of opportunity or equality of condition in Canada.[12]

Clement also illustrates the changing class origins of the Canadian economic elite. In 1951, 50 percent of the economic elite had class origins in the upper class. By 1972, this figure had increased to 59.4 percent. In 1951, 32 percent of the economic elite had class origins in the middle class. This figure had increased to 34.8 percent by 1972 changing only slightly over the period. The most significant change is in the proportion of the economic elite with working-class origins decreasing from 18 percent to 5.8 percent from 1951 to 1972.

Social Background

Social background is an important determinant of education levels. Working-class men and women have lower levels of postsecondary schooling than their counterparts in the middle class. This occurs for at least two reasons: (1) lower incomes mean working-class students have more difficulty financing the cost of postsecondary education; (2) education and studying are not a major part of working-class life and culture. Middle-class parents are more likely to have books around the house and encourage their children's schooling. Working-class parents, on the other hand, are less involved in educational concerns, preferring to defer to the knowledge of the teacher and the school system.

Sex

Men and women in Canada have similar levels of education. Only at the highest levels of the education system (master and doctorate) do men outnumber women, although this is changing. Despite similar education levels, however, men have higher average incomes

CRITICAL THINKING BOX 3.2
Should wealthy families in Canada be made to redistribute their wealth?

than women, even in cases where they perform the same jobs. Despite evidence of an increase in weekly wages of full-time female workers in the early 1990s and a levelling of the earnings of male workers,[13] in 2000 women working full-time in Canada made roughly 72 percent of the average earnings of employed males.[14] There are several reasons for this discrepancy.

First, men occupy most management and ownership positions, and they make the decisions on the appropriate incomes for employees. Traditionally regarded as the second income-earner, out to supplement the main income of the husband, women have generally been paid substantially less. This discriminatory practice, although slowly changing, makes it difficult for many women, in particular single mothers, to provide a satisfactory standard of living for themselves and their families.

Second, and related, women are disproportionately employed in occupations viewed as an extension of their household duties, such as teaching, nursing, and social work. The cultural perceptions surrounding such occupations reflect those of the "homemaker," someone who is nurturing, caring, sensitive, efficient, and emotional. Although such stereotypes are also slowly being modified, many women remain in occupations traditionally regarded as "women's work." Elementary-school teachers and nurses, for example, are overwhelmingly women. Furthermore, women who strive to build a career are often treated and viewed differently from men with similar ambitions. Women who prefer not to marry or have children in the hope of pursuing a career are frequently depicted as selfish, greedy, and "unfeminine."

Third, many women find themselves in what is called a "double ghetto." This occurs when a woman both works for a wage and undertakes housework, including looking after the children and the needs of her husband.

Fourth, over the past 25 years the number of single-parent families in Canada headed by women has almost doubled. However, the proportion of single mothers in the workforce has fallen slightly, from 32 to 26 percent. As Susan Crompton, a sociologist, noted, this shift has occurred for several reasons.[15] First, the tremendous increase in the number of working wives has essentially displaced single mothers from certain occupations. Second, as the educational requirements of jobs have increased, single mothers have found it increasingly difficult to obtain these necessary qualifications. The difficulties and expense of raising a child, or children, frequently make it financially infeasible for many single mothers to seek employment (see Box 3.2). Many mothers receive no child support from the fathers and so are forced to live almost wholly on government assistance. In 2002, 30.1 percent of lone-parent families (headed overwhelmingly by women) were considered low-income families. This is in contrast with 5.4 percent of two-parent families with children. The situation is partly responsible for the **feminization of poverty** over the past two decades.

> **BOX 3.2**
>
> *Left Behind: Lone Mothers in the Labour Market*
>
> The stagnating employment situation of lone mothers is not for lack of willingness to work. Many lone mothers currently outside the labour force want to work; those who are working are more likely than wives to be employed full time, and a substantial proportion of those working part time would rather have full-time jobs. However, wives are older and better educated and have more work experience. Moreover, having another adult to help with childcare arrangements can only make it easier for married mothers to look for and retain a job. Faced with competition from a large pool of better-educated women, it is not surprising that many lone mothers have difficulty establishing themselves in the job market.
>
> But a "hierarchy of success" can also be found in the population of lone mothers. Separated or divorced mothers are "ex-wives" who occupy a more advantageous position in the labour market than never-married women because they have more education and more work experience. Much of the labour market disadvantage of never-married lone mothers may be attributable to their lower educational attainment, which raises the question of whether pregnancy outside marriage increases the likelihood of interrupting formal education and delaying the acquisition of work experience. It is certainly clear that two distinct types of women, with considerably different demographic and socioeconomic characteristics, are merged under the rubric "lone mother." It seems a disservice to both groups to ignore the differences between them.
>
> **Source:** Statistics Canada, *Perspectives on Labour and Income*, Catalogue No. 75-001, Summer 1994.

Age

Increasingly, attention is being paid to the importance of age in promoting inequality. Many observers have focused particularly on the financial plight of our senior citizens. It has been pointed out that, over the course of the 20th century, the number of people age 65 and over has steadily risen and will rise even more markedly this century. Improved medical technologies, healthier lifestyles, and the large number of births between the years 1945 and 1965 are responsible for this increase.

> ### CRITICAL THINKING BOX 3.3
>
> Should lone, or single, mothers be given greater government financial assistance in light of the difficulties they face?

As individuals, living well past the traditional retirement age of 65 may appear to be a happy prospect; but as a society, it is becoming apparent that there are many difficulties associated with a greying population. First, partly because most elderly people do not do paid work, they are stereotyped as lazy, mentally slow, traditionally minded, and incapable of contributing to society. Second, as more and more people live well into their retirement years, many quickly find themselves spending whatever savings they may have accumulated over their working lives. This forces many seniors to live in poverty. Indeed, outside of children and youth, the elderly—most of whom are women—constitute the largest group in poverty in North America. For example, according to Statistics Canada,[16] in 1997, an estimated 49.1 percent of unattached women and 33.3 percent of unattached men over the age of 65 lived below the poverty line. In those instances when seniors are physically able to work, they generally find themselves in low-prestige, minimum-wage jobs.

With a growing senior population, the Canadian government faces increased pressure to improve the financial situation of those entering retirement. At the same time, however, government cutbacks and continued public demand to reduce spending on the unemployed, welfare recipients, and seniors (in the form of an income security system), suggest that the situation of the elderly in this century may deteriorate further.

Disability

Historically, people with physical or mental disabilities have generally been excluded from participating in mainstream society, particularly in the workforce. More recently, however, through the lobbying efforts of organizations representing the disabled, as well as a variety of employment equity programs, many people with disabilities now find themselves participating actively and fully in work and social activities. Nevertheless, people with disabilities continue to confront stereotyping, prejudice, and discrimination. People with disabilities, as a group, suffer severe social inequality, experiencing high levels of unemployment and welfare.

VIEWS OF INEQUALITY

The existence of inequality does not tell us whether it is good or bad, positive or negative, moral or immoral. Does inequality of income and wealth, for example, contribute to a more peaceful and stable society? Does it provide greater financial rewards to those who work the hardest? Or, does it serve to further the exploitation and oppression of one section of society by another? Responses to these types of questions have been numerous, but it is possible to classify them into two basic positions.

Structural Functionalism: Inequality Is Good, Necessary, and Inevitable

The first position, **structural functionalism**, argues that inequality is positive and necessary for the proper functioning of society. Writers adopting this position view society as operating in a manner similar to the human body. Just as the different parts of the body—skin, muscles, bones, organs, and so on—collectively function to maintain the operation and survival of the whole body, the different parts—the structure—of society together function to promote the overall peace and stability of society.

Everything that exists in society serves a function. Prostitution, for example, although illegal and commonly regarded as a major problem, is seen as serving the important function of allowing men, and women, to vent their sexual frustrations. Without prostitution, according to structural functionalists, it is likely that rape and other sexual assaults would increase, along with the rate of divorce. Prostitution, then, according to this view, simultaneously functions to reduce violent sexual assaults and maintain intact the important role of the family unit.

Inequality is viewed in a similar manner. Structural functionalists consider inequality both unavoidable and desirable. The existence of inequality in all societies, past and present, indicates that it must serve a positive function. But how does inequality contribute to the stability and functioning of society? Inequality, in the form of different income levels, in this view, provides an incentive for the most able people in society to work the hardest to attain those jobs considered by society to be the most functionally important. Doctors, lawyers, dentists, and so on, earn substantial incomes because of their great importance to the stability and continuing operation of society.

If a doctor and a bus driver, for example, received similar incomes, few people would want to endure the years of education and hard work required to qualify as a doctor. Not only would there arise a critical shortage of doctors, but it is also likely this shortage would lead to the acceptance into medical school of many unsuitable applicants. It is necessary, therefore, that a differential reward system exist, whereby those performing the most important jobs receive higher incomes and more prestige than those in occupations considered less important. Without differential rewards, the most talented would have no incentive to pursue the functionally most important jobs. Inequality, then, as an essential and inherent feature of the economic structure of society, is vital to the proper functioning of the whole society. For this reason, those making this argument are referred to as structural functionalists.

Some Problems with Structural Functionalism

To their credit, structural functionalists emphasize the need for some form of incentive to ensure that the most talented attain those jobs requiring extensive knowledge and training. It is unlikely that in a society in which everyone receives the same level of income, regardless of occupation, the most talented would seek to undertake the most important occupations.

There are several problems, however, with the structural functionalist position. First, the emphasis on individual talents and abilities suggests that those with the most talent and ability will succeed and that those less able and less intelligent will be less successful. To structural functionalists, a person's degree of success should be determined solely by individual effort. Consequently, everyone should have an equal chance to succeed—there should be **equality of opportunity**. This involves equal access to education and the elimination of sexual and ethnic discrimination in employment and hiring practices. This type of argument fails to satisfactorily analyze the impact of ethnicity, class, and gender on the creation and continuation of inequalities in society. Many ethnic groups, the poor, and women face severe impediments to educational and occupational success. "The point to stress," as sociologist Edward Grabb noted,[17] "is that established structures tend to define the prospects and life chances of people in most instances." As a society, we may want everyone to be given an equal opportunity, but it is apparent that many people cannot overcome major obstacles that have been put in their way, thus preventing them from moving ahead.

Second, are occupations financially rewarded based on their functional importance to society? Not always, it seems. For example, what is the functional importance of professional athletes and entertainers? Jaromir Jagr, Alex Rodriguez, and Shania Twain may annually earn millions of dollars, but are they crucial to the functioning of society?

A less obvious example relates to the functional importance of doctors. Are they more important than, say, garbage collectors? According to structural functionalists, garbage collectors receive lower incomes than doctors do because of their lesser importance to the overall functioning of society. But there is little evidence to back this up. Indeed, it could strongly be argued that as a society we could do without doctors more readily than without garbage collectors. On the one hand, the elimination of garbage collectors would result in mountains of rotting garbage and the epidemic outbreak of potentially fatal diseases. Without doctors, on the other hand, more people would undoubtedly die and at an earlier age, but it is unlikely that society would suffer serious damage.

Another example is provided by the recent arguments made by feminists and many others regarding the value of housework. Although unpaid in financial terms, such work is obviously important, if not crucial, to the overall functioning of society.

Third, some people are motivated by factors other than money. Bryan S. Turner, in his book *Equality*, noted the nonmonetary motives that are traditionally associated with the nursing profession: "A number of social roles such as nursing may be regarded as socially important despite their low income and persons who occupy such roles are typically motivated by moral or religious arguments where a direct monetary reward is absent."[18]

Fourth, structural functionalists emphasize the importance of individual hard work and perseverance in determining occupational and financial success. This assumes that movement from one class to another, both up and down, is largely a consequence of individual strengths or weaknesses and pays little attention to the role of inherited wealth in preserving economic inequalities and class differences over time. The majority of Canada's wealthiest families, as already noted, are of Anglo-Saxon origin and the beneficiaries of inherited wealth.

Conflict Theory: Inequality Is Bad, Avoidable, and Unnecessary

The second position, **conflict theory**, relates inequality to conflict in society. Conflict theorists view inequality as both a cause and an effect of exploitation, conflict, and oppression. There are many different types of conflict theorists, with disagreement centring on the precise form conflict takes and how it can be reduced or eliminated. Some conflict theorists view class inequality and conflict as central, whereas for others gender or race is more important.

Karl Marx

One of the most important conflict theorists was Karl Marx (1818–1883). Marx argued that all societies are divided between those who possess wealth and power and those who

---

CRITICAL THINKING BOX 3.4

Are some workers, such as nurses and teachers, less motivated by money and more by a desire to help people?

do not. In modern society, the major division, or inequality, is between the **capitalist class**, or bourgeoisie, and the **working class**, or proletariat. Capitalists are those who own the factories, land, machinery, and other materials used for the production of goods and services. Marx termed these materials the **means of production**—they are, simply, the means to produce goods and services. Workers, on the other hand, do not own property, and to survive they must sell their labour power to capitalists.

To Marx, the relationship between capitalists and workers is based on a fundamental conflict of interest. Capitalists are motivated by profit, and workers are interested in obtaining as much income as possible. But because higher wages mean less profit, workers and capitalists have different, and conflicting, interests and objectives. According to Marx, capitalists exploit workers by paying them less than the real value of their work. For example, suppose I work as a bartender for eight hours one day, at $6 an hour. At the end of the day, I receive $48 ($6 × $8). However, the amount of money I have in the cash register is considerably more, say $98. It is apparent, then, that I am paid $50 ($98 − $48) less than the amount I collected. Of this $50, my boss, the owner of the bar, will be required to pay business expenses, such as hydro, food, mortgage, and so on. Of the $50, then, my boss may make a profit of only $20. Nevertheless, this is money that I worked for, not my boss. To Marx, this $20, the difference between what I am paid and the value of my work, represents the degree to which I am exploited. Marx termed this difference—in this case, $20—surplus value, or profit.

The greater the amount of surplus value, the greater the level of **exploitation**. For example, if my boss's profits increase every year and my wages remain unchanged, the profit, and hence the level of exploitation, increases. During Marx's lifetime, this is indeed what occurred, as most workers were unable to survive without the aid of charity of some kind. Many starved to death as capitalists increased their share of income and wealth.

Marx referred to the process of increasing inequality and poverty as the "immiseration" of the working class. He hoped it would force workers into recognizing the exploitative nature of the system in which they lived and the need to overthrow the capitalist class and institute a new, more equal, society.

But Marx was unsure whether workers would recognize their real interests and revolt. He observed that the capitalist class, as a consequence of its ownership of all institutions in society, including the media, was informing workers that their poverty resulted not from exploitation, but from their own personal failings: they simply did not work hard enough. This made most workers feel responsible for their own poverty. Marx referred to this outlook on the part of the working class as *false consciousness*: workers essentially believed it was not the system of capitalist exploitation that was to blame, but themselves.

The Many Faces of Diversity

The increasing exploitation of the working class, Marx believed, would lead eventually to conflict between capitalists and workers (see Figure 3.1). Workers would rid themselves of their false consciousness and recognize their shared experience: exploitation. Marx referred to this shared experience as *class consciousness*. It would eventually result in an organizational effort to overthrow the capitalist class and to the establishment of a new communist society, based on equality and freedom.

In communism, exploitation would no longer exist, as all goods and services would be produced by and for the whole population. There would no longer be a division between

Figure 3.1   From Class to Cash Consciousness

Source: Eddie Grattan and Mark Galante.

owners and workers; all property would be owned by the whole population, not by one class or group of people. Goods and services would be distributed on the basis of, not a person's ability to pay, but instead a person's need: "From each according to his ability, to each according to his need" (Marx).[19]

Some Problems with Marx's Theory

Marx presented an accurate depiction of 19th-century society; however, at the time of his death in 1883, major changes were occurring in the societies of Europe and North America. Consequently, today we consider many of his predictions about the development of Western society to be somewhat misplaced or greatly in need of revision.

First, Marx's assumption of property ownership as constituting the central source of conflict within society has not proven wholly correct. Many observers argue that other factors, such as gender, race, ethnicity, and national identity have generated greater conflict and tension.

Second, Marx predicted increasing income disparities between capitalists and workers; however, two developments have occurred to prevent this. The first is the growth, starting about 1880, of a large middle class. The growth of government services over the past hundred years has created a large group of clerical (white-collar) workers who tend to view themselves as distinct, in prestige and status, from the working class. Also, and this is related, in the 20th century all workers experienced significant increases in their standard of living. Compared with the 19th century, poverty levels have declined sharply, and average incomes have risen dramatically. Consequently, many workers feel little animosity toward their employers.

Third, governments over the past century or so have assumed control of increasingly more areas of society. Although this has arguably lessened individual freedom, as will be discussed shortly, it has also enabled governments to maintain greater control of the economy in an effort to avoid major economic crises, such as the Great Depression of the 1930s, that provide the preconditions for worker unrest.

CRITICAL THINKING BOX 3.5

Most workers may be exploited, as Marx argued. Do you believe that most workers feel exploited?

Although they have lessened over the course of the 20th century, substantial income and wealth inequalities still exist in Canada. This suggests that Marx's analysis of society as comprising the haves and have-nots still provides insight into the workings of modern society.

Max Weber

Max Weber (pronounced VAY-ber) (1864–1921) accepted Marx's assertion of the importance of the ownership or nonownership of property as a major source of inequality and as a potentially important source of conflict. However, Weber highlighted the other sources of potential conflicts, as well as offering a more complex system of stratification.

Weber agreed with Marx that class, ownership or nonownership of property, is a central feature of modern society. Capitalists possess great amounts of wealth, power, and prestige. However, argued Weber, often the possession of business property is not accompanied by power and prestige. Social prestige, or status, as the degree of positive evaluation by members of society, often derives from a person's level of education, income, and occupation. Doctors, lawyers, dentists, university professors, and so on, enjoy high status in large part because of their income level and the education they possess. Wait staff, shop attendants, and bartenders experience low status or prestige because of their low incomes and the limited education requirements of their jobs.

Political power is the degree of political influence a group has. Some unions, for example, acting on behalf of workers, often exert extensive pressure on employers, including the government, in an effort to increase salaries and improve working conditions. But because not all unions have the same degree of influence, some workers are more politically influential than others. Furthermore, other groups in society, such as those sharing a common lifestyle, similar income levels, and so on, possess greater prestige and political power than many business property owners. To Weber then, differences in social prestige (status) and access to and control of the political system have important effects on the degree of income and wealth inequalities and on the nature of the stratification system.

Because Weber viewed inequality as stemming from a variety of sources and conceived of stratification as a complex phenomenon, he talked about social class and status groups rather than simply about class, the ownership or nonownership of property as defined by Marx. Furthermore, Weber contended that frequently there is no necessary relationship between economic (class), prestige (status), and political (power) rankings. For example, university professors, who have long rejected unionization, tend to have a low power ranking but a high status ranking. In addition, the economic ranking of professors tends to be significantly lower than suggested by their high status. Some manual workers, on the other hand, ranking low in terms of status, earn higher incomes (have higher economic ranking) and often have greater power than professors.

The aristocracy in Britain also serves to illustrate the often-wide discrepancies among economic, status, and political rankings. The British aristocracy, based on heredity, for centuries has enjoyed great prestige. Politically, however, its strength has been declining and no longer matches its high status ranking. The British aristocracy has also been declining economically, with many families forced to sell valuable assets in an effort to maintain their traditional lifestyle and status ranking.

Although recognizing the importance of the ownership or nonownership of property, Weber believed society to be more profoundly affected by the growth of bureaucratic institutions, such as government and large businesses. He believed these limited individual freedom. Workers may indeed be exploited, conceded Weber, but it is unlikely that this will provoke conflict with the capitalist class, and it will certainly not bring about a new, communist, society. Furthermore, even if workers were to overthrow the capitalist class and establish a new society, communism would require an even larger bureaucracy and so further restrict individual freedom. The new leaders would establish themselves as a bureaucratic ruling class, and the workers would simply be trading in one set of rulers for another.

Weber's emphasis on status as an important indicator of social class provides a more accurate representation of the class structure of modern society. The growth of the middle class, and its desire to purchase consumer products such as cars, houses, televisions, and designer clothes, symbolizes its concern with status and prestige. Also, it is apparent that growth in the number of government departments has reduced individual freedom: more and more areas of our life are controlled by rules and regulations, requiring little individual expression or decision making.

THE SOCIAL CLASS STRUCTURE OF CANADA

Depending on the theoretical perspective you adopt, you can define the nature and number of social classes in Canada in several ways. Nevertheless, this section presents a brief description of what are commonly viewed as the major social classes in Canada: the upper class, middle class, working class, and subworking class.

The Upper Class

The **upper class** in Canada comprises those who own substantial amounts of wealth, about 4 or 5 percent of the population. Within the upper class a distinction is often made between those possessing wealth passed down from generation to generation (inherited wealth) and those acquiring wealth through recent business successes ("new money"), such as Bill Gates and Donald Trump. The Eaton's department stores, for example, established in the late 19th century, were until recently owned continuously by the Eaton family of Toronto. Having inherited wealth over several generations, this type of family is often referred to as *bluebloods*, or the upper-upper class. It is estimated that at least half the wealthiest families have benefited to some degree from inheritance.[20]

Members of the upper-upper class characteristically attend expensive private schools and universities, at home and abroad. These institutions serve to teach the values, beliefs, manners, and ways of looking at the world that are exclusive to this class (see Figure 3.2).

The other members of the upper class earn income from well-paying occupations or investments. These **"new money"** capitalists make up the bulk, between 70 and 80 percent, of the upper class. They are often prohibited from socializing with "old money" families through the many private clubs and associations, as well as social events, established exclusively by and for those with inherited wealth. Bluebloods also prefer avoiding the publicity and media attention frequently craved by new money capitalists such as Donald Trump.

Figure 3.2   Lifestyles of the Wealthy

Source: Eddie Grattan and Mark Galante.

The Middle Class

Members of the **middle class** in Canada own some property—usually a house, one or two cars, and perhaps a cottage. They have relatively high-paying, secure occupations, providing a degree of satisfaction and feeling of accomplishment. Many of these occupations are in the public sector, and often offer generous benefit allowances, for example, sick, dental, and maternity benefits, privately established pensions, and, long-term job security.

The Working Class

The working class comprises those who own little or no wealth and are employed in low-paying and generally insecure occupations, such as most of those within the service sector. It is largely because of limited education—most working-class people have not progressed beyond high school or community college—that members of the working class face limited, and typically low-paying, employment opportunities.

The Subworking Class

The subworking class is made up of those, around 20 percent of the population, with the lowest or no incomes—the homeless, welfare recipients, the unemployed, single-parent families, the aged, and those in extremely low-paying occupations. Life for this class tends to be very unstable, both financially and emotionally. Families have limited, or no, savings as nearly all income is required for the purchase of necessities.

Members of the subworking class often live in a separate area of a town or city, and their cultural environment—values, beliefs, attitudes, behaviour, and social activities—tends to be distinctive.

SOCIAL MOBILITY IN CANADA

In North America, especially the United States, it is commonly suggested, particularly by structural functionalists, that economic success is achieved through individual hard work and determination. This idea implies that movement up and down the social class structure occurs frequently and primarily because of personal effort and initiative. This view of social mobility is ritually portrayed and celebrated in television programs and Hollywood movies: obstacles to personal success reside solely within the individual.

> **CRITICAL THINKING BOX 3.6**
>
> Can you provide an example of a middle-class family experiencing upward social mobility?

How accurate is this portrayal? How many people actually move up the class ladder? How many fall down? What is the pattern of social mobility in Canada? What is the likelihood, for example, of the daughter or son of a steelworker eventually becoming a member of the middle class?

The social class structure of Canada has remained relatively stable for several decades. The number of people within each class has remained constant, although many of those within each class have changed. A study conducted in 1986 revealed that close to 40 percent of men and women experienced some form of upward mobility, and a similar figure moved downward.[21]

The degree of social mobility in Canada is greater than that of most other Western countries. Nevertheless, as pointed out earlier, ascribed statuses continue to affect movement up and down the social and occupational hierarchy. In Canada, as Joanne Naiman concluded, "the very top and the very bottom of the status hierarchy remain relatively closed, and most occupational mobility is mainly small movement in the middle."[22]

An important element of upward mobility is level of education. As already mentioned, studies indicate a close relationship between a person's first job and his or her education level. However, it would appear that this may be changing as the educational requirements of jobs increase and the Canadian economy, like economies in the rest of the industrialized world, experiences major structural changes. Free trade, the growth of service industries, and the shrinking of industrial production suggest that the mobility and social class patterns of the past forty or fifty years may be undergoing a fundamental restructuring.

CHAPTER SUMMARY

This chapter has examined several aspects of social inequality in Canada. (1) There is inequality of income and wealth in our society. Although both income and wealth are unequally distributed, historically inequality of wealth has exhibited greater extremes.

(2) Two major theories, structural functionalism and conflict theory, each provide insight into the nature of social inequality, although neither is without problems. (3) Social inequality pervades nearly all aspects of our lives. Social class, race and ethnicity, sex, age, and physical and mental ability all have an impact on social inequality. (4) Canada has four major social classes. (5) In recent years structural changes in the global economy have increased social inequality in Canada and elsewhere, and in the future, this is unlikely to change.

KEY TERMS

ascribed status, p. 74
capitalist class, p. 83
conflict theory, p. 82
disposable income, p. 72
equality of opportunity, p. 81
exploitation, p. 83
feminization of poverty, p. 77
income, p. 68
means of production, p. 83

middle class, p. 89
political power, p. 86
social inequality, p. 67
stratification, p. 67
structural functionalism, p. 80
upper class, p. 88
wealth, p. 72
working class, p. 83

DISCUSSION QUESTIONS

1. To explain social inequality, which of the ascribed statuses (social class, race and ethnicity, sex, age, disability) do you consider the most important? Provide evidence.
2. Which theory do you believe more adequately explains social inequality: conflict theory or structural functionalism?
3. Why are single women more likely to be living below the poverty line?
4. Provide reasons why you believe the Canadian government should or should not attempt to reduce the degree of income and wealth inequality in Canada.
5. As a person grows older, which ascribed statuses do you think become more significant in his or her life?

NOTES

1. John J. Macionis, Juanne Nancarrow Clarke, and Linda M. Gerber, *Sociology* (New Jersey: Prentice-Hall, 1993), p. 243.
2. G. Picot and A. Heisz, "The Labour Market in the 1990s," in *Canadian Economic Observer*, February 2000, Statistics Canada, Catalogue No. 11-010-XPB (Ottawa: Statistics Canada, 2000), pp. 312, 315.
3. The number of social assistance recipients dropped from 3.1 million to under 2 million by 2000, with benefit payments decreasing from $14.3 billion in 1994 to $10.4 billion in 2001 (see F. Roy, "Social Assistance by Province, 1993–2003" in Statistics Canada, *Canadian Economic Observer*, November 2004, Catalogue No. 11-010), p. 3.1.
4. Ibid.
5. Alfred A. Hunter, *Class Tells: On Social Inequality in Canada* (Toronto: Butterworths, 1981), p. 99.
6. "Likelihood of Saving Increase with Income," *Infomat, A Weekly Review*, 20 July 2001, Catalogue No. 11-002E.
7. Ibid., p. 132; Macionis et al., *Sociology*, p. 337.
8. Jane Badets and Tina W.L. Chui, *Focus on Canada: Canada's Changing Immigrant Population*, Statistics Canada, Catalogue No. 96-311E (Ottawa and Scarborough: Statistics Canada and Prentice-Hall, 1994), p. 79.
9. Anton L. Allahar and James E. Cote, *The Structure of Inequality in Canada* (Toronto: James Lorimer and Company Ltd., 1998), p. 63. Using data for 1990, Feng Hou and T.R. Balakrishnan reach the same conclusion in "The Economic Integration of Visible Minorities in Contemporary Canadian Society," in James Curtis, Edward Grabb, and Neil Guppy, eds., *Social Inequality in Canada: Patterns, Problems, Policies*, 3rd ed. (Scarborough: Prentice Hall Allyn and Bacon Canada, 1999), p. 223.
10. Ellen M. Gee and Steven G. Prus, "Income Inequality in Canada: A 'Racial Divide'" in Madeline A. Kalbach and Warren E. Kalbach, eds., *Perspectives on Ethnicity in Canada: A Reader* (Toronto: Harcourt Brace, 2000).
11. Macionis et al., *Sociology*, p. 337.
12. Wallace Clement, *The Canadian Corporate Elite: An Analysis of Economic Power* (Ottawa: Carleton University Press, 1986), pp. 192, 364–65.
13. Picot and Heisz, "The Labour Market," p. 312.
14. Melissa Cooke-Reynolds and Nancy Zukewich, "The Feminization of Work," *Canadian Social Trends* 72 (Spring 2004): 27. Statistics Canada, Catalogue No. 11-008.
15. Susan Crompton, "Left Behind: Lone Mothers in the Labour Market." *Perspectives*, Summer 1994: 23.
16. Statistics Canada, *Income Distributions by Size in Canada, 1997*, Catalogue No. 13-207-XPB, 1999.
17. Edward G. Grabb, *Theories of Social Inequality: Classical and Contemporary Perspectives* (Toronto: Holt, Rinehart, and Winston, 1990), p. 190.
18. Bryan S. Turner, *Equality* (London: Tavistock Publications, 1986), pp. 40–41.
19. This quotation appears in Karl Marx, *Critique of the Gotha Programme* (Moscow: Progress Publishers, 1970).

20. James B. Davies, "The Distribution of Wealth and Economic Inequality," in James Curtis, Edward Grabb, and Neil Guppy, eds., *Social Inequality in Canada: Patterns, Problems, Policies*, 3rd ed. (Scarborough: Prentice Hall Allyn and Bacon Canada, 1999), p. 72.
21. Gillian Creese, Neil Guppy, and Martin Meissner, *Ups and Downs on the Ladder of Success* (Ottawa: Statistics Canada, 1991).
22. Joanne Naiman, *How Societies Work: Class, Power, and Change in a Canadian Context*, 2nd ed. (Toronto: Irwin Publishing, 2000), p. 224.

CHAPTER 4

Race and Ethnicity: The Obvious Diversity

Paul U. Angelini and Michelle Broderick

We kept the blacks out and we did it in a peculiarly Canadian fashion—by pretending publicly that our immigration laws did not discriminate against anyone by reason of race, creed or colour while at the same time preventing all Negroes from crossing the border into Canada.
— Pierre Berton, Why We Act Like Canadians

We are all immigrants.
— common Canadian saying

Objectives

After reading this chapter, you should be able to

- sketch the diverse history of Canadian immigration

- outline the major determinants of personal and group identity and clarify the meaning and implications of the term *multiculturalism*

- analyze the most important factors that influence societal interaction

- appreciate the nature and seriousness of the problems and issues in race and ethnic relations in Canada

- talk about who you are in a manner that develops a positive self-image and acceptance of racial and ethnic differences

INTRODUCTION

Few issues can raise temperatures to such heights, with relative ease, as the issue of race and ethnic relations in Canadian society. Discussions inevitably lead to debates concerning what it means to be Canadian and how one becomes Canadian. From its very beginnings, Canada has been a multicultural and multiracial society. Sadly, this fact is frequently omitted from school curricula. As a result, Canadians tend to be largely ignorant of the history of immigration in Canada. It is assumed that British and French settlers "founded" Canada, irrespective of the Native Peoples who already occupied much of what we call North America. In fact, people from many parts of the world arrived on Canadian shores expecting to begin a prosperous life in a new land. People arrived from all over Europe, Asia, and the United States—people from all walks of life, including slaves and free people of colour. With so many different people arriving from so many different parts of the world, it is little wonder that Canada has experienced and is experiencing all the growing pains of so many different people living together. This does not mean we should accept some of the problems as "normal"; it simply means we should acknowledge them (past and present) and deal with them directly in hopes of not repeating what we have done in the past or are doing now. History does not repeat itself—human beings do!

The purpose of this chapter is to introduce students to the history of immigration to Canada, multiculturalism, identity, and the factors that affect interaction in Canadian society. The hope is to help students come to understand the challenges facing them in developing a society tolerant and accepting of cultural and racial differences.

A BRIEF HISTORY OF IMMIGRATION TO CANADA

Discussions concerning the merits of immigration to Canada are never far from debates on Canadian identity—what is a Canadian, who is a Canadian? One of the most widely held myths in Canada is the belief that Canada has always been a "white" British and French country; from its very beginnings, Canada has been multiethnic (see Box 4.1). A historical analysis of Canadian immigration patterns is essential if we are to have a thorough understanding of race and ethnic diversity in Canada.

In 1970 the Royal Commission on Bilingualism and Biculturalism outlined four distinct stages to Canada's immigration history.[1] Let us look briefly at each of these. We will then discuss a fifth stage resulting from a later change in policy.

Stage 1: The Beginnings to 1901

Immigration to Canada before 1901 was slow. People came to Canada possessing many different skills and with quite different experiences: fishermen, farmers, merchants,

> **BOX 4.1**
>
> *An Insidious Myth*
>
> ....
>
> The myth that Canada was and is a "white" country has no foundation in fact, yet it is part of our national consciousness. Writer Adrienne Shadd refers to this as an "insidious" myth. People tend to forget that Native Peoples were here first: black slaves were among the earliest to arrive in Canada during the 17th and 18th centuries; the Chinese presence in Canada dates back to the 19th century. A wide variety of races and nationalities helped to build Canada. This truth, however, is not properly reflected in school curricula.
>
> *Source:* Carl E. James, *Seeing Ourselves: Exploring Race, Ethnicity and Culture*, 2nd ed. (Toronto: TEP, 1999), p. 160.

traders, soldiers, adventurers, slaves, and fugitives. French and British immigrants were dominant in terms of their numbers, cultural influences, and power, but they were not the only ones to arrive. Before 1800 or so, only 10 percent of Canada's population was not British, French, or Native. More than half of all immigrants who arrived in Canada during the 19th century, however, were of German origins. Germans settled in New France and Nova Scotia, where 1500 of them founded the Lunenburg settlement between 1750 and 1753. Germans also settled in other parts of the Maritimes and in what is today Quebec. German sectarians, including Mennonites, Moravians, and Tunkers, predominantly from the United States, came to Canada from about 1780 until well into the 1800s. In the middle of the 19th century, some of them settled in Ontario, especially in Waterloo County.

About 10 percent of the United Empire Loyalists who came to Canada after the American Revolution were black.[2] Some of these people were slaves that white owners brought with them. Contrary to popular myth, slavery did indeed exist in Canada (see Box 4.2). Moreover, between 1815 and 1860, 40 000 to 60 000 fugitive slaves and free people of colour arrived.[3] The Dutch and the Scandinavians were the only other large groups who came, accounting for less than 1 percent of the total immigration population at this time.

The extinction of the Native inhabitants of Newfoundland, the Beothuk, is almost universally forgotten. The tribe died out by the 1820s due to a combination of starvation (not having free access to their traditional fishing grounds), tuberculosis (which took a heavy toll), and being hunted for sport by the European settlers![4]

> **BOX 4.2**
>
> *Slavery in Canada, Part I*
>
> ---
>
> **TO BE SOLD,**
> A BLACK WOMAN, named PEGGY, aged about forty years; and a Black boy her son, named JUPITER, aged about fifteen years, both of them the property of the Subscriber.
>
> The Woman is a tolerable Cook and washer woman and perfectly understands making Soap and Candles.
>
> The Boy is tall and strong of his age, and has been employed in Country business, but brought up principally as a House Servant—They are each of them Servants for life. The Price for the Woman is one hundred and fifty Dollars—for the Boy two hundred Dollars, payable in three years with Interest from the day of Sale and to be properly secured by Bond &c.—But one fourth less will be taken in ready Money.
>
> PETER RUSSELL.
>
> York, Feb. 10th 1806.
>
> *Source:* Daniel G. Hill, *Human Rights in Canada: A Focus on Racism* (Ottawa: Canadian Labour Congress, 1977), p. 3.

Stage 2: 1880–1918

Stage 2 is characterized by Europeans leaving Europe. So many left for the New World that these years have been described as "the mightiest movement of people in modern history."

As European society continued its evolution from a rural one to an urban one and as its population continued to grow, millions and millions decided to leave, hoping to find better opportunities for themselves elsewhere. Many went to the United States, South

The Many Faces of Diversity

> **CRITICAL THINKING BOX 4.1**
>
> Why is the past existence of slavery in Canada not a well-known fact? Why is it not a standard part of high-school curricula? Did you know slavery once existed in Canada? If so, where did you find out? If not, why?

America (especially Argentina and Brazil), and, beginning in the late 1890s, Canada. By this time the federal government had decided to settle the empty Canadian West and was encouraging migration there.

Other factors, too, brought immigrants here: the closing of the American frontier, the Yukon gold rush, huge construction projects like the transcontinental railway, and new developments in farming technology all made Canada more attractive.

In 1913 alone more than 400 000 immigrants arrived, the largest number in any single year. Between 1896 and 1914 more than 3 000 000 came to Canada. Of these, about 1 250 000 were from the United Kingdom and approximately 1 000 000 from the United States. The important feature of this period was the arrival of people from Central and Eastern Europe, including Ukrainians, Poles, Hungarians, Romanians, and Russians. By 1921, 15 percent of Canada's population was of neither French nor British origin.

The increased numbers for those groups already settled in Western Canada were nothing short of amazing. The population of German origin went from 46 800 in 1901 to 148 000 in 1911, and by 1931, 242 000 Germans were living in the Prairies. The Scandinavian population went from 17 300 in 1901 to 130 000 in 1921. The Ukrainian population increased almost twentyfold, from 5600 in 1901 to 96 000 in 1921, and the population of Polish origin experienced similar growth, from 2800 in 1901 to 32 000 in 1921.

In the rest of Canada other groups also saw their numbers rise dramatically. The Italian population went from 11 000 in 1901 to 67 000 in 1921, and the Jewish from 16 100 to 126 000 in 1921. Other arrivals included Greeks, Syrians, Lebanese, and Armenians. Like the Italians and Jews, these immigrants preferred to settle in the cities of central Canada rather than the farming communities of the West.

Stage 2 also saw the imposition of the head tax on Chinese immigrants. (The racist treatment of Asian immigrants was to be repeated during World War II; the only difference was the target—this time it was Japanese-Canadians.) The head tax was intended to slow the arrival of Chinese. In 1885 it was set at $50. In 1900 it was set at $100. This tax, however, did not slow the influx of Chinese, so the tax was increased to $500 per person, or head, in 1903.

The racist workings of the Canadian immigration policy did not stop there. In 1908 the government attempted to limit East Indian immigration by requiring that anyone arriving from India had to do so by coming to Canada directly without stopping. This became known as the "direct passage" stipulation. In 1914, 376 Sikhs arrived from India directly—they did not stop at any other site during their voyage. Their ship, the *Komagata Maru*, was made to wait in Vancouver harbour for three months—before it was shamelessly turned away.[5]

In 1907, more than 8000 Japanese immigrants arrived in Canada, particularly in British Columbia. This heightened anti-Asian sentiment in British Columbia, and led to tighter restrictions on immigration. These feelings culminated in racial riots in September of that year. By 1921, 16 000 Japanese were living in Canada, 15 000 of them in British Columbia.

Chinese immigrants are always credited with having made a huge contribution to the building of Canada's railways. This accolade, however, masks the exploitative and often brutal treatment they received during the building itself. Pierre Berton and others have remarked that there is one dead Chinese worker for every mile of rail laid. By 1921, there were 40 000 Chinese in Canada, 24 000 of them in British Columbia.

Stage 3: 1918–1945

Between World War I and II, immigration continued, but it never reached the explosive numbers of Stage 2. Between 1914 and 1939 the United States severely reduced the number of immigrants it would take, and this made Canada the favoured destination. But Canada, too, began to put restrictions on immigration. Whereas the United States chose

---

CRITICAL THINKING BOX 4.2

Every immigrant entering Canada since 1995 must pay an "administrative fee" of $975. Since this is a flat tax, all must pay regardless of their ability to pay it. A refugee family of two adults and two children must pay $1950 in landing fees. If this same family makes their own way to Canada, they will have to pay an additional nonrefundable processing fee of $500 per adult and $100 per child. Is this fee really a head tax? Is it fair to ask immigrants to pay such a fee? What are the benefits and weaknesses of asking immigrants to pay such a fee? List three benefits and three weaknesses.

Source: Andrew Brouwer, "Protection with a Price Tag: The Head Tax for Refugees and Their Families Must Go," *Caledon Commentary* (Caledon Institute of Social Policy), June 1999, available <www.caledoninst.org>, accessed 30 March 2006.

> **BOX 4.3**
>
> *Slavery in Canada, Part II*
>
> Slavery did indeed exist in Canada. The first slave was brought to New France in 1628 from Madagascar. During and after the American Revolution, United Empire Loyalists brought slaves with them. At least six of sixteen members of the first Parliament of Upper Canada owned slaves.
>
> **Source:** Daniel G. Hill, *Human Rights in Canada: A Focus on Racism* (Ottawa: Canadian Labour Congress, 1977), p. 7.
>
> *Slavery in Canada, Part III*
>
> In 1734, Marie-Joseph Angelique, a black female slave, burned down part of the city of Montreal after she was informed of her owner's intention to sell her.
>
> **Source:** Carl E. James, *Seeing Ourselves: Exploring Race, Ethnicity and Culture*, 2nd ed. (Toronto: TEP, 1999), p. 161.

to adopt a system based on quotas, Canada decided to make lists of countries that were "preferred" or "nonpreferred." These categories usually excluded Chinese immigrants, for example, and severely limited others from Asia.

Stage 3 saw the campaign against Japanese and Chinese immigrants known as the campaign against the "Yellow Peril." A Chinese exclusion law was passed banning poor Chinese immigrants: it was not repealed until 1947.[6] By 1931 there was an increase of 19 000 Asians in Canada, bringing the total to 85 600, but the census of 1941 showed the number had declined to 74 000.

During the 1920s blacks also suffered from the racist nature of Canada's immigration policy. In the middle of the decade the government decided that a "British subject" would be defined as a citizen of a Commonwealth country whose population was predominantly white.

In the meantime, other ethnic groups continued to arrive. Between 1923 and 1930, 20 000 Swedes, 19 500 Norwegians, and 17 000 Danes immigrated to Canada.

The Great Depression essentially stopped immigration to Canada. Between 1931 and 1941 only 140 000 immigrants arrived. Some groups saw their numbers decline. The populations of German, Russian, and Asian origin in Canada declined by 9000, 4000, and 10 500, respectively. Throughout the war, Canada was reluctant to accept the victims of terror from Nazi Germany.

> **BOX 4.4**
>
> *Maroons*
>
> Maroons were slaves originally stolen from Africa who escaped slavery and their British masters in Jamaica. The Maroons established settlements in the central mountains and hills of Jamaica and subsequently fought two wars against the white colonists. At the conclusion of the second war in 1795, many Maroons were deported to Nova Scotia and Sierra Leone.
>
> *Source:* Harry Harmer, *The Longman Companion to Slavery, Emancipation and Civil Rights* (Toronto: Pearson Education Ltd., 2001), pp. 158, 168, and 203.

Stage 4: 1945–1974

Large numbers of people came to Canada after World War II. Between 1945 and 1961, 2 100 000 arrived. This was the most prolonged period of immigration in Canadian history. It was also the most diverse in terms of social class, ethnicity, and occupation. Large numbers of Italians, Germans, Poles, Jews, and Dutch came. Immigrants from Britain, however, accounted for the largest number—one-third of the total.

Canada at this time began to experience all the benefits and problems associated with an advanced, industrialized, urban country. Not surprisingly, immigrants began to settle in cities and towns. Toronto became the favourite destination. By 1961, 42 percent of Toronto's population and one-third of Metro Toronto's were not born in Canada. In fact, 29 percent of those living in Toronto and one-third of those living in Metro Toronto arrived there between 1946 and 1961.

The largest wave of immigration in Stage 4 occurred between 1951 and 1960. In just nine years more than 250 000 Italians, for example, entered Canada. Like the Greeks and Portuguese, many of them left rural areas to settle in Canadian cities. These immigrants shared many other characteristics. They were largely unskilled, had no experience with city living, and had low levels of formal education. These three groups headed mostly for Toronto, where today they live in large recognizable communities.

The case of Italians living in Toronto is especially interesting. The Italian population of Metro Toronto now is estimated at more than 400 000, making it the largest Italian community outside of Italy.

This period is also remembered for its racist and brutal treatment of Japanese-Canadians. At the end of World War II, 4000 Japanese-Canadians were forced to leave

Canada under a "repatriation" program. More than half had been born here, and more than two-thirds were Canadian citizens! How would it feel to know that your birthplace and citizenship mean nothing if the government decides to deprive you of your basic human rights? The story of Japanese-Canadians sent to internment camps, most of them in British Columbia—with only what they could carry and the remainder of their belongings being confiscated and sold by the government—now form an integral part of discussions of Canada and the war years. When released after the war was over, most of the internees left British Columbia, and, by 1961, 8000 Japanese-Canadians were living in Metro Toronto.

Although the Bilingualism and Biculturalism Report on Immigration ended its analysis in 1974, immigration to Canada did not. We can add a fifth stage to Canadian immigration history, largely as a result of the 1974 change in Canadian immigration policy away from seeking those with specific skills to satisfying the demands of the Canadian economy. As the economy changes, so do the skills demanded of immigrants.

Stage 5: 1975 to the Present

The dominant feature of Stage 5 is emigration from countries in the developing world. Low levels of economic development characterize these countries and the overwhelming majority of their populations are not white. The arrival of these people, who form racial minorities in Canada, has had a profound effect. With new immigrants arrive new challenges; this is especially true for their preferred destination, Metro Toronto.

The most reported origins of immigrants between 1991 and 1996 were Chinese, East Indian, Filipino, Sri Lankan, Polish, and Vietnamese.[7] In fact, between 1991 and 1996, almost 1 039 000 immigrants arrived in Canada: 57 percent were born in Asia.[8] Nearly a quarter of all immigrants arrived from Eastern Asia (Hong Kong, People's Republic of China). The next largest group was Southern Asia (Indian subcontinent). European immigrants accounted for 19 percent of the 1991–96 total, down significantly from 90 percent in the years before 1961. Before 1961, European immigrants arrived mainly from the United Kingdom, Italy, Germany, and the Netherlands (see Table 4.1). Between 1991 and 1996, most European immigrants came from East European countries such as Poland, Romania, and Russia. In fact, Poles are the only European immigrants between 1996 and 2001 to be in the top 10 of source countries; Canada's mosaic is definitely becoming more colourful and educated (see Box 4.5).

Since the 1951 Census, immigrants have consistently composed approximately 16 percent of the Canadian population. This number grew to 18 percent in 2001 and is expected to reach between 19 and 23 percent in 2017.[9] By 2017, the racially visible

Table 4.1 Top 10 Countries of Birth, Canada, 2001

	Immigrated Before 1961			Immigrated 1991–2001*	
	Number	Percent		Number	Percent
Total immigrants	894 465	100.0		1 830 680	100.0
United Kingdom	217 175	24.3	China, People's Republic of	197 360	10.8
Italy	147 320	16.5	India	156 120	8.5
Germany	96 770	10.8	Philippines	122 010	6.7
Netherlands	79 170	8.9	Hong Kong, Special Administrative Region	118 385	6.5
Poland	44 340	5.0			
United States	34 810	3.9			
Hungary	27 425	3.1	Sri Lanka	62 590	3.4
Ukraine	21 240	2.4	Pakistan	57 990	3.2
Greece	20 755	2.3	Taiwan	53 755	2.9
China, People's Republic of	15 850	1.8	United States	51 440	2.8
			Iran	47 080	2.6
			Poland	43 370	2.4

*Includes data up to 15 May 2001.

Source: Statistics Canada, "Canada's Ethnocultural Portrait: The Changing Mosaic," *2001 Census: Analysis Series,* 21 January 2003, Catalogue No. 96F0030XIE2001008, available <www12.statcan.ca/english/census01/products/analytic/companion/etoimm/canada.cfm>, accessed 30 March 2006.

BOX 4.5

The Brain Gain

Independent immigrants arriving in Canada must meet education and occupation-related criteria. In 1998, 72 percent of the skilled workers selected had university degrees. Even when their dependents 15 years of age and over are included, the number drops only slightly to 59.6 percent. This rate is more than four times the rate of university graduates among

(cont'd)

> Canadian-born households (13.3 percent). In the wake of 11 September 2001, the new *Immigration and Refugee Protection Act 2002* and provisions set out in Bill C-36 reaffirmed the kind of immigrant Canada wishes to attract: highly skilled, experienced, well educated, and fluent in French or English.
>
> **Sources:** Andrew Brouwer, "Immigrants Need Not Apply," Caledon Institute of Social Policy site, October 1999 <www.caledoninst.org>, accessed 30 March 2006; Yasmin Abu-Laban and Christina Gabriel, "Security, Immigration, and Post-September 11 Canada," in Janine Brodie and Linda Trimble, eds., *Reinventing Canada: Politics of the 21st Century* (Toronto: Prentice Hall, 2003), p. 299.

population of six urban areas will exceed the national average: Toronto, Vancouver, Ottawa-Gatineau (Ontario side), Abbotsford, Calgary, and Windsor. Toronto and Vancouver are projected to have the highest numbers at 50.6 and 49.2 respectively.[10]

ETHNICITY, RACE, AND SOCIOLOGICAL THEORY

We will now examine three dominant theoretical approaches to society and how they differ with respect to what constitutes society, its development, and the various issues and problems about society that need to be discussed and debated.

Symbolic Interaction Theory

Interactionists believe that relations in society can be viewed by examining the communication and manipulation of symbols. Understanding race and ethnic relations and dealing with conflicts in society must be done by examining how each individual defines the situation he or she is in and how this definition is influenced by culture, race, and ethnicity. Interactionists ask questions such as: How do members of different groups define the world around them? What role does language play in helping define the situation? Why do symbols play such an important part in race and ethnic relations?

Conflict Theory

Conflict theorists believe that power is the key to understanding interaction in society. Power, they argue, comes primarily from the ownership of those things necessary to produce goods in society, such as land, resources, buildings, machines, technology, and knowledge. Different groups in society, those who own things and those who do not, compete for power. Some groups attempt to change the status quo, whereas others work to preserve it. Conflict theorists ask questions such as why do some ethnic groups have more power than others? Do government policies such as multiculturalism and employment equity (affirmative action in the United States) really address the power imbalance in society, or, are they simply superficial modifications to prevent real changes to the distribution of power? Generally, conflict theorists believe a radical reorganization of the power structure in society is a prerequisite for stable race and ethnic relations.

Structural–Functionalist Theory

Structural–functionalists emphasize the importance of maintaining social order. In their analysis of race and ethnicity, they concentrate on the manner in which policies such as multiculturalism and affirmative action contribute to social order. Do stereotypes and prejudice have a negative effect on society? How do race and ethnic relations in Canada promote or destroy social order? These are common questions that a structural–functionalist might ask.

THE SOCIAL MEANING OF CULTURAL IDENTITY

Socially, a person's cultural identity comprises many different elements, including race, ethnicity, class, sex and gender, religion, region, occupation, language, country of origin, and sexual orientation. We will discuss two of the most crucial of these elements of identity: ethnicity and race.

Ethnicity

Your perception of your ethnicity is one of the most important elements of your identity. Many definitions of **ethnic groups** exist, and most of them have three common elements: (1) an ethnic group shares a common ancestry and history, (2) an ethnic group shares many norms, values, and traditions, and (3) an ethnic group is considered a group by those others who do not share the first two elements. We can define ethnic groups, therefore, as those who share several norms, values, and traditions; have a common ancestry and history; and are considered distinct by the rest of society because they share these elements.

Language is essential to the identity of any ethnic group, although not all ethnic groups require their members to share a language. Without question language is the most crucial aspect of identity. It is the symbolic mode of communication we use to transmit ideas, information, and history from one generation to the next and to socialize the young. As the well-known sociologists Peter and Brigitte Berger put it: "Language provides the lifelong context of our experience of others, of self, of the world . . . it provides the most powerful hold that society has over us."[11] It is not surprising that language is such an important component of ethnicity.

Race

Race is the second major component of identity. Race is also one of the most misused and misunderstood words in the field of diversity. It has two separate meanings: (1) a biological meaning and (2) a sociological one.

From a biological perspective, we all belong to the same race—the human race. This simply means that all humans belong to one species. Therefore, all humans, regardless of which population they belong to, are genetically compatible. This in turn means that they can reproduce and produce viable (i.e., fertile) offspring. As an example, a person of Inuit descent can produce fertile offspring with an individual of Australian Aboriginal descent. If humans were not compatible genetically, we would belong to different species and, therefore, be unable to reproduce fertile offspring. A horse and a donkey belong to different species and their offspring, known as a mule, is sterile and cannot reproduce. Because we all belong to the same species, we can reproduce regardless of what other humans we decide to have children with.

The meaning of the term "race" has three problems. First, race refers to subspecies. Most traits used to define subspecies, or races, vary along a continuum. But when we attempt to divide people into different races, we do so in a completely subjective and arbitrary way. Suppose we divide people according to skin colour and type of earwax (wet versus dry), inevitably some people would share both these traits and some would not. So how would we be able to divide people into specific racial categories?

The second problem with the question of race concerns interbreeding between populations. Humans are and always have been, even before the development of mass transportation, a very mobile species. This means it is virtually impossible to accurately assess the amount of interbreeding that has taken place between human populations. As a consequence of this, there are no true "pure" human races, and there probably never have been.

Finally, the third problem with the question of race has to do with its origins—where did this term come from? The concept of race is associated with a history of exploitation. When Europeans started to travel to new lands, the economic benefit of exploiting others became the driving force of European colonization. It was easier to exploit and abuse a population if they were viewed as less than human. Systems of classifying people that were developed hundreds of years ago arranged populations into a hierarchy with the best ("closest to God") being white and the less human ("furthest from God") being darker skinned—the darker the skin the further away from God. Sadly, this kind of thinking

CRITICAL THINKING BOX 4.3

Which sociological theory do you believe best explains race and ethnic relations in Canada? List at least three reasons for your choice and explain each in detail.

> **BOX 4.6**
>
> *Protection of the French Language: Bill 101 (1977) and Bill 178 (1988)*
>
> With the intention of protecting Francophone identity, successive Quebec governments have gone to great lengths to ensure the use and survival of the French language in Quebec and, ultimately, in Canada.
>
> In 1977 the Parti Québécois government introduced Bill 101. Bill 101 reaffirmed that French would be the language of business and labour and restricted English-language schooling to two groups of people: (1) those whose parents were educated in English and (2) those who were already attending English-language schools. The target of Bill 101 was the immigrants arriving in Quebec who were sending their children to English schools and not to French ones.
>
> In 1988 the Liberal government of Quebec introduced its own language bill, Bill 178. Known as the "sign law," it restricted the use of English-language signs in two ways: first, by not allowing English on outside signs and second, by legislating that the letters on French signs inside stores be three times larger than English ones.

persists, even in the field of education, where researchers assert that some populations are not as intelligent as others are by virtue of their genes (genetic makeup).

Sociologically, our differing physical features are symbols that are accompanied by emotionally charged meanings. In short, physical features influence the way people see themselves and the way people interact. As Carl E. James has written: "Race is significant as long as groups are determined by selected physical traits, and as long as people act upon these meanings. However, we must bear in mind that race is largely based on its social meaning."[12]

It is important to understand the emotionally charged nature of physical symbols. Sociologically, race is part of everyone's life. Whites in Canada tend to take their skin colour for granted while at the same time they identify others by the colour of their skin.[13] Indeed, white people in Canada are in large part socially ignorant of the benefits of being white in a white-dominated society. Whites in Canada (the dominant group) define what is good, acceptable, excellent, and successful. In short, they have the power to define the standards by which all members of society are judged. "Whiteness" is generally associated with good and "blackness," with few exceptions, is associated with negativity, gloominess, pessimism, hostility, and evil[14] (see Box 4.7).

BOX 4.7

Racism in Cultural Symbols

- black magic is evil (and white magic?)
- blackmail (why not whitemail?)
- a black cat is bad luck
- the black prince is the evil prince
- a black mark on your record is a negative thing
- a black heart has no love
- black humour (what about white humour?)
- black leather is worn by social and sexual deviants
- black lingerie is worn by sexually deviant women
- the black sheep of the family
- devil's food cake (chocolate)
- angel food cake (vanilla)

Can you think of any others?

Over the years, many non-white students have identified incidents and situations where they have experienced different treatment when compared with white students of the same age. Students have sarcastically referred to these experiences as "special" treatment (see Box 4.8). Some common experiences of non-white students include being watched more closely when shopping in malls, being accused of public loitering when

BOX 4.8

Other Examples of "Special" Treatment

- Schools teach the history of British and French people in Canada but ignore the history and contributions of other people.
- The few minority members who are successful in the field of television news are concentrated in sports, weather, and traffic reporting. Few minorities are primetime anchors for national news programs.

> - Politicians in Canada are usually white.
> - White people tend to believe that there is a "black" culture but no such thing as a "white" culture.
> - Many people believe that all blacks in Canada are from Jamaica.
> - White people generally believe that people from the subcontinent (India, Pakistan, and Bangladesh) are poor.
>
> For a similar but more focused discussion see Peggy McIntosh, "White Privilege: Unpacking the Invisible Knapsack," in *Peace and Freedom*, July/August 1989, pp. 10–12.

conversing in groups, having change slapped down on the counter rather than being handed the change, being accused of being drug dealers when carrying pagers, and being served last in restaurants and bars.

To conclude, when we speak of identity, the social meaning of race takes precedence over the biological meaning. Biologically speaking, race is an outmoded and archaic concept.

ATTITUDES AND BEHAVIOURS THAT AFFECT SOCIAL INTERACTION

Many factors play a negative role in social interaction, including prejudice, discrimination, racism, and ethnocentrism.

Prejudice

Prejudice is the attitude of prejudging people on the basis of statements and beliefs that do not hold up to rational or critical scrutiny. Prejudice occurs when, in spite of evidence to the contrary, a person still holds negative feelings and opinions toward other people and groups. Prejudgments are seldom based on experience. Carl E. James has written, "The tendency to make prejudgments may be seen as necessary, as the human mind needs to organize the stimuli with which it is bombarded."[15] In many respects prejudging is accepted and expected in a world where the amount of knowledge understood by the human race is said to double every five years. Prejudging seems to make a complicated world easier to deal with. It is not surprising, therefore, that prejudice is a universal phenomenon and a common problem in social interaction.

The Many Faces of Diversity

> **CRITICAL THINKING BOX 4.4**
>
> How many events of "white privilege" have you witnessed or experienced recently? Why are whites largely ignorant of such "privileges"? Give at least three reasons.

The human tendency to put things into groups or categories is clearly visible when we speak of **stereotypes**. A stereotype can be defined as a collection of generalizations about a group of people that are negative, exaggerated, and cannot be maintained when subjected to critical analysis. Some common stereotypes include the following:

- All Italians belong to the Mafia.
- All blacks are criminals.
- All Aboriginals are drunks.
- All Sikhs drive taxis.
- All Pakistanis own corner stores.
- All Jews are cheap.

The key feature that explains the attractiveness of stereotypes is that they are overly simplistic. Minority groups have continually complained of their stereotypical treatment by the mass media.

Discrimination

Prejudice is the attitude, and **discrimination** is the action. Discrimination can be defined as the unequal or unfavourable treatment of people because of their perceived or actual membership in a particular ethnic group that restricts their full participation in Canadian society. When discrimination is carried out on the basis of race (people who share physical characteristics) and when common behaviour is assumed (that people who share certain physical characteristics behave in a certain way), we call this **racism** (see Box 4.9). Many regard the denial of Aboriginal land claims and self-government as a form of racism. Land claims and self-government are crucial to Aboriginal identity and survival—to regain control over their traditional land, including access to all resources, especially hunting and fishing rights. Others point to the 1999 Nisga'a agreement as a model for future agreements with Aboriginal people. This agreement met Nisga'a demands in the areas of fishing and hunting rights, land ownership, law, and financial compensation. The treatment of the Nisga'a contrasts sharply with Ottawa's treatment of the Mi'kmaq.

> **BOX 4.9**
>
> *Racism in the Canadian Justice System*
>
> The existence of racism in the Canadian Justice System was acknowledged by the minister of state for multiculturalism and citizenship in 1990, the Law Reform Commission of Canada in 1992, and the Law Reform Commission on Systemic Racism in the Ontario Criminal Justice System in 1995. The report of the latter concluded that blacks constitute just less than 3 percent of the population of Ontario but 15 percent of the prison population. The *Toronto Star* reported in 2001 that analysis of police data from 1996 to 2001 showed people with black skin are twice as likely to be stopped by police as people with white skin. A similar study released in Kingston in May 2005 found that black-skinned people were 3.7 times more likely to be stopped in that city than white-skinned people.
>
> The same is true for Aboriginal People, who make up about 2.8 percent of the Canadian population yet comprise 18 percent of the federal prison population in 2003. In 2002, Aboriginal males composed 17 percent of the federal male prison population while Aboriginal females composed 20 percent of the female population.
>
> **Sources:** Frances Henry and Carol Tator, *The Colour of Democracy: Racism in Canadian Society*, 2nd ed. (Toronto: Harcourt Brace, 2000), p. 147; John Sewell, "Same Old Diversionary Game," *Eye Weekly*, 7 November 2002, available <www.eye.net/eye/issue/issue_11.07.02/news/citystate.html>, accessed 30 March 2006; "Police Stop More Blacks, Ont. Study Finds," *CBC News*, 27 May 2005, available <www.cbc.ca/news/viewpoint/yourspace/racial_profiling.html>, accessed 30 March 2006; Aboriginal Initiatives Branch, "Facts and Figures: Aboriginal Offenders—Overview," Correctional Service Canada site <www.csc-scc.gc.ca/text/prgrm/correctional/abissues/know/7_e.shtml>, accessed 30 March 2006; Mark Lachmann, "Human Immunodeficiency Virus: Emerging Epidemic in Aboriginal People," *Canadian Family Physician*, October 2002, available <www.cfpc.ca/cfp/2002/Oct/vol48-oct-editorials-1.asp>, accessed 30 March 2006.

There are two types of discrimination, individual and institutional. **Individual discrimination** is perhaps the most common type, but, as we will see, it may not be the most damaging. Examples of such individual actions include refusing to sit next to or associate with members of minority groups while in public places; giving or receiving poor or slow service in restaurants and stores; refusing to date people from outside your ethnic or racial group; and, finally, discouraging your children from developing friendships with people from outside your ethnic or racial group. Sometimes individual discrimination takes the form of what Daniel Hill called "nice guy" discrimination.[16] This means people will act

> ### CRITICAL THINKING BOX 4.5
>
> *What About Minority Experiences?*
>
> "When discussions of Canadian diversity begin and end with the ongoing struggles between English and French Canada, the accomplishments, experiences, desires, and troubles of minority groups are pushed to the margins of Canada's cultural discourse. As a result, consciously or unconsciously, politicians, professors, teachers and community leaders erect another barrier to minority inclusion in society." Is this viewpoint true? If so, why? If not, why?

in a discriminatory manner and justify it according to the potential for negative reactions from others. So, for example, you don't hire members of minority groups for fear that other workers may "rebel" and not accept them.

The second type, **institutional discrimination**, occurs daily and limits the full participation of minority groups in the political, economic, and educational institutions in Canada. Some debate exists about the racist nature of this form of discrimination. The consensus seems to be that the intent is not racist but the outcomes certainly are. This is especially true of the people who work in these institutions—they are usually unaware of the outcomes of their institutional policies and practices.

An example of this form of discrimination is hiring practices. In 1989, racial minorities composed only 4 percent of the Metro Toronto Police Force.[17] These officers were highly concentrated in the ranks of cadets, constables, or in training; only three held the rank of inspector.[18] By 1998, gains by racial minorities were insignificant: only three held the rank of staff inspector, only three were senior police officers, and only 7.4 percent of all uniformed employees were racial minorities.[19] Yet, in 2001 only 15 percent (51 of 332) new recruits represented racial minorities; this number dropped to 11 percent in 2004.[20]

> ### CRITICAL THINKING BOX 4.6
>
> Are jokes based on racial and ethnic stereotypes "just jokes," or do they reinforce negative attitudes and beliefs and in the process destroy self and group esteem? Why are these jokes so popular? Give at least three reasons and explain them.

This does not auger well for one of the most multicultural cities in the world where the racial minority population in 2017 is expected to surpass 50 percent.[21] The city of Montreal faces similar challenges. Between May 1999 and April 2000, only 42 of 278 new recruits of the Montreal police force were racial minorities.[22]

Another example of institutional discrimination includes textbook selection in schools, marketing, advertising, minimum levels of education for hiring, and cultural biases in aptitude or qualification tests. During the 1970s, the reading texts for the primary grades in Ontario schools were the *Mr. Mugs* series. The series centred entirely on a white, middle-class family. Small wonder that in 1971 a report on the study of 400 textbooks in Ontario, entitled *Teaching Prejudice*,[23] painted an unflattering picture of textbooks used in Ontario schools (see Box 4.10).

Ethnocentrism

Ethnocentrism is the attitude whereby an individual views the world from the point of view of his or her own culture. There are two variations of ethnocentrism: (1) the assumption that what is true of your culture is true of other cultures and (2) the belief in the superiority of your culture in comparison with other cultures. The most common form of ethnocentrism is the first. It is embodied in statements that question why certain cultures do not behave or do things in the same manner as your own culture does. Statements of this nature would include variations of: "Why don't they do it like this [meaning, like we do]?" "That's a strange way of doing things." "They're weird." Ethnocentric attitudes definitely influence

---

BOX 4.10

Teaching Prejudice in 1971

Non-white groups were frequently referred to as bloodthirsty, primitive, cruel, and savage, in contrast with saintly and refined Europeans. With only a few passing exceptions, Aboriginals, blacks, and Asians who contributed to Canadian development in significant and positive ways were omitted from reference. In addition, major events in the sad history of Canada's mistreatment of minorities—the extinction of the Beothuk, the treatment of Japanese-Canadians during World War II, the abuse of Métis and Indians throughout Canada—were barely touched upon, if at all.

Source: *Teaching Prejudice*, quoted by Daniel G. Hill, *Human Rights in Canada: A Focus on Racism* (Ottawa: Canadian Labour Congress, 1977), p. 15.

social interaction, and, furthermore, many ethnocentric attitudes and beliefs are held unconsciously. An example that vividly demonstrates how racism, prejudice, discrimination and ethnocentrism affect social behaviour is the issue of racial profiling.

September 11 and the Rise of Racial Profiling

The Ontario Human Rights Commission defines **racial profiling** as any action undertaken for reasons of safety, security, or public protection, that relies on stereotypes about race, colour, ethnicity, ancestry, religion, or place of origin, or a combination of these, rather than on reasonable suspicion, to single out an individual for greater scrutiny or different treatment.[24] Racial profiling equates physical characteristics with particular negative behaviour. This makes the thinking that surrounds it prejudiced and the behaviour that it leads to discriminatory and racist.

According to the Ontario Human Rights Commission, racial profiling is considered widespread in the United States, United Kingdom, and Canada. Police and immigration officials have been singled out as the most consistent practitioners. The Ontario Court of Appeal found racial profiling exists in the province of Ontario when it upheld a lower court decision in the Dee Brown case.[25] The former Toronto Raptor had his drunk driving conviction overturned when the appeal court ruled the lower court failed to appreciate the fact that racial profiling can be a "subconscious factor" when exercising "discretionary power."[26] In recent years community policing has focused on combating and eliminating the use of racial profiling. In October 2002, on the basis of police arrest records, the *Toronto Star* published a series of articles dealing with racial profiling in the city.[27] Some 10 000 cases between 1996 and early 2002 dealing with "simple" drug possession were analyzed, and they showed 63.8 percent of those arrested were designated as white while 23.6 percent were designated as Black. Blacks were more likely than whites to experience being denied bail, held in jail overnight, and charged with driving offences (failing to update a driver's licence or driving without insurance). The use of racial profiling as an instrument of policing is not conducive to establishing good relations with different cultural groups. This is especially true in Toronto, Vancouver, and Montreal, Canada's largest and most diverse cities.

Since the terrorist attacks of 11 September 2001, racial minorities leaving Canada or returning to Canada have been subjected to racial profiling on a regular basis. During the same period, the United States pressured Canada to spend more on border/airport security and screening. American talk of "harmonizing" national regulations in these areas created a sense of uneasiness in many Canadians. These fears were legitimized and racial profiling received national attention when the case of Canadian citizen, Maher Arar, was made public. A 34-year-old, Syrian-born Canadian (in Canada since the age of 17) Arar was vacationing in Tunisia in 2002 with his wife and children. While changing planes in

> **BOX 4.11**
>
> *Racial/Ethnic Groups Are Not Monolithic Entities*
>
> No racial/ethnic group in Canadian society is a monolithic entity. There is huge variation within each group in every defining characteristic, especially income. Not all members of the dominant group earn the same or similar yearly income. Statistics Canada shows that earnings of Greeks, Portuguese, and other members of the "white" dominant group are far below British and French earnings. The same is true of racial minorities: the earnings of some groups rival and surpass those of the dominant group. Canadians of Japanese origin have average earnings that exceed those of most members of "white" ethnic groups—they are overrepresented in business, the arts, and universities considering their percentage of the Canadian population.
>
> **Sources:** Statistics Canada, "Ethnocultural Portrait of Canada," Tables 15 and 18; Walter Johnson, *The Challenge of Diversity* (Montreal: Black Rose Books, 2006), p. 43.

New York's John F. Kennedy airport, he was detained and questioned by American officials about supposed links to the terrorist group, al-Qaeda. Against his consent, he was flown to Jordan, then to Syria where he was jailed, beaten, tortured, and forced to sign a false confession. American officials knew details of Arar's life that could have only been provided by Canadian authorities. His case got so much attention in Canada that the federal government established a Commission of Inquiry on 28 January 2004 whose purpose was to investigate the role played by Canadian authorities in the affair, including the use of racial profiling.

THE MEANING AND IMPLICATIONS OF MULTICULTURALISM

Multiculturalism is one of the most used and most misunderstood words in the field of ethnicity. It has many definitions and explanations; some focus on the concept of culture, some on the ideology on which it is based, and still others refer to it in terms of government policy.

Multiculturalism is commonly associated with the idea of the "Canadian mosaic"—the view that different cultural groups come to Canada, retain their language and culture, and still become Canadian. It is not uncommon to consider yourself a hyphenated Canadian: French-Canadian, Greek-Canadian, Ukrainian-Canadian, Jamaican-Canadian, and the like. The concept of hyphenated citizenship is at odds with what has prevailed in the United States, where people are expected to abandon their heritage culture and assimilate into the dominant one. Thus, U.S. society is termed a "melting pot."

Perhaps one of the best attempts to define this complicated word is provided by Fleras and Elliott. They combine the cultural and ideological components with the reality of government policy for an all-inclusive definition. According to Fleras and Elliott, multiculturalism is "a doctrine that provides a political framework for the official promotion of cultural differences and social inequality as an integral component of social order."[28]

Multiculturalism is based on the ideology of **pluralism**, which is the belief that ethnic diversity and conflict remain central features of modern industrial societies and that ethnicity continues to be an essential aspect of individual identity and of group behaviour. The implications of the acceptance of pluralism are threefold. First, pluralism recognizes that cultural and racial conflict is inevitable and unavoidable; at best we can only hope to manage and mitigate this conflict as opposed to pursuing some fantasy of eliminating it. Second, pluralism recognizes that people hold on to their cultural heritage with much passion and commitment; therefore, to better manage cultural conflict, we must openly acknowledge and accept cultural differences. Third, pluralism is the belief that pursuing a multicultural policy is beneficial to Canadian unity, for it strengthens Canada and enriches the Canadian experience, especially in the fields of literature and the arts. In addition, pluralism lets Canadians know that they are part of a larger global community where we all must communicate and cooperate with each other if we are to survive.

Pluralism is the opposite of **assimilation**. Assimilation is the process by which immigrants adopt the language, values, norms, and worldview of the host culture. All immigrants experience some degree of assimilation. Immigrant cultures cannot totally insulate or isolate themselves from the influences of the host society—this is especially true of immigrant children, because children must attend school. The education

---

CRITICAL THINKING BOX 4.7

How do, or did, your high-school and postsecondary school texts compare with the standards of *Teaching Prejudice*? What about the newspaper you read, the news programs you watch, or the movies you see at the theatre and rent at the video store?

> **CRITICAL THINKING BOX 4.8**
>
> In 2005, the Ontario government officially rejected formal calls from the Islamic community that Muslims be given the right to use the body of Shariah law for divorce proceedings for those Muslims who ask for it. In other words, some Muslims were asking for the right to forgo the application of Canadian divorce laws in favour of Muslim ones. Should the Islamic community be accommodated? If yes, why? If not, why? Give three reasons for each.

system is perhaps the single most persuasive assimilationist force that immigrants are exposed to. While in school, immigrant children are socialized to the complete worldview of the host culture: its history, accomplishments, values, norms, customs, biases, and prejudices. Some assimilation on the part of immigrants is an inevitable fact of immigration and, thus, never completely avoidable.

The Purpose of Multiculturalism

The response of the federal government to the history of immigrant experiences with prejudice, racism, discrimination, stereotyping, and the like has been to officially recognize the racial and ethnic diversity in Canada and to support and encourage Canadians to learn more about the people of Canada and the country itself. For decades the federal government has encouraged the acceptance of different cultures through various policies and initiatives broadly labelled as "multicultural." In 1988, Parliament officially enacted the **Multiculturalism Act**, which pledges federal government assistance "in bringing about equal access and participation for all Canadians in the economic, social, cultural and political life of the nation."[29]

To achieve these goals the Act outlines 10 specific objectives. The objectives have much consistency; we can narrow the list to three general goals:

1. The promotion of both official languages. The federal government declares that multiculturalism will be pursued within a bilingual, English and French, framework.

2. A commitment to help all members of cultural groups to overcome barriers that limit their full participation in Canadian society.

3. The promotion of understanding among different groups and the acceptance of cultural differences.

Multiculturalism is not the only government initiative designed to promote equality and equity. Other initiatives of the federal and provincial governments include the introduction of a *Charter of Rights and Freedoms* with the repatriated Constitution of 1982, human rights commissions, and employment equity programs (see Box 4.12).

With these ideas in mind, Canadians need to come to a consensus concerning the purpose of multiculturalism. Too many people believe, with encouragement from the media, that multiculturalism is supposed to cure all problems in society that are a result of numerous different cultures living in close proximity to each other. Nothing could be further from the truth. The reality of multiculturalism, as government policy, is about the *management* of racial and ethnic conflict. It is not about the *elimination* of such conflict, which is unrealistic. Cultural conflict can be better managed if all groups have the same degree of cultural freedom. The more all groups feel they are free to live in and promote their culture; the more they will respect the culture of other members of society.

BOX 4.12

Employment Equity and Toronto Firefighters

In the 1990s, more than 90 percent of the employees of the Toronto Fire Department were white males. To help the department better reflect the community they serve, the personnel committee of the Toronto City Council recommended that the department give preference to 13 *qualified* women and racial-minority candidates, ahead of white men who scored slightly higher during the qualification process. The chief of the department protested vigorously even though the 13 candidates were drawn from a total pool of 140 fully qualified candidates selected from 4000 applicants; all candidates had passed the physical, health, and aptitude tests. The motion to hire these candidates was defeated by city council. The Ontario Human Rights Commission filed a complaint, claiming that the Toronto firefighters' union was blocking the proposal and creating a "poisoned work environment."

Source: Frances Henry and Carol Tator, *The Colour of Democracy: Racism in Canadian Society*, 2nd ed. (Toronto: Harcourt Brace, 2000), p. 364.

THE TYRANNY OF MULTICULTURALISM OR MULTICULTURAL TYRANNY?

The Tyranny of Multiculturalism: The Experience of Minority Groups

If language is the key to the survival of culture, why does Canada only guarantee the use of two official languages? Language is the most important element of culture. Without it, there is no culture. It provides the symbols, tools, perceptions, and definitions for each and every culture. In fact, language provides worldviews that make cultures distinct. In all these ways, language provides the foundation for identity. Without guaranteeing language, multiculturalism is reduced to a "feel-good" policy designed to give people a false sense of cultural security in a country that officially and loudly professes a commitment to cultural acceptance but really prohibits and limits important cultural behaviours on a daily basis.

Official multiculturalism within a "bilingual framework" is problematic in many ways. As Carl James and others[30] point out, immigrant and minority full participation in society is limited without knowledge of the English and/or French languages. Minority ethnic groups are promised government help to "overcome barriers that limit their full participation in Canadian society" (the second purpose of multicultural policy—see number 2 on page 117). Again, as James points out, at the same time, the government openly declares that "there is no official culture, nor does any ethnic group take precedence over any other," while publicly acknowledging "the undisputable role played by Canadians of French and British origins in 1867, and long before Confederation."[31] The dominance of the two dominant groups is strengthened by institutionalizing their worldviews. The English and French languages dominate the operation of government, the bureaucracy, and the most influential assimilating institution of all, public schools. With political, economic and social dominance, the pressures on minority groups to conform are inescapable. Structural problems with the unequal distribution of power cannot be corrected with multicultural solutions.[32] In fact, official multiculturalism marginalizes minorities by blocking their access to power and resources.[33] The official policy of multiculturalism is really "monoculturalism in disguise."[34] As such, minority groups can realistically regard official multiculturalism as a tyrannical entity intent on destroying their various cultures.

Multicultural Tyranny: Majority Group Experiences

Does official multiculturalism help or hinder the acceptance of cultural diversity? There is a huge difference between cultural diversity and multiculturalism. As an idea, diversity means "difference" and refers to the different languages, religions, food, dress, sport, social customs, and leisure activities that make Canadians different from each other. Accepting the idea of diversity means accepting the fact that people are different. Generally, this is not regarded as a problem. Multiculturalism as official policy, however, means accepting all cultures as equal. This is simply not true in Canada at the beginning of the 21st century. The policy gives minority groups the illusion of being a part of Canadian society yet many of their norms, customs, and folkways are treated with contempt. Countless attitudes and behaviours that various minority groups have brought and continue to bring to Canada are deemed unacceptable to the vast majority of Canadians and in some cases, illegal (see Critical Thinking Box 4.9). In fact, when real cultural differences challenge established Canadian traditions and Canadian law itself, the line between right and wrong becomes blurred. Most Canadians do not believe multiculturalism means the acceptance of cultural differences that challenge established norms, mores, and laws. Yet many immigrants believe that it does. Canadians are simply not prepared to accept polygamy, female circumcision, eating dogs, and using an assortment of hallucinogenic drugs. In some cultures, these behaviours are equally as acceptable as monogamy, male circumcision, eating cows, and drinking alcohol are in Canadian culture. Yet the former set of behaviours is illegal and the latter is not. Once some cultural behaviours are criminalized, the visible and "safe" differences of dress, music, faith, leisure activities, and cultural festivals are the ones that remain. In the end, individual tolerance of visibly mundane cultural behaviours is equated with societal acceptance of "multiculturalism."

There is a difference between multiculturalism as official government policy and multiculturalism as reality. Accepting our "hyphenated" labels (e.g., Jamaican-Canadian, Polish-Canadian, Philippino-Canadian) tends to take focus away from our shared "Canadian-ness" and highlight our minority status.[35] As Neil Bissoondath has pointed out, official multiculturalism stymies what the designers of the policy were looking to encourage,

---

CRITICAL THINKING BOX 4.9

Official Multiculturalism or Official Hypocrisy?

East African immigrants principally from Somalia and Sudan practise female circumcision. This procedure involves the removal of some, or in extreme cases all, of the female genitalia. Some negatively refer to this process as "female genital mutilation." The Canadian government has officially prohibited this practice, but it continues to allow male circumcision. Why is it permissible to eat cows, pigs, and chickens, but not dogs? How can we justify one set of practices and criminalize others? Can you think of other double standards?

the acceptance of different people into Canadian society and the Canadian way of life. In fact, it has aggressively encouraged immigrants to demand that Canada adopt their laws, customs, mores, folkways, and worldviews and not the other way around.[36] The recent attempts by Muslims in Ontario to have Shariah law (dealing with divorce) implemented alongside established secular law and the continued debate over female circumcision are two such examples. The ongoing debates dealing with use of religious dress/symbols (e.g., Turban, Hijab, and Kirpan) in everyday life continue to be contentious. Many Canadians believe official multiculturalism is at the root of continued minority demands for cultural and religious accommodation in the areas of law, language, dress, and attitudes. They view these neverending demands as a form of multicultural tyranny (see Box 4.13). In fact, it is not uncommon to hear emphatic variations of the phrase, "If that's what they want, they can go back to where they came from—this is Canada!"

Quebec and Multiculturalism

Most provinces have accepted the principles of multiculturalism. In 1974, Saskatchewan was the first province to accept the principles of multiculturalism as a basis for defining majority–minority relations. Since 1974, most of the remaining provinces have done the same. Officially Quebec is designated as a multicultural province. For a variety of reasons, however, federal multiculturalism is officially rejected. The centre of this rejection deals with the belief that the federal policy challenges Quebecois authority to determine the direction of majority–minority relations in

---

BOX 4.13

Multicultural Tyranny?

In 1993, a *Globe and Mail* article suggested hockey examples used in school textbooks be removed since new immigrants are unaccustomed to the game. A doctoral student at the Ontario Institute for Studies in Education (OISE), Pushpa Seevaratnam, cited the case of a 10-year-old Sri Lankan student who couldn't conclude how long it would take a Bobby Hull slapshot travelling at 52.9 metres per second to travel 25 metres. Seevaratnam referred to the example as "ethnocentric." Does this situation meet the criteria for being "ethnocentric"?

Source: Daniel Stoffman, *Who Gets In? What's Wrong with Canada's Immigration Program and How to Fix It* (Toronto: Macfarlane Walter and Ross, 2002), p. 147.

their own province. In short, federal multiculturalism does the following: it contravenes Quebec's special status as one of the founder or charter members of Canada; it undermines Quebec's right to manage its own diversity as it sees fit; and finally, it violates the notion that Canadian federalism was constructed on a bicultural foundation.[37]

Employment Equity in a Multicultural Society

In 1986 the federal government introduced the *Employment Equity Act*. It affects all employees of the federal government and federally regulated industries. The purpose of the *Act* is to promote the equality of opportunity for four groups of workers: racial minorities, Native Peoples, women, and people with disabilities. The *Act* is based on the reality that these groups have historically been discriminated against in the labour market. Section 15.2 of the *Charter of Rights and Freedoms* protects employment equity.

Employment equity programs are considered by many to be the most effective way of reversing at least a century of discriminatory hiring practices that favoured white males. The dominant group has vigorously opposed such programs. Some of the more popular reasons cited for resisting such programs include that equity programs are reverse discrimination, that they ignore the merit principle, and that they ignore the belief that fairness is best achieved by treating everyone the same[38] (see Box 4.12). Yet, as Henry and Tator point out, each of these reasons cannot stand up to critical scrutiny.[39] First, are such programs really reverse discrimination or urgently necessary to correct at least a century of preferential treatment of the dominant group, especially white males? Are white males the *only* qualified applicants? Second, do they really ignore the merit principle or simply choose to disregard personal characteristics that do not affect job performance, such as cultural background, skin colour, gender, and family/friendship networks? Third, does treating everyone fairly translate into equal treatment? Or does ignoring and refusing to accommodate ultimately lead to discrimination?[40]

Although employment equity programs are important, it is highly unlikely that they can achieve their goals without the implementation of other measures, such as institutions making individuals accountable for discriminatory and racist behaviour; sanctioning those who behave in unacceptable ways; and being committed to devoting significant resources to combating discrimination and racism even in difficult economic times.[41]

CONCLUSION

Race and ethnicity are powerful determinants of identity in Canadian society, and it is important to understand the nature and depths of the problems associated with developing stable relations in a society characterized by racial and ethnic difference. The nature and composition of Canada is changing. New immigrants are arriving from the developing world, especially from Asia and the Indian subcontinent. Like those who came between 1991 and 2001, they settle primarily in the urban areas of Ontario, Quebec, Alberta, and British Columbia. There is no evidence to suggest this trend will change anytime soon.

Canada is not the same country today that it was fifty years ago—nor will it be the same fifty years from now. The challenge is to ensure that the social institutions that serve our society reflect the people they serve. This must be done in a manner that accepts the cultural, linguistic, and racial diversity of all Canadians—it will be difficult. Prejudice, racism, and discrimination seem to be part of every society, and it may be naive to think that we can eliminate them. We must, however, work to manage them in a productive manner so that all Canadians can develop a positive self-image and a more complete understanding of our country.

CHAPTER SUMMARY

It is almost impossible to overstate the importance of race and ethnic relations in Canada. An analysis of Canadian immigration history is crucial for a complete understanding of Canadian society and to destroy prevalent myths. Canada has never been just a purely British and French, white society—Aboriginal peoples were here, and other people arrived and continue to come from all over the world. In the early 19th century this included black slaves and in later decades Chinese and Japanese people as well. Immigrants from the developing, non-white world continue to arrive in Canada in large numbers—the Canadian mosaic is becoming ever more colourful.

Race and ethnicity are important components of individual and group identity. Meanwhile, prejudice, racism, stereotyping, ethnocentrism, and discrimination continue to have a strong negative impact on social interaction. Therefore, a policy of official multiculturalism has been enacted to help manage social interaction in Canada.

KEY TERMS

assimilation, p. 116
discrimination, p. 110
ethnic groups, p. 105
ethnocentrism, p. 113
individual discrimination, p. 111
institutional discrimination, p. 112
multiculturalism, p. 115

Multiculturalism Act, p. 117
pluralism, p. 116
prejudice, p. 109
race, p. 105
racial profiling, p. 114
racism, p. 110
stereotypes, p. 110

DISCUSSION QUESTIONS

1. What stage of immigration is known as "the mightiest movement of people" in modern history? What accounted for this mass movement of people?
2. What is the significance of the fifth stage (1975 to the present) of Canadian immigration history?
3. Define prejudice and discrimination. What is the difference between them?
4. What do we mean by race and ethnicity? Why are these ideas so important to Canadians? Provide some examples.
5. How does conflict theory account for the existence of racism and discrimination?
6. How would structural–functionalism interpret Canada's policy of multiculturalism?
7. Provide three (3) criticisms of official multiculturalism. Give an example for each.
8. Describe what is meant by the term *racial profiling*? Why is racial profiling always detrimental to community cohesiveness?

NOTES

1. "Report of the Royal Commission on Bilingualism and Biculturalism, Book IV, The Contributions of Other Ethnic Groups," in Howard Palmer, ed., *Immigration and the Rise of Multiculturalism* (Toronto: Copp Clark, 1975), pp. 1–16. Unless otherwise noted, all statistics and examples of historical events are drawn from these pages.
2. Adrienne Shadd, "Institutionalized Racism and Canadian History: Notes of a Black Canadian," in Carl E. James, ed., *Seeing Ourselves: Exploring Race, Ethnicity and Culture*, 3rd ed. (Toronto: TEP, 2003), pp. 165–68.
3. Ibid.
4. Ninette Kelly and Michael Trebilcock, *The Making of the Mosaic: A History of Canadian Immigration Policy* (Toronto: UTP, 1998), p. 36.
5. Daniel G. Hill, *Human Rights in Canada: A Focus on Racism* (Ottawa: Canadian Labour Congress, 1977), p. 10.
6. Harry Hiller, *Canadian Society: A Macro Analysis* (Toronto: Prentice Hall, 2000), p. 174.
7. Statistics Canada, "1996 Census: Immigration and Citizenship," *The Daily*, 4 November 1997, available <www.statscan.ca/Daily/English/971104/d971104.htm>, accessed 31 March 2006.
8. Ibid.
9. Canadian Policy Research Network, "Populations Projections for 2017," CPRN site, 31 March 2006 <www.cprn.org/en/diversity-2017.cfm>, accessed 10 December 2005.
10. Andrew Heisz, "Ten Things to Know About Canadian Metropolitan Areas: A Synthesis of Statistics Canada's Trends and Conditions in Census Metropolitan Areas Series," Statistics Canada (Ministry of Industry, 2005), p. 14, available <www.statcan.ca/bsolc/english/bsolc?catno=89-613-MIE2005009>, accessed 31 March 2006.
11. Peter Berger and Brigitte Berger, *Sociology: A Biographical Approach* (New York: Basic Books, 1971), p. 75.

12. Carl E. James, *Seeing Ourselves: Exploring Race, Ethnicity and Culture*, 3rd ed. (Toronto: TEP, 2003), pp. 41–42.
13. Ibid., p. 42.
14. P. Essed, *Everyday Racism: Reports from Women of Two Cultures* (Claremont, CA: Hunter House, 1990).
15. James, op. cit., pp. 134–135.
16. Hill, op. cit., p. 13.
17. Frances Henry and Carol Tator, *The Colour of Democracy: Racism in Canadian Society*, 2nd ed. (Toronto: Harcourt Brace, 2000), p. 105.
18. Ibid.
19. Ibid.
20. Walter Johnson, *The Challenge of Diversity* (Montreal: Black Rose Books, 2006), p. 43.
21. Heisz, op. cit., p. 14.
22. Johnson, op. cit., p. 90.
23. Hill, op. cit., p. 13.
24. Ontario Human Rights Commission, *Paying the Price: The Human Cost of Racial Profiling*, Inquiry Report 2004, p. 6. See also Charles C. Smith, "Crisis, Conflict and Accountability: The Impact and Implications of Police Racial Profiling" (Toronto: African Canadian Community Coalition on Racial Profiling, March 2004), and Tom Wise, "Racial Profiling and Its Apologists," *Z Magazine*, March 2002.
25. Johnson, op. cit., p. 91.
26. Ibid., p. 87.
27. James, op. cit., p. 150.
28. Augie Fleras and Jean Leonard Elliott, *Multiculturalism in Canada: The Challenge of Diversity* (Scarborough: Nelson, 1992), p. 272.
29. *The Multiculturalism Act*, 21 July 1988.
30. James, op. cit., p. 210.
31. Ibid.
32. Augie Fleras and Jean Leonard Elliot, *Engaging Diversity: Multiculturalism in Canada* (Scarborough: Nelson, 2002), p. 100.
33. Ibid.
34. Ibid., p. 48.
35. Ibid., p. 100.
36. Neil Bissoondath, *Selling Illusions: The Cult of Multiculturalism in Canada* (Markham, ON: Penguin Books, 2003), and Daniel Stoffman, *Who Gets In? What's Wrong with Canada's Immigration Program and How to Fix It* (Toronto: Macfarlane Walter and Ross, 2002).
37. Fleras and Elliot (2002), op. cit., p. 76.
38. Henry and Tator, op. cit., pp. 364–65.
39. Ibid., pp. 364–66.
40. R. Abella, *Report of the Commission on Equality in Employment* (Ottawa: Supply and Services Canada, 1984), p. 3.
41. Henry and Tator, op. cit., pp. 366–75. An excellent discussion.

CHAPTER 5

Aboriginal Peoples

John Steckley

Several years ago in a sociology class on social problems, I recall wondering if anyone else was poor, because the professor repeatedly referred to Native people as statistical examples of poverty. . . . Not for one moment would I make light of the ugly effects of poverty. But if classroom groups must talk about Indians and poverty, then they must also point out the ways in which Native people are operating on this cancer. To be sure, the operations are always struggles and sometimes failures, but each new operation is faced with more experience, more skill, more confidence and more success.

— Métis writer Emma LaRoque, "Three Convention Approaches to Native People"

Objectives

After reading this chapter, you should be able to

- understand better the diversity of Canadian Native culture
- understand the historical background and development of Native issues and contemporary circumstances
- understand the extent to which Natives are in different legal circumstances than are other Canadians
- understand the negative impact on Natives of official policies in the areas of education, religion, and the justice system
- understand some of the strengths and challenges of a Native-run justice system

INTRODUCTION

Knowing about Aboriginal Canadian people takes years. There is much to learn, so little of which can be presented in a short chapter. First, you need to appreciate the incredible length of time that Aboriginal people have been in Canada. Second, you need to know that not all Aboriginal people are the same, be that in terms of traditions and such cultural features as language, housing, and food preferences, or in other ways. There are rich Aboriginal people and poor ones. There are Aboriginal alcoholics, but also those who have never had a drink or who just have a couple of beers "with the guys," like most other Canadians. Some Aboriginal people adhere to traditional religious ways, and others make Christianity a "traditional" Aboriginal belief system. There are Aboriginal hunters and trappers, and computer programmers, too.

Much, however, is held in common. Aboriginal people share a genetic heritage, an identity, and a complex set of legal regulations that both serve and restrict them. Neither abolishing those laws—basically one law, the **Indian Act**—nor keeping them as they are will create solutions. Redefining the *Act* is the only answer, but it is difficult. The question is not one of giving Aboriginal people a separate status, but of taking the separate status that now exists and changing it for the good of Aboriginal people—and for all Canadians.

BEGINNINGS: IT ALL STARTED LONG AGO

No one knows precisely how long Aboriginal people have been in Canada. Aboriginal traditions tell us that the Creator placed them here and that they lived nowhere else before. Anthropologists say that the people came from Asia. Current speculations, based on archaeology, biological anthropology (especially the study of DNA, dental patterns, and skull shapes) and linguistics speak of four waves of people. The first (and most controversial) wave may have come from Southeast Asia, taking boats up the Asian coast and down the west coast of the Americas. The next and main wave involved "just about everybody." They walked across what is now the Bering Strait (width 90 kilometres) on a 1600-kilometre north-to-south land bridge formed during the last Ice Age (12 to 15 thousand years ago), when the water level went down with the increase in ice. The third wave may have been of speakers of Athabaskan languages. The fourth is the Inuit. Aboriginal traditions often tell a different story. Natives and anthropologists agree that Aboriginal people are the First Nations of Canada.

So what date can we use? The North American archaeological site with the best evidence for early occupation is the Meadowcroft Rockshelter in Pennsylvania.[1] Two levels at this site have very early claims to human occupation: 20 000 and 15 000 years ago. If we take the second date as valid, we can say that Aboriginal people came to Canada by at least 15 000 years ago. The Vikings came for a short stay in Newfoundland about 1000 years ago. This means roughly 93.3 percent of Canadian history is Aboriginal history before Europeans came. In terms of the analogy of a day of history, Aboriginal people came here at 12:00 midnight, and Europeans didn't even visit here until slightly after 10:30 at night.

The great length of time that Aboriginal people have been here means that they have very deep roots in this country. From these roots have grown their sense of the sacredness of the land, their feeling of being its primary caretakers.

THEORY

What is the place of Aboriginal people in Canadian history since contact? This question can be answered from several different theoretical positions. The structural functionalist position looks just at how Aboriginal people helped this country develop. During the first two centuries of postcontact history (the 17th and 18th centuries), the fur trade was Canada's biggest industry. Aboriginal people played a key role as both suppliers and intermediaries. Furthermore, the newcomers needed to learn how to survive here. Aboriginal people enabled them to travel by providing canoes, dog sleds, kayaks, snowshoes, toboggans, and information for maps. Aboriginal people taught the newcomers how to feed themselves, instructing them in how to make maple syrup and pemmican and how to grow Native crops that were new to Europeans—corn, beans, and squash—and by showing which berries and other wild foods were edible. In the wars of the 17th and 18th centuries, Aboriginal allies helped to ensure that whoever governed Canada, first the French and then the British, could stave off the forces from what is now the United States. In the War of 1812, the role of Aboriginal nations such as the Mohawk, Ojibwa, and Wyandot was critical to the survival of the colony that was to become Canada. And in the 20th century, in both world wars, Aboriginal people were proportionately among the highest groups represented in our armed forces.

To stop here would be to paint too rosy a picture of the interaction of Natives and newcomers. A conflict theory approach would emphasize that although the role of Natives in the fur trade often started between equals, exploitation of the Natives eventually developed, particularly with the introduction of alcohol in the trade. At the Hudson's Bay Company (HBC) trading post of York Factory, on the west side of Hudson Bay, the amount of rum traded per year hit a peak of 3928 litres by 1753. Between 1720 and 1774, the HBC traded 98 346 litres of rum from that post alone. Conflict theorists would denounce the clear manipulation of the treaties and the other "legal" but morally questionable means by which Natives were removed from their lands, the cultural prejudice that led to the banning of traditional ceremonies from late in the 19th century until 1951, and the destructive effects of the residential schools in the 20th century.

DIVERSITY: THE MANY TONGUES OF NATIVE CULTURES

At the time of first contact with Europeans, Aboriginal cultures in Canada were very diverse. This diversity can be seen in the foods the different peoples ate and in the size of their houses. Although all Aboriginal peoples hunted, fished, and gathered foods, some, along the St. Lawrence River and in southern Ontario, also engaged in agriculture. They cultivated and had as the greater part of their diet corn, beans, and squash. Some Aboriginal peoples lived in single-family dwellings such as the familiar wigwams, tepees, and igloos. Others, however, in British Columbia and Ontario, lived in long, narrow buildings, some of which stretched more than 100 metres long and served as multihousehold "apartment buildings" for very extended families. Maquinna, a powerful Nootka (a west coast nation) leader early in the 19th century, had a house that was 45 metres long, perhaps nearly 12 metres wide, and about 3 to 4 metres high. Around one hundred people called that house their home.

Language illustrates well the diversity of Native culture in Canada. Today roughly fifty Native languages still have speakers, although this number is decreasing. Native languages in Canada have a greater diversity than in Europe: Canadian Native languages form 11 separate groups; eight are **language families** or groupings of related languages, and three are **language isolates**, with no known relatives. Europe has only two language families (Indo-European and Finno-Ugric) and one language isolate (Basque).

The largest language family is Algonquian, which has languages in every province. The family includes two widespread languages with many speakers. Ojibwa is spoken by thousands of speakers from Quebec to British Columbia. Cree, from which Saskatchewan ("fast-flowing water") and Manitoba ("spirit strait") take their names, likewise has thousands of speakers, living from Quebec to Alberta. The other languages are more regionally restricted, with Mi'kmaq spoken in the Atlantic provinces and Quebec (meaning "It narrows" in Mi'kmaq), Maliseet in New Brunswick, Innu (Montagnais and Naskapi) in Labrador and Quebec, Abenaki in Quebec, Delaware in Ontario, and Blackfoot (Blackfoot, Blood, and Peigan) in Alberta.

Athabaskan languages are found in the four western provinces, the Northwest Territories and Yukon Territory; they include Beaver, Carrier, Chilcotin, Chipewyan, Dogrib, Han, Hare, Kaska, Gwich'in, Sarcee, Sekani, Slavey, Tagish (now extinct), Tahltan, and Tutchone.

The Many Faces of Diversity

The Eskimo-Aleut family, which includes the various distinctive dialects of Inuktitut in Canada, is found in all three territories, and in Labrador and Quebec.

The Iroquoian family, whose languages gave Canada its name (meaning "village" in the extinct language of St. Lawrence Iroquoian), as well as the names for its most populous province (Ontario, "It is a large lake" in Huron) and city (Toronto, "poles in water" in Mohawk) comprises the languages of the six nations of the Iroquois Confederacy (Mohawk, Oneida, Onondaga, Cayuga, Seneca, and Tuscarora), which are spoken in Ontario and Quebec.

The Siouan family, distantly related to Iroquoian, includes two languages spoken in the Prairies, Nakota or Assiniboine and Dakota.

British Columbia has the greatest Native language diversity in Canada. Found in that province, but no other, are the Salishan languages (Bella Coola, Comox, Halkomelem, Lillooet, Okanagan, Pentlatch, Sechelt, Semiahmo, Shuswap, Squamish, Straits, and Thompson), the Wakashan languages (Haisla, Heiltsuk, Kwakwala, Nitinat, and Nootka), the two Tsimshian languages, and language isolates Haida, Kutenai, and Tlingit (which is spoken also in Yukon).

DEMOGRAPHICS

Historical Picture

The number of Native people in Canada at the time of first contact is estimated to have been between 500 000 and 2 000 000. The areas of greatest population were where Canada's three largest cities of Vancouver, Toronto, and Montreal are now. The Pacific coast, with its rich forests and marine resources of fish, seals, sea otters, shellfish, and whales, had the largest groups, perhaps totalling some 200 000 people in British Columbia. Groups in southern Ontario and the area along the St. Lawrence River in Quebec, where Native people grew crops, fished, and hunted deer, were next-largest, numbering probably more than 60 000.

Why, then, did so many 18th- and 19th-century European writers refer to the great open "empty" lands of Canada? Much of the country had been emptied by disease. Europeans who came to Canada carried with them, unknowingly, diseases they had endured for centuries. But Natives had not experienced these afflictions before. Their bodies had not built up immunities with which to combat them. Native people became easy victims of smallpox, influenza, scarlet fever, whooping cough, typhus, measles, and tuberculosis. With deadly efficient regularity, most of a Native nation would die within a few years of first contact with Europeans.

The greatest killer was smallpox. It swept west gradually, but relentlessly, killing people in high numbers. In the 1630s it hit the Montagnais, Algonkin, and Huron. The Huron lost perhaps two-thirds of their population in just four years. During the 1770s, Prairie Natives were mowed down like so much wheat. In 1776–77 the victims included the Cree and the Assiniboine. The explorer Simon Hearne claimed in 1781 that 90 percent of the Chipewyan had fallen prey to disease. In 1862–64 smallpox found British Columbia. Coastal groups such as the Tsimshian, and interior peoples such as the Chilcotin, suffered high casualties. In 1870 smallpox returned to the Prairies and killed as many as 3500 Natives.

The Contemporary Picture

How many Native people are in Canada today? This is difficult to determine. There are registered Indians, Inuit, and Métis. Who is "Indian" and who is not is a complicated legal matter. The *Indian Act* of 1876 enshrined a sexist definition by stating that an "Indian" was any man of "Indian blood" reputed to belong to a particular band, any child of such a man, or any woman who is or was married to such a man. Under this act, a man would keep his status, no matter whom he married; a woman, however, would lose her status if she married someone not legally an Indian and her children would share that fate. Adding insult to injury, a white woman who married an Indian man would gain Indian status. This discriminatory law was in force until 1985.

In 1881 there were 108 547 "status Indians." By 1984, before the change in the *Indian Act*, this number had risen to roughly 349 000. **Bill C-31** was passed in 1985, enabling people who had lost their Indian status through marriage or through the marriage of their mother to apply to be reinstated. Currently, there are more than 700 000 **registered Indians** (the term that replaced "status").

"Registered Indian" cards are one reason that people have applied to regain Indian status. Perhaps you have worked at a store when someone presented you with an "Indian card," which meant that the person did not have to pay provincial or federal taxes on what you were selling. This is because the land most Aboriginal people live on is technically "outside the province," federal land. Some non-Natives resent this, believing that Aboriginal people are receiving special privileges. The peoples' response to this is that they have prepaid these taxes many times over with the treaties that cost them their lands.

CRITICAL THINKING BOX 5.1

There has been a "sudden rise" in the number of registered Indians over the past 10 years. Why is this?

> ### BOX 5.1
>
> *Bill C-31 Double Standard*
>
> In Bonita Lawrence's recent study of people of mixed heritage living in Toronto, she noted a double standard held by the people she studied concerning Bill C-31 Indians. They generally believed that individual intent for asking for band membership was more important than collective entitlement:
>
>> It is clear that the forced estrangement of so many children from their mothers' home communities has been individualized in the minds of most Native community members. . . . While most of the participants expressed in an abstract manner that it was an injustice not to reinstate people to their bands, most of them implied that Bill C-31 Indians must demonstrate the *right* reasons for wanting reinstatement (selfless devotion to community) rather than the *wrong* reason (looking for education funding or other financial benefits from the band). . . . A selfless desire to put the wishes of the community before one's own educational or other needs is, in fact, demanded of nobody but individuals seeking reinstatement after Bill C-31.[2]

Inuit

Canadians often ask why the people are called Inuit now and not "Eskimo." The answer is that the Inuit never called themselves "Eskimo," which comes from the languages of their Algonquian neighbours. It refers to eating food (probably fat and meat) raw. Their being called "Eskimo" is rather like French people being called "frogs," or Germans "sausage-eaters." It lacks respect. The word "Inuit" is a plural noun in the Inuktitut language, and literally means "men." The singular is Inuk. There is no such word as "Inuits." Inuit differ from "Indians" in being in Canada for a shorter time, less than 10 000 years. There are Inuktitut speakers in Siberia, something that cannot be said for other Aboriginal languages. Culturally, the Inuit developed items, such as the kayak and the igloo, unlike anything found among other Native people in Canada.

Not until 1939, when Canada asserted territorial claims in the Arctic, did the federal government take official responsibility for the Inuit. Each Inuk was given a metal disk with a number that was used as a token of their status. Today, about 60 percent of Inuit have disk numbers.

Since "joining Canada," much has happened that has been harmful to the Inuit. Southerners involved with various projects in the Arctic in the 1940s and early 1950s killed off sufficient caribou to make the Inuit who lived off that animal disappear as a people. In 1953 the federal government forced the people of Inukjuak (Port Harrison) in northern Quebec to move some 3200 kilometres north to uninhabited Ellesmere Island. Ottawa claimed that this was for the betterment of the people, as local natural resources were being depleted and Ellesmere Island was untouched. A darker interpretation is that this forced move was made primarily to guarantee Crown rights to that contested territory, and that the government officials did not know or care that the kinds of resources in the two areas were very different. In the 1950s and 1960s the Inuit had the highest rate of tuberculosis in the world.

Inuit status has taken on a new nature for some since the territory of Nunavut ("our land") came into being on 1 April 1999 in the eastern two-thirds of the Northwest Territories. More than 80 percent of Nunavut's population of about 25 000 is Inuit. They own 18 percent of the land (almost all the rest being Crown land), have subsurface rights to oil, gas, and other minerals for about 2 percent of Nunavut, and will receive royalties from the extraction of those minerals from the rest of the territory. They do not require a licence to hunt or fish to meet their basic needs.

Métis

The term **Métis** is used in two ways. When written with a lowercase m, it typically refers to anyone of mixed genetic heritage: part Native and part white. When written with an uppercase M, it usually refers to the descendants of French fur traders and Cree women. Beginning in the late 18th century, these latter, the Métis, developed a culture that effectively combined European and Native features. They also developed the Michif language, which has nouns that are usually French and verbs that are usually Cree. They lived for part of the year in river-lot farms in the Winnipeg area, but their main annual activity was the buffalo hunt. From the hunt came pemmican, a key Métis contribution to the fur trade. It supplied much-needed, well-preserved food for the men who paddled the canoes and carried the huge loads transported in the fur trade.

The Métis came to think of themselves as a nation. They had earned this sense through battle, and in legal struggle with the Hudson's Bay Company (HBC). "The Bay" owned most of the Prairies and about half of present-day Canada, through a 1670 charter given by English King Charles II, who little knew what he was signing away. Charles II also granted the HBC a trading monopoly. In 1811, a senior HBC official placed settlers in the middle of Métis territory, not concerned that people already lived there. In 1814 the governor of these settlers declared that the Métis could not trade in pemmican, as the HBC had a monopoly. This led in 1816 to a struggle at Seven Oaks. The resulting Métis victory became a symbol of their nationhood, celebrated in song.

In 1849 the Métis apparently beat the HBC again. Pierre-Guillaume Sayer, charged with illicitly trafficking in furs, was released, even though he had been convicted. It is thought that the judge had been influenced by the armed presence of two to three hundred Métis outside the courtroom.

In 1867 the HBC started negotiating with Canada to sell its land, and the federal government moved to set up a colony in Manitoba. Métis rights to the land were not considered. The Métis took action. Led by 25-year-old, college-educated Louis Riel, the Métis achieved a military takeover in 1869 and set up an independent government to negotiate with Ottawa. Initially, things went well for them.

In 1870 the *Manitoba Act* of 1870 established the province of that name with a legal recognition of Métis rights. This took the form of "scrip," a certificate declaring that the bearer could receive payment in land, cash, or goods. But government officials and land speculators ensured that the Métis were "legally" cheated—the laws changed 11 times in 12 years—of their land. Only a few remained. Most Métis simply moved west. But in 1885, with their rights ignored again, and again led by Louis Riel, the Métis set up an independent government in Saskatchewan. They were defeated by federal Canadian forces. Riel was hanged for "treason."

As a separate people, the Métis almost disappeared. However, during the 1930s, Alberta Métis pushed for the creation of communal settlements similar to reserves. In the words of Métis leader Adrian Hope: "We've had enough of negotiable scrip . . . to buy booze. . . . What we are asking for is land we cannot sell, cannot mortgage, but land to which we can belong."[3] In 1938 eleven Métis "colonies" were formed (eight still exist). Unfortunately, the Alberta government controlled them. The Métis could only advise or recommend. From 1969 to the present, the colony Métis have been engaged in a legal fight for two basic rights:

1. They want the colony to have the political power of a municipality. For the most part, that was granted during the 1980s.
2. They want more say concerning economic development. In this they have not been so successful.

The colony Métis have tried to obtain royalty payments for gas and oil extracted from their land, but this has been a bitter and unrewarding fight. Any economic plans that they make are still subject to provincial veto. Beyond the colonies, there are the Métis National Council and provincial organizations in Ontario and in the Prairie provinces; these suffer from difficulties of legal definition and lack of recognition.

How many Métis are there in Canada? Estimates vary from 500 000 to 1 000 000.[4]

How many Natives, then, are there in Canada? That depends on how you define "Native." If you go by legal definition, you start with more than 700 000 registered Indians and add some 20 000 disk-carrying Inuit and 4000 Métis "colonists" to that. That is one definition. Self-identification is another. If every Canadian with Native genetic background were to be counted, they would number more than 2 000 000. French-Canadian biologist Jacques Rousseau established that more than 40 percent of the French-speaking population in Canada has at least one Native ancestor.[5]

BOX 5.2

The Powley Case: Métis Hunting Rights

In 1993, Steve Powley and his son killed a bull moose in the area of Sault Ste. Marie, in northwest Ontario, without a provincial licence. They were charged under the Ontario Game and Fish Act with unlawfully hunting and possessing a moose. Their defence was that Powley was Métis, and that hunting for food was his Aboriginal right. On September 23, 2003 a unanimous decision of the Supreme Court of Canada upheld his right to hunt for food. It established what has been called the **Powley test** for Métis hunting rights. The test has 10 main components:

1. Characterization of the Right
The right involved is the right to hunt for food (not sale) in designated territories, not confined by provincially set hunting season or licence.

2. Identification of the Historic Rights-Bearing Community
A Métis community has existed from the 18th century in the Sault Ste. Marie area.

3. Identification of the Contemporary Rights-Bearing Community
The historic Métis community (which has a population of roughly 900 people) had persisted into contemporary times.

4. Verification of the Claimant's Membership in the Relevant Contemporary Community
This involves three elements: (a) long-term self-identification as Métis; (b) evidence of an ancestral connection to a historic Métis community; and (c) acceptance by the contemporary community that the individual is a member.

(cont'd)

5. Identification of the Relevant Time Frame
This entailed shifting the time frame for Aboriginal rights from the precontact "time immemorial" of status or registered "Indians" to one that was relevant to the creation of Métis communities after contact.

6. Determination of Whether the Practice Is Integral to the Claimant's Distinctive Culture
Hunting for food was a key element of the traditional culture of the community.

7. Establishment of Continuity Between the Historic Practice and the Contemporary Right Asserted
Hunting for food has continued from the historic period to contemporary time.

8. Determination of Whether the Right Was Extinguished
When an Aboriginal right is extinguished it means that there has been specific legislation that has taken the right away. No such legislation had been passed.

9. If There Is a Right, Determination of Whether There Is an Infringement of That Right
It was determined in this case that arresting the Powleys for hunting for food was an infringement of their Aboriginal right.

10. Determination of Whether the Infringement Is Justified
It was *not* determined that the infringement of the Powleys' Aboriginal right was justified.

This has had implications for Métis elsewhere in Canada. By the end of 2005 the Powley case had been used as justification for Métis hunting rights in parts of British Columbia, in Alberta, and in Saskatchewan. It was still being contested in Manitoba.

TREATIES

What is a **treaty**? It is difficult to determine the original meaning of any treaty for those on either side of the negotiating table. The government felt that Natives would eventually disappear. Precise wording might not have been so important to them. Native people had no precedent for giving away land forever. Today's Aboriginal people tend to feel that these treaties are international agreements between "nations." They consider treaties to be statements of recognition of their sovereign or independent status, made necessary by the **Royal Proclamation of 1763**. This key document contained two statements of significance to Native issues in Canada.

First, land not part of New France and not owned by the Hudson's Bay Company was declared to be "Indian land." Second, Indian land could only be taken from them through "public purchase," that is, treaties. Most of Canada is covered by such treaties, but significant exceptions exist. British Columbia has a few treaties, but the province removed Natives from most of their land without public purchase. Quebec received its northern half on the understanding that Native land rights would be dealt with, but nothing was done until the James Bay Agreement. Newfoundland and Labrador, separate from Canada until 1949, has no such treaties.

Generally, in treaties, the Natives involved would agree to give up their rights to a certain area of land that they traditionally used. In return they would have a smaller area or areas reserved for their use, hence the name **reserves** for these blocks of land. The calculation was typically a certain amount of land per family of five, say 65 hectares. The Natives would also receive a certain amount of money. In that sense, treaties were like land sales. They were different, however, in that Natives could not get their hands on the money. It was held for them "in trust" by the federal government. From 1818 on, Natives would also receive ***annuities***, or annual payments, usually involving a small sum per person, such as three dollars a year.

Treaties generally moved westward with non-Native settlement. The oldest treaties (the "Peace and Friendship Treaties") were with the Aboriginal peoples of the Atlantic provinces. These treaties have generated a great deal of controversy concerning their promises of hunting and fishing rights. In 1713, the Treaty of Portsmouth, New Hampshire included the Maliseet and the Mi'kmaq in Canada. It involved not only "peace and friendship," but also "free liberty for Hunting, Fishing, Fowling, and all other of their Lawful Liberties and Privileges."[6] Similar statements of hunting and fishing rights for the Mi'kmaq occurred in two treaties of 1725. In 1752 another treaty was signed, which reaffirmed the rights promised earlier. In the case of *Regina v. James Matthew Simon* (1985), the Chief Justice recognized that the 1752 treaty had supremacy over provincial game laws, saying, "It is an enforceable obligation between the Indians and the white man."[7] Conflict arose in 1988 and 1989 between the Nova Scotia Ministry of Lands and Forests and Mi'kmaq hunters concerning their rights to hunt moose to provide their families with food. The Mi'kmaq were granted special one-week extensions one week before and after the two-week hunting season for non-Natives.

Greater conflict occurred in 1999 and 2000 concerning a treaty of 1760. After a Mi'kmaq man was convicted in 1996 for catching eels out of season, the court's decision was overturned on 17 September 1999 by the Supreme Court of Canada. The Supreme

Court's ruling was that the Mi'kmaq could hunt or fish not for "economic gain" but to obtain a "moderate livelihood" for a family's sustenance and survival. As this decision was handed down around the beginning of lobster season, the Mi'kmaq began to set up lobster traps without obtaining licences. Non-Natives possessing those licences were angry. Some responded with violence to Mi'kmaq lobster traps, the boats they used, and processing plants thought to be handling Mi'kmaq-caught lobster. The state of tension increased by the clumsy response of the federal government. A 30-day Mi'kmaq lobster fishing moratorium was declared, which was accepted by all but two affected Mi'kmaq communities. One in particular, Burnt Church, in New Brunswick, became the scene of several confrontations. Eventually the Mi'kmaq were awarded the right to use about 13 000 traps, in an Atlantic fishery that generally involved about 2 000 000 traps.

During the last part of the 18th century and the early part of the 19th, a series of treaties covered most of southern Ontario. Following these were the Robinson Treaties of 1850, involving the land immediately north of Lakes Superior and Huron. Then came the "numbered treaties," 1 to 11, which covered most of northern Ontario, across the Prairie provinces to the eastern part of British Columbia, and parts of Yukon Territory and the Northwest Territories. Almost all of these treaties can be connected with the federal government wanting something, particularly mineral resources. The Robinson Treaties were signed after the discovery of metals in northern Ontario. Treaty Number 8 was occasioned by the Klondike Gold Rush near the end of the 19th century. Oil found in 1920 led to Treaty Number 11.

Why do Aboriginal people make such a big deal about the treaties? First, they hold the view that in the treaties they were recognized as sovereign nations. Second, there are few treaties in which the Native groups involved have not had some grievance concerning either verbal promises that were not written or written promises that were not fulfilled. For example, in 1818 in partial payment for giving up the "Mississauga Tract," land including in part what is now the city of Mississauga, the King's representative promised "to pay to the said Nation of Indians inhabiting as above mentioned, yearly and every year for ever the said sum of five hundred and twenty two pounds ten shillings currency in goods at the Montreal price."[8] Forever was brief. These annuities were paid for only a few years.

Two years later, the Mississauga reluctantly agreed to give up their exclusive fishing rights to a few local creeks and rivers, as well as all their remaining land, "[s]aving and reserving, nevertheless, always to . . . the people of the Mississagua [sic] Nation of Indians and their posterity for ever a certain parcel or tract of land containning [sic] two hundred acres."[9]

The Mississauga never signed away those 200 acres (about 80 hectares), but they soon came to understand that without a land deed of the type that the settlers had, they had

> **CRITICAL THINKING BOX 5.2**
>
> Why is it difficult to determine how many Native people there are in Canada?

no hope of keeping that land, despite the written treaty promise. They left the area in the 1840s. Fortunately, the Iroquois gave them land on which to live.

LAND CLAIMS

Typically, land claims are put forward by Aboriginal groups not covered by any treaties. They are based on the principle that people who have not signed away their rights to their land have Aboriginal rights that have not been extinguished. Sometimes, land claims involve dispute over whether a people is included in a treaty. The Temagami Anishnabe (Ojibwa) of northeastern Ontario asserted that they never signed the Robinson–Huron Treaty. For more than a century, they pushed to sign an agreement so that they could obtain a reserve where they worked and lived. The position of the provincial governments over the same period remained that someone had signed for them. Governments were reluctant to allow them a reserve in that pine-and-tourism-rich area of the province. During the 1980s, the courts ruled against the people. The provincial government then made an offer that divided the band (only part of the community accepted the offer), creating a rift that has lasted into the 21st century.

More often, a land claim is made where no treaty applies. One famous case involves the Cree of east James Bay. When the federal government handed over northern Quebec to the province by the federal government in 1898 and 1912, it was stipulated that the Cree, Inuit, and Innu would have their rights dealt with through treaty. No Quebec government did anything until 1971, when Premier Robert Bourassa announced his plans for the James Bay Project, a grand scheme of constructing power dams on four large rivers. These dams would have flooded the homes and the hunting and trapping territories of the Cree. Bourassa had not talked with the Cree, nor thought of negotiating with them concerning their rights to the land they had lived on for perhaps thousands of years. After nearly three years of legal and public relations battles, the James Bay and Northern Quebec Agreement of 1975 was signed, with extensions north and east added in 1978 and 1984.

This agreement has been a mixed blessing. The Cree received a lot of money, although not as much as was promised. The Cree School Board has achieved a great deal more than their critics thought possible for a Native-run, Native-language, and culture-based system. Financially, some individuals have prospered, and some businesses have done well. Politically, the Cree have become a force to be reckoned with. They were able to stop the second phase of the Project. During the Quebec Referendum of 1995, the Cree were prepared with legal documentation that could very well have won them the right to withdraw from a separate Quebec.

On the negative side, mercury, raised by the flooding, exists in dangerously high amounts in fish, a major item in the local Cree diet. Communities that in the early 1970s were made up mainly of financially independent hunters and trappers have seen a good number of people forced onto welfare. Social problems such as substance abuse and suicide, unheard of before the James Bay Project, have emerged.

The Nisga'a Treaty

The Nisga'a treaty of 2000 has prompted much discussion, particularly in British Columbia, where a great deal of land "suddenly" became "Indian land." This is not a new land claim. The Nisga'a, a Tsimshian-speaking people, worked for more than a century to have their land claims resolved. They sent delegations to Victoria in 1881 and 1887. The premier told them that "when the whites first came among you, you were little better than the wild beasts of the field."[10] The Nisga'a spoke to prime ministers in 1885 and in 1910. In the latter year, Prime Minister Wilfrid Laurier promised to resolve the land issue. His government was defeated in 1911. In 1909 and 1913, the Nisga'a sent representatives to Britain, with no success.

In 1927, the federal government made it difficult for the Nisga'a to get their land claims settled. Ottawa took away from Natives the democratic right to organize to discuss land claims. The Nisga'a did not give up. They and other British Columbia Native groups formed the Native Brotherhood of British Columbia in 1931, which discussed in secret how they could get land claims settled.

In 1967, Nisga'a leader Frank Calder took the land question to court, seeking a declaration that his people had held Aboriginal title to the land before colonization and that their title had never been extinguished. The Nisga'a lost the decision in the B.C. court but appealed the decision, taking it to the Supreme Court of Canada. In 1973, the appeal was lost in a split decision: the judges agreed that Aboriginal title had existed but disagreed as to whether that title continued to exist.

In 1976, the federal government and the Nisga'a began negotiations to settle their land claims under the new "comprehensive land claims policy." British Columbia did not enter into the discussion until 1990, eight years after the Canadian Constitution recognized and affirmed Aboriginal title.

On 22 March 1996, the Minister of Indian Affairs, the B.C. Aboriginal Affairs minister, and Nisga'a Tribal Council President Joseph Gosnell Sr. signed an Agreement-in-Principle for the first modern treaty in British Columbia. In 1999 Parliament, and on 13 April 2000, the Senate, passed the bill.

The 5500 Nisga'a received title to 1930 square kilometres in their homeland and $487.1 million in benefits and cash. They obtained the right to make laws concerning land use, employment, and cultural preservation (resembling municipal, provincial, and federal governments in that lawmaking capacity). They own the forest and mineral resources on their land (as a private owner would) and have to manage them according to British Columbia's laws and standards.

It was not all gains. The Nisga'a lessened their original claim by 25 percent, and they gave up future claims on more than 80 percent of their traditional territory. They gave up their tax exemptions.

Although the non-Native media, particularly in British Columbia, have presented this treaty as a step toward separation, it is more accurate to say that it puts the Nisga'a in a legal position similar to that of other Canadians.

BANDS

Typically, an Indian **band** is made up of people of one cultural tradition with some historical connection with each other. They own land and funds in common and are governed in part by a band council headed by a chief. There are some 592 bands across Canada. The largest is the Six Nations, an Iroquois band in southern Ontario whose website declares that they have more than 20 000 members.[11] Bands can be created or done away with by the federal **Department of Indian Affairs (DIA)**.

Bands now face a problem that is dividing some communities. Membership is sharply rising, with no compensating increase in land or funds. The band council handles housing and postsecondary education money. With the influx of "Bill C-31 Indians" the competition for such programs is tight. Many reserves are already crowded, and the problem is getting worse. There is tension in many communities.

A great frustration for chiefs and band councillors has long been their lack of power. They have less power than a municipality, which can raise funds through taxation (a band cannot), and which possesses an independence that a band would envy. The DIA has the power to veto any decision that a band council may make. Much of what bands do has to be sent to DIA for approval.

Native self-government is something that non-Natives tend to fear, thinking that the country will be divided into little separate nations. Native groups differ as to what they think

Native self-government is. Some just want the power that a comparably sized town or village has. Others desire the right to set up the structure of their choice, one that instead of having an elected chief and band council, has more-traditional-style leaders and spokespersons for clans. For most Canadian Native groups, "the chief" was more a foreign creation than a traditional form of leadership. All bands want to be less controlled by the DIA. Native government is already distinct. Natives just want that distinction to be less of a powerless one.

RESERVES

Students often ask me, "What is it like on a reserve?" It is as if they expected me to tell them stories of bizarre mystical rites and shape-shifting shamans. They are often disappointed when I say that many reserves look just like other, comparable communities. They wonder, too, why Native people seem to cling to reserves even when some are overcrowded and violent. One answer is that for many Native people the reserve is their "home and Native land." There they are not a minority, subject to stares and discrimination.

The uncertainty that sometimes surrounds reserve status creates problems. One major reason that the 1990 confrontation at Oka (in Quebec) took place was because the people there did not have a reserve that could protect their burial grounds from a local golf course. They still don't. The tragedy behind the story at Davis Inlet in 1993, with Innu children suffering from serious substance abuse and attempting suicide, has a lot to do with the federal and Newfoundland governments forcing the people to move the reserve from the Labrador mainland to an island. This was done against the wisdom and the will of the local Innu, who knew they would have less access to the natural resources they valued, such as caribou. The move took a lot of the heart out of the people.

Reserves have a unique legal status. Technically, a reserve can be called federal rather than provincial land. The band owns the land, but the federal government, through the DIA, has ultimate authority. Although people have their own houses and property lines, the band holds the land in common. They do not pay municipal taxes, and housing is usually cheap. On the negative side, a family's part of the reserve cannot be used as collateral to get a loan from a bank.

Most Natives live on reserves. A little less than 60 percent of registered Indians live on reserves. Reserve patterns differ across Canada. In central and eastern Canada each band has only one reserve. Out west, a band is more likely to have more. In British Columbia, for example, there are fewer than 200 bands but more than 1600 reserves. Typically, reserves are remote. In 1989 more than 60 percent were farther than 50 kilometres from a city or major town. A remarkable 18.6 percent were not connected to a city or major town by a year-round road.

A growing number of Native people work and live in towns and cities, however, estimate that in 1991, 11 Canadian cities had an Aboriginal population of at least 10 000.[12] Edmonton, Montreal, Toronto, and Vancouver top the list, with an estimated 40 000 to 50 000 each.[13] These figures are probably a lot lower than the actual numbers. Estimates for Toronto's Aboriginal population are often given as more than 70 000; it is sometimes called "the largest Indian reserve in Canada." This move to the city has been a trend of the past 35 years. In 1971 none of these cities had a reported Native population of more than 5000.

Unfortunately the greatest urban visibility of Aboriginal people is as the stereotypical "Indian drunk." In Toronto they are otherwise pretty much an "invisible minority." An Ojibwa student of mine told me that for years, when she worked in a bank in that city, customers would guess at her identity, asking her if she were Filipina, Chinese, or Korean. No one ever guessed correctly.

As more Natives move to the city, there is a growing need for Native-run organizations to help them cope with urban life. These are appearing, usually starting with a Native Friendship Centre.

Urban Reserves

A new type of reserve is being created in Saskatchewan: **urban reserves**. These are lands located in a municipality or Northern Administrative District. Their main function is to provide central urban locations for Aboriginal businesses. To enable the creation of these reserves, the Saskatchewan Treaty Land Entitlement Framework Agreement was signed in 1992. A fund of approximately $446 million was established so that the land could be purchased. Twenty-eight of the 70 Saskatchewan First Nations signed on, initiating the development of 28 urban reserves. Nine of them are located in cities. More than 1350 people work in the businesses developed there.

Why did urban reserves come about? First of all, there were a good number of outstanding treaty land entitlements in Saskatchewan. Land promised in the treaties of over 100 years ago had yet to be delivered to the people. Secondly, an increasingly high percentage of the Aboriginal population is moving (at least temporarily) from the rural reserves and into cities. There they have been as a group markedly less successful than non-Aboriginal people. Thirdly, the economic potential of rural reserves, on average, could not be as great as reserves located in cities. Fourth, for shared and different reasons, the federal government and First Nations wanted the latter to be more economically self-sufficient, not so dependent on economic transfer payments from the federal government.

The first community to develop an urban reserve was the Muskeg Lake Cree Nation. It has about 1200 members, most of them living away from the main reserve (located 93 kilometres outside of Saskatoon). The leaders of this community first approached the federal government in 1984, and were granted 33 acres of land in Saskatoon. In 1993, they signed a services agreement with the city. This involved the band giving an annual payment for municipal services such as snow and garbage removal, with electricity and water being billed directly to individual customers on the reserve. Businesses operating there are almost all Aboriginal-owned, with some 300 employees.

NATIVES AND THE JUSTICE SYSTEM

Natives do not fare well in the Canadian justice system. They are less likely than non-Natives to be put on probation and to be released on their own recognizance. Natives are more likely to be ordered to pay fines, and much more likely to be put in jail because they cannot pay the fines. They are more likely to serve their full sentence rather than having their sentence shortened by parole. Drinking in the same bar with non-Natives, and consuming roughly the same amount of alcohol, they are more likely to be picked up for being drunk around closing time. Natives are overrepresented in our prisons. Shock-value statistics abound, often doing more harm than good (see the opening quote). Stories of what is happening and how solutions are being applied are less often presented and are more likely to improve the situation. The following are a few such stories and potential solutions.

Different Sides to Policing Natives
Starlight Tours in Saskatoon

Saskatoon has a population of about 200 000. Its Native population has been estimated at about 15 percent of that number, but the percentage of the poor in Saskatoon is almost double that at 30 percent.

In 1997, a veteran police officer wrote a column in the *Saskatoon Sun* that told about a night on the beat for two fictional officers. In the story, they pick up a loud, verbally abusive drunk and take him on what is sometimes known as a "starlight tour," driving him outside town and leaving him there to walk back.

On the night of 27 January 2000, 33-year-old Darrell Night, an unemployed bricklayer and a big, sturdy man, got drunk. In the early morning hours he was arrested outside a friend's apartment by two officers, handcuffed, and put into a police cruiser. He was then taken outside town and dropped off by Queen Elizabeth II Power Station south of the city, without a coat. The temperature had dropped to about −28°C. Darrell went to the station and got a night watchman to call him a cab. That may have saved his life.

He did not tell his story to anyone in legal authority at first. He was Cree. The officers were white. Who would be more likely to be believed? It would take the report of two tragedies to make him tell his own story.

What made him tell his story was the discovery the next night of the body of a 25-year-old Cree man, Rodney Naistus, and, a few nights later, of a 30-year-old Cree student, Lawrence Kim Wegner. Both had been drinking. On 20 September 2001, the two officers who took Darrell Night in his starlight tour were convicted of unlawful confinement and were fired by the Saskatoon Police Service. An inquest into the deaths of Rodney Naistus and Lawrence Kim Wegner found no wrongdoing.

The Positive Side of the Saskatoon Police Service

It would be wrong to simply label the Saskatoon Police Service (SPS) as racist. Since 1994, the SPS has developed several programs and events dedicated to creating a bridge between the police service and the Saskatoon Native community. In 1995, the SPS won the National Ivan Ahenakew Award for efforts in employing Aboriginal officers and in the development of cultural programs.

Perhaps the most successful elements of the SPS's efforts to connect with the Native community have come through the peacekeeper programs, which put officers together with Aboriginal at-risk youth, often under the guidance of Elders and other Native leaders. The programs include innovative cultural activities such as Project Firewood/Rocks. This involves Aboriginal youth and SPS officers travelling to northern Saskatchewan to load a semi-trailer with firewood and rocks to be taken to Saskatoon for the Elders. The Elders will use these materials in **sweat lodge** ceremonies (see below), some involving both the youth and officers as participants.

Native Policing Services

Following the Manitoba Aboriginal Justice Inquiry of 1988–89 and the 1990 Oka confrontation, the federal solicitor general initiated in 1991 the First Nations Policing Policy. It operates on a principle of partnership involving First Nations, the federal government, and the provincial or territorial governments. These three enter into tripartite agreements for police services that fit the needs of the particular Aboriginal communities involved. Fifty-two percent of the funding for the agreements comes from the federal government, with the remaining 48 percent coming from the provincial or territorial government. By the end of 1999, there were 52 such agreements across the country. The following is the story of one such agreement.

CRITICAL THINKING BOX 5.3

What problems do band leaders face that leaders of towns and cities do not?

The Stl'atl'imx Tribal Police Case Study

On 20 December 1999, British Columbia received its first Aboriginal police service with full jurisdictional authority. The Stl'atl'imx Tribal Police (STP) force is responsible for 10 First Nations communities, serving roughly 3150 people in southern British Columbia. Nine of the participating communities are Lillooet, the 10th is Shuswap, a closely related people.

Although the force dates back to 1988, it wasn't until April 1992 that a tripartite agreement was signed. The STP is guided by the Stl'atl'imx Tribal Police Board (STPB), which has 10 members, one from each of the participating communities. The STPB has more responsibility for managing and leading the STP than is found in most municipal police departments. The tripartite agreement included some dependence on the RCMP, with the latter providing assistance in the investigation of relatively serious cases.

The notion of a Native police service is new to non-Natives, easily interpreted by them as threatening—particularly, as in this case, when the Native police service is asked to give assistance to provincial police off-reserve. By May 1992, three petitions with 493 signatures were sent from the non-Native town of Lillooet to the B.C. attorney general with complaints of non-Natives being "harassed" by Native police.

The British Columbia Police Commission reviewed the STP in 1996. At that time the STP had nine policing positions: a chief constable, a supervisor, three constables based in the community of Mount Currie, three in Lillooet (the Native community), and one in Seton Lake. The problems revealed in this review were typical of the fledgling Native police services across Canada: insufficient and often late year-to-year funding, difficulties in hiring experienced staff (particularly when contracts are a year long, rather than long-term), inadequate training, the difficulties of policing a broad area of separated communities with few officers (making full-service, full-time shift work almost impossible), low morale, discontent with the authoritarian management style of the non-Native chief constable, and what might be called a clash of policing cultures.

The Police Board members were unanimous in wanting to have a police service based on the traditional Lillooet "watchman" system. The "watchman" was respected for his decision making. Traditionally, members of the community felt comfortable in approaching him to hear their cases whenever they felt that someone had committed a transgression. The majority of STPB members believed that the chief constable and the STP officers were not applying this philosophy, but were following a more mainstream-culture policing approach. The officers themselves, most of whom were Native, were divided as to which philosophy or cultural style they preferred.

The culture of policing clash is one of the major challenges faced by all Aboriginal police services. It raises several important questions. How free are Aboriginal police services to innovate following traditional values? How might that freedom be restricted by the mainstream training their officers receive and the mainstream policing experiences of those who early on assume positions of leadership in these services? To what extent can Native police services reflect their traditional culture in trying to work with and get the respect of other, more established policing services and of provincial and federal funding administrators?

Despite some of the negative aspects of the review, the B.C. Police Commission noted that the Board and the STP were conscientious and dedicated to improving the quality of the service. This must have served to good effect, for three years after the review, the STP moved ahead to assume full jurisdictional authority. As Chief Constable Harry McLaughlin noted in the B.C. attorney general news release, "This occasion instills a new level of pride in our officers and a renewed sense of commitment to the Stl'atl'imx communities."[14]

Sentencing Circles

Sentencing circles provide alternatives to incarceration, through applying traditional notions of restorative justice. The circles typically involve community members such as Elders and Native social workers, lawyers, health care workers, and those affected by the crime, including the offenders, the victims, and their families. The circles are more informal than traditional courts. People sit in a circle. No judge is involved. Consensus is stressed in deciding what the sentence will be. The following is an example.[15]

In 1994, the United Chiefs and Councils of Manitoulin Justice Project began diverting cases from the courts to sentencing or justice circles. From 1 January 1998 to the end of June 1999, 77 justice circles were created, involving 110 members of the community. The kind of cases were typically property offences (i.e., vandalism, break and enter, and theft) and relatively minor instances of assault, but they did not include more serious crimes such as murder, manslaughter, sexual or spousal abuse, or impaired driving.

Sentences are referred to as "plans of action." These are traditionally based, but also creative and new. Public apologies are made by the "clients," not only to the victims involved, but also to the police and others affected by the harmful acts. Some of the apologies have aired on a local cable TV station.

Personal healing is a significant part of the plans of action. This healing has included referrals to drug and alcohol addiction treatment centres or mental health facilities, but it has also entailed traditional activities. The intent of these activities is to heal clients by raising their self-esteem and sense of identity through learning about their culture. Examples of the traditional activities recommended are helping to gather traditional medicines in the bush, participating in an arduous canoe trip around Manitoulin Island, and researching a family tree. Some clients have done more than was required of them, as they felt rewarded by the activity.

Since healing the community is a priority, the plans of action often require some form of community service. This has involved doing such things as cleaning up the results of vandalism, painting a new detox centre, and participating in a walk held annually to raise awareness of violence against women.

Making amends is also important. In one case, a woman had assaulted another woman, a long-time neighbour. Had she decided to fight the charge, which was her initial intention, she probably would have been put on probation and told to stay away from the victim. Instead, with the apology, the two women have managed to remain on good terms, something that would have been less likely to happen in the confrontation format of the courtroom.

An easy criticism of the Justice Project would be that people get off lightly. However, the clients speak of it being more difficult to face people you know than the anonymous strangers who are judge and jury.

There are difficulties. The workload is heavy. A lot of time is taken in putting together reports and proposals for the funding necessary to continue and to grow. Still, during the period studied (January 1998 to June 1999) no one had to return to the circle for noncompliance with their plans of action.

ABORIGINAL RELIGION

What religion do Aboriginal people have? There is no simple answer. A number of First Nations have adopted and adapted forms of Catholicism that can now be considered "traditional" to their people (Angela Robinson gives a Mi'kmaq example[16]). Christian texts and hymns written in Moose Cree, and the important roles played by Aboriginal ministers in their home communities have made Anglicanism a part of Native religious traditions.

Older than these traditions are indigenous forms. From thousands of years before contact to the 21st century, these forms have flourished and adapted to changes in times. Today Aboriginal religion centres primarily around a series of practices, some unique to particular peoples, some shared across Aboriginal Canada. Some of the latter will be discussed here.

The Four Medicines

The four medicines are sweetgrass, tobacco, sage, and cedar. Each is associated with a direction. When burned, each has medicinal functions that go beyond its mere physical properties. Sweetgrass, not a drug as the name might imply, is a sweet-smelling member of the grass family. When burned, it purifies the person and the immediate setting with its sacred smoke, much as incense does for Catholics, Hindus, Jews, and Buddhists. Its directional association, the north, is associated with purification and healing. People who fan the smoke over themselves are said to be **smudging**, and it is a daily morning ritual for some.

Tobacco is often put in a fire, and the smoke is said to communicate with the spirits. Tobacco is associated with the east, from which comes enlightenment and vision. In Ojibwa culture, when people want to obtain a name in their own language, an increasing phenomenon for adults and children, they go to someone who is known to provide names, and give the person tobacco. The namer places the tobacco into a fire, and asks the Creator for help in finding the name in a vision. Generally in Aboriginal culture people approach elders with tobacco if they want the elder to share knowledge or wisdom with them.

Sage and cedar, associated with the south and west respectively, are believed to cleanse in both a physical and a spiritual sense. You can smudge with sage. Cedar is often used in the sweat lodge, a key element of both traditional and contemporary Aboriginal spirituality (see below).

The Sweat Lodge

The following is a general description of the sweat lodge as the Ojibwa (Anishnabe) use it in the context of the modern vision quest.[17] Sweat lodges differ both within the traditions of the Ojibwa people, and between them and other Aboriginal peoples.

The lodge is typically constructed of 16 overlapping willow poles. It is shaped to resemble the sky world above that covers the earth like a dome. The entrance must face the east, where the sun rises.

A pit is dug in the centre, where the "grandfathers" are received and where they meet Mother Earth. The grandfathers are stones that because of their great age hold the mysteries of the past. Before the grandfathers are brought into the lodge, they are placed at the bottom of a sacred fire in front of the lodge, joined to it by a pathway of cedar. Here they are heated until white-hot. How long the vision quest seeker plans to sweat will determine how many stones will be called on to guide the individual in the sweat lodge.

The Elder and the participants throw tobacco into the fire and pray, giving thanks to the Creator. The Elder then tells how the sweat lodge came to be, a story that describes how a little boy was sent to the seven grandfather spirits in the star world during a time of great sickness and was given the gift of the sweat lodge to bring back to his people for healing. A water drum, called the "Little Boy Drum" in remembrance of the little boy who brought back the medicine teachings, is placed inside close to the entrance. The participants greet the drum as they enter. It has seven stones that surround the top, representing the seven grandfathers who first gave the little boy the teaching. When the sweat lodge is occupied, the Elder sings traditional ceremonial songs. The grandfathers are then brought in to the lodge and placed in the pit. The Elder throws water onto the grandfathers, and the steam makes the lodge quite hot.

Medicine Dances

Medicine dances are important elements in contemporary Aboriginal spirituality. One such dance is the **jingle dress dance**, which is recognized as Ojibwa in origin. Sometime between 1918 and 1920 in the reserve community of Whitefish Bay, on the shores of Lake of the Woods, near the Ontario/Manitoba border, a seven- or eight-year-old Ojibwa girl, Maggie White, became ill. There were no signs that she was getting any better, so her father sought a vision to help cure her. In his vision, he saw both the jingle dress and the healing dance that girls and women should perform while wearing it. The "jingle" of the dress comes from the metal cones that cover it. Typically these cones number 365, one for each day of the year. Traditionally, they were made out of snuff can lids bent into the shape of cones. When the jingle dress wearer dances, the sound is an amazingly soothing, lightly chiming sound.

The original jingle dance society was initially confined to four young girls (Maggie White and three others), each of whom wore a colour signifying one of the four directions. The primary purpose was healing. People would offer tobacco to a dancer who would dance as a prayer of healing for a person who was sick.

The jingle dance continued as an Ojibwa medicine dance until the late 1960s and early 1970s, when it was picked up by other peoples and included both as a medicine dance and as a woman's dance as part of intertribal powwows, even in competitions. For many, however, it is still a medicine dance, a way for a community to contribute to the healing of their people.

Religious Oppression

Aboriginal religion has had to face a long history of oppression in Canada. This peaked in the late 19th century with the banning of the potlatch and the Sun Dance. The potlatch is the most important traditional ceremony for British Columbia Native groups such as the Bella Coola, Haida, Kwakiutl, and Nootka. It is a celebration in some ways not unlike a birthday party, bar mitzvah, christening, mass, confirmation, or marriage. It also involves the "multimedia" telling of a people's history. In a potlatch, people dance, sing, drum, and wear masks and costumes to tell stories expressing who they are as a people. Traditionally, the celebration also involved a great giving of gifts, saved up for a year or more by the family holding the ceremony and its clan. The next year they might attend a potlatch in which they are the guests and therefore the receivers of equal gifts. In the 19th century, potlatch ceremonies might take days.

Missionaries saw the potlatch as a "pagan ritual"; governmental officials saw it as a "backward practice" that would prevent the Indians from becoming "civilized." So, in 1884, a ban on the potlatch was added to the *Indian Act*. Anyone who held or attended a potlatch would be "guilty of a misdemeanour, and liable to imprisonment for a term of not more than six nor less than two months in any gaol or other place of confinement, and any Indian or other person who encourages, either directly or indirectly, an Indian or Indians to get up such a festival or dance, or to celebrate the same, or who shall assist in the celebration of the same is guilty of like offense, and shall be liable to the same punishment."

The people resisted in several ways, of which the most successful was to take it underground, to hold the ceremonies where no missionary or official could know what was going on. Less productive was the following petition, signed by Elders who showed more wisdom and knowledge of the spirit of justice in Canadian law than those who had banned their ceremony. It appeared in *The Daily Colonist*, a Victoria newspaper:

> If we wish to perform an act moral in its nature with no injury or damage, and pay for it, no law in equity can divest us of such right.
>
> We see the Salvation Army parade through the streets of your town with music and drum, enchanting the town. . . . We are puzzled to know whether in the estimation of civilization we are human or fish on the tributaries of the Na'as River, that the felicities of our ancestors should be denied us.[18]

In 1895, this ban spread to the equally significant Sun Dance of Prairie peoples such as the Blackfoot, Blood, Piegan, Sarcee, Assiniboine, and Plains Cree. In 1906, the ban was extended to all Aboriginal dancing. Aboriginal people could dance in a traditional way only with the permission of a minor DIA official. Not until 1951 was this part of the *Indian Act* repealed—17 years after the American government had repealed similar legislation in their country.

NATIVE EDUCATION

To understand Native society in Canada today you must learn about **residential schools**, which were initiated in 1910. Unfortunately, the negative effects of this education are its prominent features. One tragically mistaken idea in these schools was the notion that to educate Native children, it was necessary to separate them from the "corrupting influence" of their language, their culture, and ultimately, their parents. About half the schools were Roman Catholic, the rest primarily Anglican and United Church. The religious groups received a grant of land and money and were left more or less to run the schools on their own. Although intentions were good, the results were horrifying. Schools were frequently hundreds of kilometres away from the homes of the students, who were forced to board in prison-like buildings, often for 10 months of the year. Students were sometimes dragged from their homes by police.

Residential schools were not just bad educationally, with a simplified curriculum, under-trained teachers, hand-me-down educational materials, and "learning a trade" often an excuse for exploiting students as unpaid farm and domestic labourers; these schools also harmed the Native family. Generally, strict teachers and principals were the only parental figures that the Native children experienced for most of the year. Abuse—physical, emotional, and sexual—occurred far too frequently. When the students became parents, they would often come to repeat the practices of their white role models. Brothers and sisters were kept apart in sexually segregated classes and residences, sometimes not getting an opportunity to talk with each other for months on end. One female Native student in a residential school stated: "I never did get to know my brothers. We were kept away from each other for too long. To this day I don't know much about my brothers. I just know that they are my brothers."[19]

When children left for residential school, often they knew only one language, the language of their home, community, and ancestors. But those who "spoke Indian" would be punished, usually with a severe beating.

The residential schools lasted until the 1960s. Their effects are still being felt. Many Native people are skeptical about education even with schools run by their own people. Still, increasingly Natives are taking over their own schools. In 1975–76, there were only 2842 pupils attending just 53 band-operated schools in reserves across Canada. That number shot up to more than 53 312 students in 372 band-operated schools by 1993–94.

Native students are staying in school much longer than before. The proportion of Native students living on reserve who remained in school until Grade 12 was only 15 percent in 1970–71. This reached 47 percent by 1990–91. Similar figures appear for postsecondary schooling. In 1973 fewer than 1000 college and university students were registered Indians. By 1987 that number had reached 14 000.

Taking charge of their own education is not an easy task for Native people. Good Native material is still in short supply, especially regarding language. Provinces carefully guard their curriculum as the standard of education, sometimes making it difficult for Native students to gain accreditation for grades and courses taken in band-run schools when it comes to applying for transfer to any level of schooling from within the Native system. And the Department of Indian Affairs still controls the purse strings.

THE FUTURE

What is the future of Native people in Canada? Politically, it does not seem very bright. The federal government, concerned about debts not promises, is looking to divest itself of its responsibilities, either passing them down to the provinces or wiping the slate clean. Both strategies are looked upon with suspicion by Natives.

> **CRITICAL THINKING BOX 5.4**
>
> Why do you think the federal government banned the potlatch, the Sun Dance, and, eventually, all forms of Native dancing?

Still, some recent legal changes have a positive potential. The creation of urban reserves, although it does involve some devolution of federal responsibility, may be a way for bands to generate income and jobs and still keep their communities intact. The recognition of Métis hunting rights promises to lessen the deprivations some communities have had to endure.

There is a brighter side. Native people are saying that the answer to their difficulties will not come from making governments face their responsibilities, but from Natives healing themselves and getting their own cultures in line. Strength, health, and purpose come from within, not without. In fact, Native philosophy, rooted in the traditional but tested in the self-help, self-healing needs of Native life today, presents an alternative viewpoint for the rest of Canada to meet its challenges. This is already being seen in attitudes toward the environment. It might well guide us in justice and in other areas of life.

CHAPTER SUMMARY

In this chapter we have looked at a broad range of topics concerning Natives in Canada, especially with regard to four main points. (1) Diversity exists as well as sameness. Not all Natives are alike. We see this especially in language and other aspects of traditional culture, but it exists as well in adaptations to contemporary society. (2) Understanding any aspect of Native culture requires some knowledge of its historical roots, which often involve a very damaging prejudice and discrimination. In particular this is true concerning treaties, education, religion, and the justice system. (3) Native people are legally as well as culturally different from other Canadians. Native status brings with it different rights and limitations. And Natives are divided themselves into different legal classes: registered Indian, treaty Indian, Bill C-31 Indian, Métis, métis, and Inuit. (4) Native culture is vibrant and adaptive, not some dust-covered museum showpiece. As many Native people put it, "We're still here, and we will continue to be here."

KEY TERMS

band, p. 141
Bill C-31, p. 131
Department of Indian Affairs (DIA), p. 141
Indian Act, p. 127
language families, p. 129
language isolates, p. 129
Métis, p. 133
Powley test, p. 135

registered Indians, p. 131
reserves, p. 137
residential schools, p. 151
sentencing circles, p. 147
smudging, p. 149
sweat lodge, p. 145
treaty, p. 136
urban reserves, p. 143

DISCUSSION QUESTIONS

1. In percentage and in terms of "the year" of Canadian history, compare the length of time Natives and other Canadians have been living in this land.
2. Compare the diversity of Native languages in Canada to that of languages spoken in Europe.
3. How are the Inuit different from "Indians"?
4. Why are treaties of so long ago such an important political issue to Natives today?
5. Will urban reserves go a long way toward raising the standard of living of Aboriginal people in Canada?
6. What concerns might municipal officials and the owners of businesses in the area of urban reserves have? Are these justified?
7. How is the Native notion of medicine different from the usual mainstream Canadian idea of medicine?
8. What challenges are faced by Native police services?

NOTES

1. J.M. Adovasio and Jake Page, *The First Americans: In Pursuit of Archaeology's Greatest Mystery* (New York: Random House, 2002).
2. Bonita Lawrence, *"Real Indians" and Others: Mixed-Blood Urban Native Peoples and Indigenous Nationhood* (Vancouver: UBC Press, 2004).
3. Donald Purich, *The Métis* (Toronto: Lorimer, 1988), p. 140.
4. James S. Frideres and René R. Gadacz, *Aboriginal Peoples in Canada: Contemporary Conflicts*, 6th ed. (Toronto: Prentice Hall, 2001), p. 37.
5. Donald Smith, *Le Sauvage* (Ottawa: National Museum of Man, 1974), p. 88.
6. Olive P. Dickason, *Canada's First Nations: A History of Founding Peoples from Earliest Times* (Toronto: McClelland and Stewart, 1997).

7. Peter Kulchyski, *Unjust Relations* (Toronto: Oxford University Press, 1994).
8. *Indian Treaties and Surrenders*, Vol. 1 (Toronto: Coles Publishing, 1971), p. 48.
9. *Indian Treaties and Surrenders*, p. 52.
10. Premier William Smithe, 1887, quoted in Alex Rose, *Spirit Dance at Meziadin: Chief Joseph Gosnell and the Nisga'a Treaty* (Medeira Park, BC: Harbour Publishing, 2000), p. 13.
11. Six Nations Council, "Six Nations of the Grand River," available <www.sixnations.ca/index.htm>, accessed 27 February 2002.
12. James S. Frideres and Rene Gadacz, *Aboriginal People in Canada: Contemporary Conflicts*, 6th ed. (Toronto: Prentice-Hall, 2001).
13. Frideres and Gadacz, *Aboriginal People in Canada*.
14. Chief Constable Harry McLaughlin, quoted in "B.C. Gets Its First Aboriginal Police Force," B.C. Attorney General news release, available <http://turtleisland.org/news/news-policing.htm>, accessed 1 April 2006.
15. Adapted from John Steckley and Bryan Cummins, *Full Circle: Canada's Native People* (Toronto: Prentice Hall, 2001), pp. 237–38.
16. Angela Robinson, *Ta'n teli-ktlamsi Tasit (Ways of Believing): Mi'kmaw Religion in Eskasoni*, Nova Scotia. (Toronto: Pearson Education Canada, 2005).
17. Adapted from Brian Rice and John Steckley, "Lifelong Learning and Cultural Identity: A Lesson from Canada's Native People," in Michael Hatton, ed., *Lifelong Learning: Policies, Programs, and Practices* (Toronto: Asian Pacific Economic Cooperation Publication, 1997), p. 226.
18. Dickason.
19. R. Bell in Jean Barman, "Aboriginal Education at the Crossroads: The Legacy of Residential Schools and the Way Ahead," in D.A. Long and O.P. Dickason, eds. *Visions of the Heart: Canadian Aboriginal Issues* (Toronto: Harcourt Brace, 1996), p. 294.

CHAPTER 6

Religion as Meaning and the Canadian Context

Mikal Austin Radford

A religion is a unified system of beliefs and practices relative to sacred things, that is to say, things set apart and forbidden—belief and practices which unite into one single moral community called a Church [sic], and all those who adhere to them.
— Emile Durkheim, *Elementary Forms of Religious Life*

*Life is bigger,
It's bigger than you,
And you are not me. . . .
That's me in the corner,
That's me in the spotlight,
Losing my religion.*
— R.E.M. (Berry/Buck/Mills/Stipe), "Losing My Religion"

Objectives

After reading this chapter, you should be able to

- list the issues involved in finding a definition of religion
- distinguish the key differences between the world religious traditions
- describe how religion operates as an important component within the "cultural project"

- explain the importance of religion as part of socio-religious identity formation
- recognize how religion played an important role in the historical formation of Canada
- appreciate some of the concerns surrounding issues of religion by contemporary Canadians

INTRODUCTION

For most of us religion was one of *those* three topics—the others being politics and sex—that our parents cautioned against adopting as part of *any* casual conversation. Perhaps the idea was to avoid heated disputes and the loss of friends. What it has meant, however, is that the discussion of one's own religion or religious belief is very often relegated to either a myopic study of religion held in the privacy of one's own church, mosque, or temple, or a branch of learning undertaken by small, underfunded departments in our colleges and universities. As a result much of the public discourse of the past few decades has tended to think of religion in terms of something in opposition to the modern world—some quaint, antiquated curiosity that predates the Western "modernist experiment." In short, religion has often been treated as a cultural fossil, an antiquity that was to be replaced by science, economics, and modern statecraft.

The events surrounding 11 September 2001 have dramatically changed the timbre of that conversation.

This chapter starts from the premise that to fail to understand the context, meaning and *ethos* underlying the religions of the world is to fail to understand ourselves as we have developed, and continue to develop, both culturally and historically. The study of religion and religious belief provides three important components to understanding human diversity. The first is a lens into the great heritage of human civilization, what Ninian Smart calls "humankind's various experiments in living."[1] It can be argued, for example, that in order to understand the present context of capitalism or globalization, one should have more than a passing familiarity with the rise of the Protestant Reformation during that great period of "human experimentation," the European Renaissance. Max Weber was the first to state, more than a century ago, that to understand the "spirit" that underlies modern capitalism one must know the relationship between the religious radicalism of Calvinism and its interrelated concepts of Divine Providence, asceticism, and salvation. This background would enable contemporary culture to understand the West's march toward a capitalistic system with its cultic focus on "individualistic economic salvation."[2]

Secondly, to understand the world's religions within their cultural contexts and that of the Canadian multicultural mosaic gives us an opportunity for a new discourse that attempts to understand the meanings and values of the plural cultures both of the world and of those who come to Canada. For example, how is one to understand the passions surrounding current events in the Middle East and Central Asia, if one does not know something about the rise of Islam, and its cultural relationship to both Judaism and Christianity? More locally, how is one to understand why a young Muslim woman, born and raised in a primarily secular country such as Canada, would happily *choose* to wear the *hijab*? Why would a young Jain born into a North American society so committed to a meat-based diet be so enthusiastic to advance the principles of vegetarianism? To understand these questions we must first understand the cultural contexts and associated sacred meaning underpinning their choices.

Thirdly, the study of world religions and their particular *ethos* gives us all the opportunity to individually reflect, shape, and articulate our own unique vision of reality. That is, in order to be adept in our own judgments about the range of philosophical and ethical choices in modern life, we need a comparative perspective.

This chapter is divided into two main sections. The first section will deal with the issues surrounding a working definition of religion, then with religion's importance within the "cultural project," and finally with religion as part of identity formation. The second section will deal with some aspects of the history of religion in Canada. This is obviously a broad topic, so there will be a focus on certain immigration patterns and the resultant political legislation. And finally, there will be a concluding statement on future challenges for religion in this country.

CRITICAL THINKING BOX 6.1

If Weber is correct in drawing the connections between capitalism, economic progress, and Divine Providence, in which religious context do you think he would place the globalization model of today?

COMING TO TERMS WITH RELIGION

In the Lecture Hall: So What Is Religion?

Picture a small, darkened seminar room in Halifax, Nova Scotia. The theme of the seminar is "Religion and Meaning, and the Meaning of Religion." As the keynote speaker approaches the podium, the stage is bathed in an ethereal white light from the data projector. Block letters, filling the screen, pose the following question: "SO WHAT IS RELIGION?" Over the speaker system of the hall, the audience of religion students, academics, and the curious are treated to a rendition of the popular R.E.M. song, "Losing my Religion."

As the song fades, the speaker approaches the microphone and asks, "Interesting song? So, who knows what the title refers to?"

An awkward moment of silence, then a single hand from the back of the room. "Give it a shot," urges the speaker.

"The song is referring to an old saying," states a young woman who, judging by her eloquent inflection, is from the southern regions of the United States, "when someone *loses their religion*, it means they are losing their footing in the world. They are losing their mind, their grip on reality."

"Well done," responds the speaker. "So how, exactly, would you define the religion that the singer of our song appears to have lost?"

An anonymous voice from the darkened room: "I think he's lost his connection with God. It is like he has lost his bond with the transcendent, that thing that is bigger than you and me."

"Could be," responds the speaker. "Any other suggestions?"

"Perhaps he's feeling disoriented in the world. Maybe he's lost his sense of community, and his identity in that community. Perhaps his community is that thing that 'is bigger than you and me,' and he feels overwhelmed."

"Again, an excellent answer. Perhaps we should spend some time answering the question, 'What is religion?' and then maybe we'll have a better idea of what it is that the singer, Michael Stipe, might be losing. Today I want to start our discussion with the following definitions." The speaker offers a second slide, and reads to the audience:

> Durkheim's primary thrust in his sociological definition of religion is that "the idea of religion is inseparable from the idea of Church, it conveys the notion that religion must be eminently a collective thing."[3]

The speaker pipes in, "Perhaps this reflects what Michael Stipe has lost in his song—that somehow he has lost his sense of relationship to the collective community? After all, he appears to be all alone while under the spotlight. That somehow losing one's religion represents incredible loneliness."

After a short moment for audience reflection, the third slide appears on the screen:

> According to Weber, the most elementary forms of behaviour motivated by religious or magical factors are oriented to *this* world. "That is, 'may go well with thee . . . and that thou mayest prolong thy days upon the earth' (Deut. 4:40) expresses the reason for the performance of actions enjoined by religion or magic. . . . Thus, religious or magical behaviour or thinking must not be set apart from the range of everyday purposive conduct, particularly since even the ends of the religious and magical actions are predominantly economic.[4]

Followed by the fourth slide:

> Max Müller states that "Religion is an effort to conceive the inconceivable and to express the inexpressible, an aspiration toward the infinite."

There's an audible murmur in the crowd as the fifth slide is put on the screen:

> According to Clifford Geertz, "A religion is (1) a system of symbols which acts to (2) establish powerful, pervasive, and long-lasting moods and motivations in men [sic] by (3) formulating conceptions of a general order of existence and (4) clothing these conceptions with such an aura of factuality that (5) the moods and motivations seem uniquely realistic."[5]

After a short pause, the speaker asks his audience, "So, are we satisfied with these definitions of religion? What is right with them, or what's wrong with them?"

A voice from the front row of the seminar theatre says, "Well, from a positive standpoint these definitions certainly appear to do their best to put all religions into a 'nutshell'; they appear to make the definition clear, concise, and to the point."

Another interjects, "I have a problem with that. As I see it, these definitions try to **essentialize** religion, to reduce all religious belief to some single common 'essence' or solitary universal definition. As I see it, by trying to search for the 'essence' of religion, we end up with a definition that in many ways is much too vague, too incomplete."

"Can you provide an example of what you mean?" asks the speaker.

"Well, for one thing, I consider myself very spiritual, but I don't belong to a particular church or religious community, and I certainly don't believe in some supreme transcendent being or the 'infinite' as your Müller slide tries to tell us. Instead, I follow what my grandmother and mother taught me. I find my spiritual connection with nature. I still view myself as being a religious person, but if I had to have a 'church' I would have to say it is being in the presence of nature."

The speaker concludes, "Obviously we're going to have to have another look at defining religion."

Attempting a Definition

Although many of us cannot provide a definitive answer as to why we think the following, most agree that Judaism, Christianity, Islam, Hinduism, Buddhism, Jainism, Taoism, Confucianism, Shinto, and the tribal belief systems around the world are part of a category we call "world religions." But that raises a question: So what makes a religion a religion? Is it a system or institution concerned only with metaphysical speculation? If this is the case, we might have a problem with including Confucianism in our list, in view of its early emphasis on social philosophy, social hierarchy, and societal harmony rather than religious conjecture and metaphysical speculation. We might have similar problems with the metaphysical atheism of Buddhism and Jainism. It is certainly a nuance that has to be taken into consideration.

From such public discussions as shown in the section above, it is clearly very difficult to come up with "one essential statement" that is true of, and defines, all religions. Taking Müller's attempt for a moment, he states that all religions are "aspirations towards the infinite." Although on the surface this seems a good starting point, it does present us with our first problem. To what "infinite" is Müller specifically referring? Is it some transcendent, Supreme Being such as the "Creator God" we have in the Judeo-Christian and Islamic traditions, a Godhead with which one can have a personal relationship? Or is he referring to some impersonal **"Ultimate Reality"** similar to the principle of Brahma as described in the mediaeval and contemporary religious texts of Hinduism? Or perhaps the *tao* and *chi* of early Taoism, in which there is something akin to "the Force" described in the film *Star Wars*, which permeates the universe, but is devoid of personality traits?

Coming to Terms with the Infinite

Certainly if we were to interpret Müller's definition, at least from one perspective, and apply it to the Jain and Buddhist traditions, we would run into some difficulties. In both these religious traditions, the concept of a **transcendent God**, an infinite Supreme Being, or some form of an **Ultimate Reality** that creates, and ultimately "judges" the universe, is absent. At best, if one did have to express an "ultimate concern" within these traditions, one would have to say that the primary focus is on the principles underpinning **karma**—that is, action-as-cause-and-effect. In a *karma*-based tradition, there is no Supreme Being judging the actions of the individual. Instead, *karma* is the accumulated sum of all actions in which one has participated, and the subsequent results of those actions that "bear fruit" in some future existence. It is the individual, therefore, who is solely responsible for his or her own future experiences, not some external divine entity. Therefore, if we define religion as something that must have a "Godhead," one has to ask: Are Jainism and Buddhism religions, or are they simply an interesting worldview that ignores metaphysical speculation and its concerns over a relationship with an "infinite Being"?

BOX 6.1

General Categories of World Religions

Although most introductory texts suggest there are eight major 'world religions' and a few major philosophies, the reality is there are more than six thousand distinct religions in the world today, each with their own rituals, practices and distinct belief systems. Yet despite these seemingly overwhelming numbers it is possible to categorize these world religions and philosophies into a few core categories.

Philosophical Category	View on Reality	View on Human Nature	Understanding of Truth	Core Values
Empiricist (the belief that there is an empirical explanation for all things) Atheism Existentialism Agnosticism (open to possibility of a god or Ultimate Reality)	There is no metaphysical universe. Reality is seen as purely empirical, and is based solely on the "laws of nature." There is no such thing as a soul or spirit, or afterlife.	Human beings are the product of both chance and the "natural process" of evolution. We are products of our biology.	Truth exists only as an existential or empirical experience. Truth is only that which can be proved primarily through the five senses (a posteriori).	Morals tend to be viewed as preferences that usually manifest as socially useful behaviours. Morals are subject to change, and the "laws of nature," including evolution.
Pantheistic (the belief that every existing entity is, in reality, only one Being; all other forms of reality are either modes, or appearances, of this Ultimate Reality.	Emphasis tends to be on the metaphysical as the primary component of existence. The physical realm is often viewed as an illusion or	Ultimately human beings are to recognize that they are truly one with the "Ultimate Reality,"	Truth is beyond all rational description and empirical experience. It can only be experienced when in unity	In the pantheistic traditions awareness of the true nature of the universe is the ultimate

Hinduism Taoism Buddhism Jainism New Age	deception from the true nature of reality (e.g., *maya* in Hinduism). The metaphysical realm such as Brahman in Hinduism, or the *tao* in Taoism, is eternal and impersonal. In general terms, everything is part of the Godhead, and the Godhead is a part of everything and everyone.	and therefore are spiritual and eternal. The idea of an individual self is the "great illusion" to be conquered.	with the "oneness" of the *universal principle*, or "Ultimate Reality." In the case of Buddhism and Jainism the *universal principle* is the law of *karma*.	moral "goodness." Failing to understand the essential unity of all things is the "illusion" or "evil" that must be conquered.
Theistic (the belief in the existence and continuance of the universe is owed to one supreme Being.) Judaism Christianity Islam Sikhism	God is both an eternal and a personal entity who created a finite, material world that has both a beginning and an end. Reality is viewed as both material and spiritual.	Humankind is the unique creation of God. Therefore, human beings are *individuals* that are spiritual and biological beings who have a personal relationship with their creator.	The true nature of the Godhead is known primarily through revelation (personal experience, sacred texts, etc.), the five senses, and rational thought.	Moral values are those put forth by "the Absolute Moral Being"—God.

(cont'd)

Spiritism (the belief that the dead communicate with the living) **and** **Polytheistic** (the belief in many deities, usually male and female) Thousands of folk religions Many new religious movements	Although there may be an initial "creator being," the universe is populated by many spirit-beings who govern and are the cause of all "natural events."	Human beings are just one of the many creatures brought about by the gods and goddesses. Often in tribal traditions, the group or clan will have a special relationship with one god or goddess, or spirit guardian (either for protection of the clan or to administer punishment).	Truth about the natural world is discovered through the shaman figure, who has visions telling him what the gods and demons are doing and how they feel.	Moral values take the form of taboos, which are things that irritate or anger various spirits. These taboos are different from the idea of "good and evil" because it is just as important to avoid irritating evil spirits as it is good ones.
Religious Postmodernism (a philosophical strategy that attempts to destabilize monolithic modernist concepts such as identity,	Reality is seen as a "social construction" that must be interpreted through our language and cultural "paradigm."	Human identity is exclusively a product of social setting. The idea that people are autonomous and free is a	Truths are mental constructs that are meaningful only to individuals within a particular cultural paradigm. They do not	Values are part of our social paradigms as well. Tolerance, freedom of expression, inclusion, and refusal

historical progress, epistemic certainty, and the univocity of meaning)	myth; they are a product of society.	apply to other paradigms. Truth is relative to one's culture.	to claim to have the answers are the only universal values.

We would run into a similar problem if we were to suggest that a religion, by its very nature, must believe in the existence of an immortal entity identified as the soul. Interestingly, the idea that something of each individual existing beyond death is very old, or at least the archaeological evidence appears to demonstrate that our earliest ancestor provided great care to their deceased some 200 000 years ago. And though we do not have a definition of what these ancestors believed existed after death, we do know they felt that something of the individual existed beyond the grave.

So, if the existence of something akin to a soul was a prerequisite to our definition of religion, many of the ancient tribal and indigenous traditions would certainly qualify. And if we continue with this as our operating definition, certainly the Judaic, Christian, and Muslim traditions would all qualify, though they have a different understanding about the nature of the soul. But despite these differences, there is also a commonality in the various traditions—they all maintain every individual has a soul, and that in some way this soul will be judged in the afterlife. That is, every soul is a uniquely individual component of personality that will be either punished or rewarded in the "afterlife" according to that individual's actions in this world. In most cases this judgment is final and eternal.

Here, however, is where we run into problems with our definition. Though the Jains believe in an entity called the soul (*jiva*), in contrast to the traditional Judeo-Christian and Muslim definitions of soul, or the soul as ancestor spirit as described by many of the folk traditions around the world, both the Jains and the Hindu sects do not understand this entity to have any individual personality associated with it. In point of fact, the Jains would view those characteristics defined as personality more in keeping with those components that are the direct result of *karma*. The Jains understand personality to be *karmic* material that acts much like a ship's anchor holding the boat firmly to the ocean floor. It is karma that keeps the soul in bondage and away from liberation (*moksha*). Personality, therefore, is all those attachments and desires that manifest themselves within the material universe as either a human being, a god, a hell-being, an animal, or a plant, and holds the "pure soul" within the eternal cycles of reincarnation (*samsara*). Personality is one of the components of **samsara** (cycles of reincarnation) from which one should free one's self to gain liberation.

BOX 6.2

World Religions by Population

The following table includes both officially recognized organized religions that primarily adhere to a single orthodoxy of belief and those less formal in their social structure, religious hierarchy, or singular orthodoxy of belief. It is also interesting to note that most of these traditions have varying degrees of representation among the Canadian population. Unfortunately it is difficult to get exact percentages of population from Statistics Canada, as many fall under "Other" in the categories; however, a cursory glance at the telephone directory of any large urban centre or similar resources (e.g., the Multifaith Council of Canada) provides insight into the number of different religions and sects represented in Canada.

World Religions by Population, 2006 (averaging of various sources)

Christianity	2 billion
Roman Catholicism	1.1 billion
Protestantism	360 million
Eastern Orthodoxy	220 million
Anglican	84 million
Other Christians	280 million
Islam	1.3 billion
Sunni	940 million
Shiite	120 million
Hinduism	900 million
Secular/nonreligious/agnostic/atheist	1 billion
Buddhism	376 million
Mahayana	185 million
Theravada	124 million
Chinese traditional religions[a]	394 million
Indigenous/folk religions[b]	285 million
African traditional and religions that trace roots to African origins	95 million[c]
Sikhism	25 million
Spiritism[d]	14 million
Judaism	14 million
Bahá'í Faith	6.8 million
Jainism	5.3 million

Shinto	4.4 million
Cao Dai[e]	3 million
Tenrikyo[f]	2.4 million
Zoroastrianism	2 million
Neopaganism	1.1 million
Unitarian Universalism	800 000
Rastafarianism[g]	700 000
Scientology	550 000

[a]Not a single organized religion; includes all Chinese religions that contain elements of Taoism, Confucianism, and traditional indigenous folk religions.

[b]Includes world folk traditions and Shamanism and Paganism.

[c]Not a single organized religion; includes several traditions such as Yoruba, and Santeria, and Vodoun.

[d]Not a single organized religion; includes a variety of belief systems.

[e]A universal faith with the principle that all religions have the same divine origin, which is God, or Allah, or the Tao, or the Nothingness, and the same ethic based on love and justice, and are just different manifestations of one single truth.

[f]A modern eastern religion based in concepts of the Japanese Shinto and Buddhist traditions.

[g]Primarily an Ethiopian-based religion; prominent within the Jamaican community.

The Buddhists, on the other hand, would agree with the Jain philosophical position of *moksha*—that one must become free from the cycles (*nirvana*) of reincarnation (birth, death, and rebirth)—but would counter the Jain position stating that there is no such object, being, or eternal entity that one could call a soul. Would we be willing to say at this juncture that Jainism is a religion because it believes in an entity called "soul," but Buddhism is not because it does not believe in such an entity? To most scholars the answer would be inclusive: of course, both are religions.

"Big R" and "Little r" Religion

Before attempting to provide a descriptive template for religion, two more issues should be addressed. The first concerns those people who feel they are deeply religious, and yet do not participate within the framework of a formal religious movement or organization. In many cases, they belong to a group that is often marginalized outside the traditional mainstream religions (some may characterize these entities as "religious cults"). Many members of the Wiccan community across Canada, for example, have often stated to me in interviews that they do not recognize the imposed hierarchies of mainstream religions, nor do they accept the concepts of "the Infinite" or "the Transcendent" as used in the traditional senses. Instead, they often view "the Ultimate" in terms of both *unity with the forces of nature* and a unifying relationship they have with other members of their coven or local communities. For them, religion is not some essentialized doctrine or dogma, nor is it the hierarchical structure often associated with mainstream religious organizations; rather, it is an individual, internalized experience in relationship to an external spiritual and material world.

Secondly, it is apparent that when people talk about their religion, they do so in two different modes. On the one hand, there is **"big R" religion**—the type of religion one often reads about in introductory texts about religion and developed by either the religious specialist or academic. Then there is "little r" religion, which often refers to the area of study anthropologists term "that part of religion that becomes messy." This is the religion that is often passed from grandmother, to mother, to daughter, or from grandfather, to father, to son. **"Little r" religion**, then, is that religious practice that occurs *on the ground*—to coin an anthropological term—or in the home of an individual family unit. It is the traditional practices of our families, and not that which was conveyed by some religious specialist, academic, or religious institution. It is the type of religious practice that tends to be impossible to essentialize, because every family has a different practice or custom.

Religion as Builder of Worlds

In the opening pages of his text *The Sacred Canopy*, Peter Berger states, "every human society is an enterprise of world-building [, and] religion occupies a distinctive place in this enterprise."[6] Extending the social theories of such notables as Karl Marx, Emile Durkheim, and Max Weber, Berger contends that the phenomenal world, and how that world is perceived, understood, and acted upon by human beings is actually a **dialectical process** that constantly "creates" and "re-creates" the world through a procedure of **externalization**, **objectivation**, and **internalization**. Berger defines these terms as follows:

> Externalization is the ongoing outpouring of human being into the world, both in the physical and the mental activity of men. Objectivation is the attainment by the products of this activity (again both physical and mental) of a reality that

---

CRITICAL THINKING BOX 6.2

Despite Canada's guarantee to its citizens of religious freedom, can you think of examples of either "big R" or "little r" that should not be practised within the Canadian context? For example, if male circumcision is allowed based on religious belief and practice, should we also allow for the practice of female circumcision if performed in a sterile, hospital setting?

confronts its original producers as a facticity external to and other than themselves. Internalization is the reappropriation by men of this same reality, transforming it once again from structures of the objective world into structures of the subjective consciousness. It is through externalization that society is a human product. It is through objectivation that society becomes a reality *sui generis*. It is through internalization that man is a product of society.[7]

To try to put these definitions into context, externalization occurs when human perceptions and understanding of the universe become externally manifest as representations in both the things that we make (objects, tools, art, music, institutions, culture, etc.) and the things we do with those "products" within the public sphere. In short, human beings project meaning into the empty vastness of the universe by creating both a material and an institutional culture that reflects that meaning.

In turn these **products of the human cultural project** become the primary objects (*objectivation*) of our attention. That is, we begin to interact with the representations we have created as if they were the universe itself. Interaction with, and the subsequent internalization of, these objects begins to change us. For example, consider the institution of marriage. The roles of husband or wife are objectively defined by the existing culture to represent models for individual conduct within a particular culture. As Berger states,

> By playing these roles, the individual comes to represent the institutional objectivities in a way that is apprehended, by himself and by others, as detached from the "mere" accidents of his individual existence. . . . Society assigns to the individual not only a set of roles but a designated identity. . . . Internalization is . . . the reabsorption into consciousness of the objectivated world in such a way that the structures of this world [culture] come to determine the subjective structures of consciousness itself.[8]

From this perspective, culture is a project meant to bring order to a universe that may appear to the individual as both chaotic and sometimes meaningless.

Religion as our Response to the Meaningless

On the surface, religion-as-dialectic-process appears to be both culturally stabilizing and self-contained. Indeed, if the individual finds resonance with these social roles and cultural institutions, he or she freely identifies with and participates within the "social project," as opposed to feeling society is forcing one to participate in a given role. This stable social environment is what Berger calls the ***nomos***.

There is a problem, however, and this begins with the realization that the "dialectic" is actually a process, rather than an end. In other words, human culture is a project that is never "finished"—culture is inherently in a constant state of flux. Social environments change, in large part, because the "institutional programs are sabotaged by individuals with conflicting interests," or the "original" meaning underpinning the socially constructed world is simply forgotten from one generation to the next. Reaction to this change by the social collective, and the subsequent reabsorption of a newly reconstructed vision by both the individual and the consciousness of the social collective, can create "new worlds" of meaning and understanding, but it also gives rise to the identification of the primary human predicament—*culture is inherently volatile, unpredictable, and unstable.*

Religion, therefore, is the ultimate response to this predicament. In order to protect the individual, and ultimately society, from the breakdown of the existing social order, myths, rituals, and orthodoxy are established to help merge the internalized, humanly constructed *nomos* with that which is perceived to have been divinely constructed. In other words, to convince the social collective that society and its roles and institutions are not simply the construction of human beings that are prone to "human error," religion is used to sacralize the existing social order—to make it sacred, to set it apart as something divinely constructed, and therefore something both eternal and immortal. In this sense religion is "world maintaining."[9]

Religion as Destroyer of Worlds

There is, however, another side to religion. To use a cliché, religion is very much a two-sided coin. One side reveals religion as the ultimate response to the volatility of the cultural project. Religion is used to make sacred the cultural project, provide its members with social stability, and, in turn, provide the universe with meaning. But what happens when the cultural project begins to alienate its own participants? What happens when members of a culture discover that, instead of meaning and constancy, their sense of the universe and the state of the culture is actually in a condition of chaos and disintegration? To a large extent, history has provided the answer; it has shown us the other side of the coin, so to speak.

In contrast to religion being an enterprise that is "world building and world maintaining," it can also be the very organism that overthrows an existing social order. While religion can be initiated by an existing social order to maintain its way of life through claims that it was modelled after the diktat of some divine source, it can also challenge the very nature of this premise. It can pull back the veil of society's religious mystery, the **sacred canopy** protecting its cultural institutions from assault, and reveal that the status quo is nothing but a human construct, a project neither immortal, nor divinely inspired, nor immune to critical errors. Religion can also be the very mechanism of challenge and change to the social order with a new "divine message" or diktat.

Religion as Cultic Project

Throughout history, there are many examples of religion being the catalyst for social change: the Buddha (Buddhism) and Mahavira (Jainism) challenging the social and metaphysical hierarchies developed within the South Asian Vedic sacrificial culture; Jesus of Nazareth testing the priestly order in the Jerusalem Temple; Mohammed casting out the pagan idols contained within the Ka'aba and reinstituting monotheism in the Arabian peninsula (and beyond); St. Francis questioning the Christian monastic orders in Europe; Confucius addressing the chaos of the Warring States period in China; and in many ways, the religio-cultic political figures of Lenin, Mao, Castro, and Che in their struggle against the excesses described within Weber's model of Protestant theology and its new morality for entrepreneurial capitalist behaviour. Although we don't often think in terms of these new religious movements as being **cults**, these examples are in fact religio-cultic responses to a preexisting cultural order that has either marginalized or alienated (***anomie***) individuals within their midst. In response, the cult is born. That is, to paraphrase both Lorne Dawson and Steven Tipton, the reason for an individual or group's conversion to an alternative cultural paradigm is that people experience "ethical contradictions of unusual intensity" in the present cultural project, and therefore will gravitate to a new paradigm that provides a coherent solution.[10]

Within this paradigm, cults can be considered both beneficial and harmful. To the existing cultural order, any cult presents a challenge to their power and the ideals of the status quo. On the other hand, for those on the margins of society, the cult—whether it pays homage to a divine being, a charismatic leader, a cultural object or institution, or, in more modern times, an abstract ideal—can provide an alternative cultural project to restore order and meaning to their lives.

Religion and Fundamentalism

> Have you heard of that madman who lit a lantern in the bright morning hours, ran to the market place, and cried incessantly, "I seek God! I seek God!" As many of those who do not believe in God were standing around just then, he provoked much laughter. . . . "Whither is God," he cried. "I shall tell you. *We have killed him*—you and I. All of us are murderers. . . . God is dead. God remains dead. And we have killed him. . . .
>
> — Friedrich Nietzsche, *The Gay Science* (1882), section 126.

Nietzsche made one of the most controversial declarations about the state of the modern humanist movement when he proclaimed that modernism's greatest accomplishment was to have finally "killed" God. This God, according to Nietzsche, was created and used as an instrument of social oppression, a cultural tool to redirect society's attention away from the potential freedoms of this world, and to point it toward some escapist otherworld

BOX 6.3

Cults and the Danger Signs

In many ways, the definition of "cult" has become so blurred it has lost any significant meaning. As shown above, cults can be the early form of life-affirming cultural movements whose intent is to replace social disintegration and *anomie* with order and meaning. In this sense, cults can be liberating, but they can also be a tool to oppress their members. Below are some of the cautionary danger signs of such cults.

Charismatic Leadership
- Most cults tend to be led by a charismatic leader, who is usually male. In many cases the leader has been married, but he or she tends to become "single" in conjunction with the growth of the cult.
- The leader dominates the membership, closely controlling them physically, sexually, and emotionally.
- The individual personalities of cult members are subsumed by that of the charismatic leader.

Apocalyptic Beliefs
- One warning sign is that the leader focuses heavily on the impending end of the world, often involving a great battle (e.g., images of Armageddon).
- Another sign is that the leader advocates suicide in order for the group to be transported to "another world" to escape the coming world devastation (e.g., the Solar Temple cult in Canada; the Heaven's Gate group).
- Members are expected to play a major, leadership role at and after the "end time."

Social Encapsulation
- Most cults tend to be small religious groups, and not an established denomination (although some may be "breakaway" groups).
- Most of the members, especially the "core members," live in communities isolated either physically or psychologically from the rest of society.
- Non-members are often demonized and considered "the enemy." There is often the sense within the cult that they are being closely monitored by the social authorities, and that these authorities are attempting to persecute the cult.
- Both information and contacts from outside the cult are severely curtailed.

Other Warning Signs
- In accordance with the social encapsulation process, many recent cults have been known to prepare defensive compounds and assemble a vast array of weapons and poisons (e.g., People's Temple, Branch Davidians, Solar Temple, Aum Shinrikyo, Heaven's Gate).
- Theologically many cults appear to be associated with a familiar religious tradition, but they often have some unique deviation from it. They often stress this as "the new Revelation" with a focus on end-time prophecies.

Important note: Though they may exhibit some of the characteristics listed above, some cult communities may not be considered dangerous. For example:

- Some of the factors may not be practised to such an intense degree, or may be absent altogether. For example, some groups may have a charismatic leader, and promote an "end times" theology, but not present a danger or advocate violence to either the "outsider" community or its own members.
- Some cults may "advocate" hatred toward the "outsiders" or minority groups (e.g., homosexuals, ethnic groups, or those with a particular political affiliation), but not call for direct or immediate violence against them. The problem, of course, is that they may impel others associated with the cult to take action.
- Some religious groups, although not a threat to the larger society, may risk the health of their own members. For example, groups such as the Jehovah's Witness recommend that their members refuse blood transfusion, and members of the Christian Science Church ask that their members refuse all medical help, and seek healing through prayer. Also, issues surrounding the treatment of children and corporal punishment have been raised in recent years.

CRITICAL THINKING BOX 6.3

Often cults are considered only within a religious context. But they can also be applied to political ideologies. Can you think of cults that might be political in nature? Using the model for cultic warning signs, how much do these political ideologies resemble religious cults?

located in heaven (or hell). Much like Karl Marx, Nietzsche viewed religion as an anathema that oppressed the "masses," an illusion created by the social order to maintain social obedience among "the herd," and to curtail the full exploration of individual freedom. To kill off God was to clear the path to the experience of true freedom for the individual (even if that freedom includes the freedom to feel pain as well as joy).

The problem of secularizing the social order—and this concept was not really developed by Nietzsche and other existentialists—was that it took away the *sacred canopy* of society. On a surface reading, "killing God" was seen as an opportunity to be truly free, an opportunity to break the yoke of an outdated social construct that no longer provided meaning and order to the world. But for those who valued the role of the sacred canopy, its removal exposed secular society as a "Godless society" with a "Godless morality," a society constructed by human beings alone, very much prone to the "evils" of human error. For this group "killing God" meant killing a divine order and, more importantly, a disintegration of its divine morality. One response to the failures of the modernist state—viewed as a morally bankrupt political system—was the rise of religious **fundamentalism** during the 19th, the 20th, and now the 21st, centuries.

In North America this renewed religio-cultural project tends to be associated with a wide spectrum of traditional evangelical (both Protestant and Catholic) and Christian fundamentalisms, many of which are associated with the political conservative right-wing of both American and Canadian politics. Interestingly, this development in North America has come in response to what has been perceived as the steady decline in the influence of the more traditional religious denominations, and the increase of the control of the public sphere by the modernist state. In Quebec, for example, the "Quiet Revolution" took place during the 1960s with the election of the Liberal Party. The result of this election was that the church no longer held sway over education, social services, and health care (Newfoundland was to follow this path a bit later)—the *curé* in Quebec was replaced by the modernist state. In Ontario we see a similar "revolution." The traditional domination of Anglo-Protestantism, particularly in the urban centres such as Toronto, was quickly losing its grip as it found itself adapting to the federal government's policy of multiculturalism (1971). For many, religious conservatism was losing out to the "revolutionary" ideals of social liberalism.

> ### CRITICAL THINKING BOX 6.4
>
> Do you see a Canadian politician's declaration of being a Christian Fundamentalist a concern for our multicultural political and social stability?

Fundamentalism in the World

These movements are not confined, however, to either Europe or North America. On an international level we see the rise of fundamentalist forms of Hinduism in South Asia, new religions in China, evangelical movements in South Korea, the Orthodox church in both the former Soviet Union and its subject states, and various fundamentalist factions of Islam throughout the world—most a response to the corruption, social chaos, and meaninglessness created by the human secularization of the cultural project. In other words, the fundamentalist perceives the modernist ideal of the secular nation-state as a system of **failed states**—a system that is morally bankrupt because of the secularization project. In response then, religion and its "sacred canopy" is very much reentering the global cultural discourse. And to some extent, understanding this trend, however distasteful we may view the extremist outcome produced by these sentiments, may provide some insight into the events leading up to "9/11."

When I first heard this joke about thirty years ago, I didn't realize the importance of the statement: "God is dead—Nietzsche / Nietzsche is dead—God." In the light of recent events around the world, I wonder if this statement of humour has even more relevance today, as it raises the important question: How is the modernist cultural project to deal with religious groups that understand God to be, despite Nietzsche's claim, very much

---

CRITICAL THINKING BOX 6.5

Canada's Solution—The Multiculturalism Act of 1985

Using the material excerpted from the Act quoted below, choose one of the following areas of discussion:

- With an eye to the *Multiculturalism Act*, discuss or write your views on the best way to integrate religion with the secular.
- Is Canada's multicultural policy a good solution to issues of religious conflict? Why or why not?
- Canada was the first country in the world to actually legislate multiculturalism. Should all countries in the world follow Canada's lead? Why or why not?
- Should students/citizens be given provision/locations in the public arena in which to conduct prayers or religious services?

(cont'd)

From the *Multiculturalism Act*:

An Act for the preservation and enhancement of multiculturalism in Canada
[1988, c. 31, assented to 21st July, 1988]
Preamble

- WHEREAS the Constitution of Canada provides that every individual is equal before and under the law and has the right to the equal protection and benefit of the law without discrimination and that everyone has the freedom of conscience, religion, thought, belief, opinion, expression, peaceful assembly and association and guarantees those rights and freedoms equally to male and female persons;
- AND WHEREAS the Constitution of Canada recognizes the importance of preserving and enhancing the multicultural heritage of Canadians;
- AND WHEREAS Canada is a party to the International Convention on the Elimination of All Forms of Racial Discrimination, which Convention recognizes that all human beings are equal before the law and are entitled to equal protection of the law against any discrimination and against any incitement to discrimination, and to the International Covenant on Civil and Political Rights, which Covenant provides that persons belonging to ethnic, religious or linguistic minorities shall not be denied the right to enjoy their own culture, to profess and practise their own religion or to use their own language;
- AND WHEREAS the Government of Canada recognizes the diversity of Canadians as regards race, national or ethnic origin, colour and religion as a fundamental characteristic of Canadian society and is committed to a policy of multiculturalism designed to preserve and enhance the multicultural heritage of Canadians while working to achieve the equality of all Canadians in the economic, social, cultural and political life of Canada; . . .[11]

alive and meaningfully guiding their lives? One response might be Canada's *Multiculturalism Act*, an act meant to provide equality to all religious points of view, yet ensuring no domination by one group over another.

CANADA AND ITS RELIGIOUS EXPERIENCE

It has been four centuries since the arrival of the Europeans and the establishment of their permanent settlements in Canada. And although the arrival of the French and its Roman Catholic faith may initially have been on good terms with the indigenous First Nations peoples—a focus more on trade and commerce than religious conversion—this peace was not to last. Native spirituality as the dominant religion in Canada was, by the early 1600s, supplanted by the Franco-Europeans, whose focus shifted from trading commodities to imposing their cultural values on the religious landscape. The template for Canadian history had been formed.

As we look back at Canada's rich cultural tapestry, the threads of religion have always been located front and centre. Whatever the origin of its peoples—First Nations, French, British, American, Irish, Scottish, Asiatic, South Asian, Ukrainian, Italian, African, Arab, or Russian—religion has always provided these new Canadians with both a sense of meaning and an awareness of order and justice. Like a double-edged sword, however, religious belief and practice has also been one of the leading causes of conflict throughout Canadian history. As Robert Choquette states,

> [T]he arrival of the Europeans in Canada [has] often been characterized by rivalry, acrimony, and conflict among the leading religions in place. European Christians worked to eradicate Amerindian spiritualities, Protestants fought Catholics and vice versa, Jews were frequently the victims of discrimination by Christians, and Asiatic Chinese and Japanese were the targets of repressive legislation. That has been because each religion considered that it had the monopoly of truth; all others were in error. Therefore, the religious group that was dominant usually made life difficult for all religion minorities.[12]

For the most part, religious conflict in the first 250 years of the European presence in Canada was primarily fuelled by the religious and political hostility centred in Europe—the fighting between Roman Catholics (French, Spanish, and later the Irish) and Protestants (British and Dutch). These religious wars continued in the lands the Europeans colonized, and in the case of Canada and North America they meant a battle for souls by "evangelizing" the peoples of the First Nations. North America was seen by the Europeans as a land not only of economic opportunity (politically and economically

The Many Faces of Diversity

based colonialism), but also one in which to develop their religious vision of a "New Jerusalem"—a religious experiment in which a religious community could practise their particular tradition freely.

Between the years 1600 and 1760, the British and French fought for the sovereignty of Canada; however, with the end of what several historians call the "first true world war," France was to cede the Canadian territories to the British in 1763 with the signing of "the Treaty of Paris." For the next six years a series of treaties were signed with one of the major components being the establishment of a religious policy in the *new* Canada—a policy that was to last more than 150 years (but one not to be severely challenged on many fronts). For the most part these treaties spoke of a policy of religious freedom for Canada's Catholics, an unusual policy by the British-as-conqueror, and a unique policy experiment for its time. This was, however, only a surface reading of the British intent in Canada. The "unofficial" message given the governors of Quebec was quite different. For the most part they were directed to "hold the church on a tight leash, to restrict its freedom as much as possible, and to do everything they could to promote the interests of the Protestant religion."[13] Between the 1760s treaty-years and the 1800s both the British Crown and its

---

BOX 6.4

Top 10 Religions in Canada, 2001

Group	Identified Membership	Percentage of Population
Roman Catholic	12.8 million	43.2%
No religion	4.8 million	16.2%
United Church of Canada	2.8 million	9.6%
Anglican	2.0 million	6.9%
Apostolic, Evangelical	780 thousand	2.6%
Baptist	729 thousand	2.5%
Lutheran	607 thousand	2.0%
Muslim	580 thousand	2.0%
Other Protestant	549 thousand	1.9%
Presbyterian	410 thousand	1.4%

Source: Adapted from Statistics Canada, "Top Ten Religions in Canada", adapted from the Statistics Canada publication "Highlight tables, 2001 Census," Catalogue 97F0024XIE2001015, Released May 13, 2003, available at: <http://www12.statcan.ca/english/census01/products/highlight/religion/Index.cfm?Lang=E>

Protestant émigré population undertook a project to make Protestantism the official religion of Canada. The British largely succeeded in this venture—Upper Canadian and Maritime politics was to be dominated by Protestantism, but Lower Canada and parts of Newfoundland were to be dominated by Roman Catholicism (a power that was not seriously challenged in Quebec until the rise of the sovereigntist movement in the 1960s). As the British pushed westward, so did the domination of its pro-Protestant political and social policies.

The First Challenge to British Canada and Protestant Domination

Despite travelling a rather "rocky road," it appears that by the 1850s there was a relative religio-political equilibrium between the dominant Protestant ruling class and Roman Catholicism. Perhaps one of the major factors contributing to this was the relative singular ethnicity of its peoples. That is, most new Canadians were arriving from the land of western and northern Europe. By the 1860s, however, the constancy of this "European ethno-culture" was to be challenged, and for the next 100 years the ethnic, cultural, and religious composition of Canada was to become one of the most diverse in the world. As space is too short to thoroughly outline the details of this diversity, we will focus on two non-European immigrant groups, the Chinese and South Asians, as an example of this new immigrant pattern and the dominant culture's reaction to it.

The First Major Wave of Non-European Immigrants to Canada: The Chinese

Although Captain John Meares, a retired British naval officer, noted the presence of people of Chinese origin on the western shores of Canada (British North America) as early as 1788, most sources indicate that it wasn't until 1858, with the beginning of the Fraser Valley Gold Rush, that we see large-scale immigration to Canada of the Chinese from both California and China—in the period of 1858 to 1884 entry into Canada as either labourer or prospector was unrestricted. By the 1860s, and particularly after Confederation (1867), pressure from a land-hungry United States forced an exacting effort by the newly formed Canadian government to push into the Canadian West in order to maintain sovereignty from coast to coast. This required building a national railroad. Despite considerable efforts by the Canadian government of its day to increase immigration from Europe, labour shortages in the underpopulated West, especially for the building of the railroad, continued to plague the government.

In China during this period, increases in both population and taxes forced many off the land. Industrialists in Canada saw an opportunity—Canada needed labour, and those Chinese who chose to leave their homeland saw economic opportunities in Canada. Between the years 1881 and 1884, 17 000 Chinese arrived in Canada, most under contract to the Canadian Pacific Railway to complete the connection between British Columbia and the rest of Canada.[14]

With the completion of the railway in 1885, the West was finally opened to "the rails" and the need for Chinese labourers had all but disappeared. The "white society" of British Columbia looked upon members of the Chinese community as competition for the few jobs remaining in the region. The government's response was to impose a $50 "head tax" on all Chinese immigrants. The "head tax" policy, continued well past World War I, in time had increased from $50 to $500 per person. Despite this enormous tax, and the fact that many Chinese were excluded from industry and commerce in the West and forced to move eastward as far as Toronto, immigration from China still continued until the early 1920s. However, compounding the demise of the Chinese labour market was the return of the Canadian soldiers from the war. The situation became so tense between the "white community" and the Chinese community that the government was "forced" to bring two acts before Parliament: the *Dominion Elections Act* of 1920, which took away the federal franchise rights of both the Chinese and South Asian Immigrants and the *Chinese Immigration Act* of 1923, which virtually prohibited any Chinese immigrant from entering Canada.[15]

South Asian Immigration to Canada: The Problems of Identification

In many ways the immigration pattern of the South Asian community parallels that of the Chinese immigrant population, as did the response by the dominant "white" culture in Canada to this group. It was Canada's relationship with Great Britain and the other Commonwealth nations, however, that was profoundly different from the Chinese experience and facilitated the first wave of South Asian immigrants to continental North America.

Members of the British Commonwealth were initially granted special access to other Commonwealth countries. South Asians, especially those who served in the British Army (i.e., the Sikhs), were considered "full members" of the Commonwealth nations and given

---

CRITICAL THINKING BOX 6.6

Both the Chinese and Japanese had private property either destroyed or confiscated during different periods of their Canadian experience. In the light of Canada's present policy of multiculturalism, is it the duty of its current citizens to make reparations for these past transgressions?

limited preference in consideration for immigration. However, the host countries did not necessarily appreciate this special status, and many supported laws that discriminated again non-white members of the Commonwealth. In time, this movement to ban South Asian entry into Canada grew exponentially. The Chinese, on the other hand, were never members of the Commonwealth (with just a few colonial exceptions); they were never considered for special status, and therefore were the first to feel the effects of discriminatory laws in Canada. Eventually, these laws were passed to include both Asians and South Asians.

Unfortunately, existing documentation in the early days of Confederation (1867) can provide only a meagre glimpse into South Asian immigration and visitation patterns between 1867 and 1899.[16] For example, according to Roger Daniels there were a total of 491 "Indian entries between 1871 and 1899 . . . but this data reflects place of birth as well as race," and in many cases, unlike in the Chinese example, race and place of birth were often confused. Immigrants might be of British, French, Dutch, Persian, or African ancestry, but if they were born within the colonial jurisdiction of British India they might be incorrectly classified as "Indian" by birth—or vice versa.[17]

Keeping these descriptions in mind, there are some Canadian studies that attempt to paint a slightly more definitive picture of the early South Asian immigrant and visiting patterns to Canada. For example, C.I. Petros concludes that although specific names of the first South Asian immigrants to Canada are not known, it is clear from the early records and reports that there "had been visits of Indian dignitaries and Indian seamen, to Vancouver, in the west coast of Canada, especially after the opening of the trans-Canada rail route to the eastern ports."[18]

Patterns of South Asian Immigration to Canada: The First Wave

Although the historical context of the story varies, the first wave of South Asians to establish themselves in Canada were largely members of the Sikh community both from India (retired army officers who had served under the British Raj) and "from established Indian communities in Hong Kong."[19] According to George Kurian (1993) and web publications by both the Government of Canada and members of the Canadian Sikh community,[20] the event that marked the genesis of South Asian immigration from the Indian subcontinent to North America was the visit of Sikh soldiers to British Columbia on their return journey (to Hong Kong and India) after participating in Queen Victoria's Diamond Jubilee celebrations in 1897. The visiting group initially received a friendly welcome from their "official" Canadian hosts, particularly from prospective employers. As Buchignani states, highly placed officials within the Canadian government did not share the local concerns that South Asian immigrants might not be "warmly received" and cited "the beneficial effect of South Asians in British Guyana, Trinidad, and elsewhere.[21] Thanks in large part to the various rumours of job availability (particularly with regard to the expansion of the national railway, the west coast lumber mills/logging camps, and farming), offers of Canadian farmland to any British subject, and compelling sales tactics by shipping companies hoping to cash in on any migrants willing to book passage on their ships, the soldiers returned to India (particularly the Punjab) to "spread the news."

A variation on this story states that the single incident that fuelled immigration of Indians to Canada was the visit of a group of soldiers under Sergeant Major Kadir Khan Bahdur who were on their way back after attending the coronation of Edward VII in 1902. Although both versions of the story are acknowledged by the Sikh community and the Canadian governmental sources, a greater emphasis is placed on the first (with Captain Kesur Singh being the first Sikh soldier to arrive in Canada). Records with Citizenship and Immigration Canada mention both versions, but again, they lean toward the "Queen Victoria version." What all sources agree upon is that the first permanent South Asian immigrants arrived in British Columbia in 1903–04, and totalled 45 "Hindoos"[22] (an obvious misnomer, even though a few were in fact members of the Hindu religious tradition) who had mortgaged their land at 10 to 12 percent interest to raise the $65 they needed for their ship fare and the amount dedicated for the immigration (head) tax.

BOX 6.5

Religious Trends in Canada by Percentage of Total Population

Religion	1981 Census	1991 Census	2001 Census
No religion	7.4%	12.3%	16.2%
Buddhism	0.2%	0.6%	1.0%
Catholic	47.5%	45.2%	43.2%
Protestant	41.2%	34.9%	29.2%
Eastern Orthodox	1.5%	1.4%	1.6%
Other Christian	—	1.3%	2.6%
Hinduism	0.3%	0.6%	1.0%
Judaism	1.2%	1.2%	1.1%
Islam	0.4%	0.9%	2.0%
Sikhism	0.3%	0.5%	0.9%

Source: "Religious Trends in Canada by Percentage of Total Population," Adapted from Statistics Canada, 1981, 1991, and 2001 Census of Canada.

In contrast to racial issues surrounding the earlier "wave" of Chinese and Japanese indentured labourers brought to the west coast of Canada, the initial reception of the South Asia migrants (who were already British subjects, and in many cases either educated professionals or retired members of the British army) was relatively "warm." This "warmth" was particularly demonstrated by one segment of Canadian society, the industrialists and railway company executives who had lobbied the government to increase the immigration numbers of inexpensive, unskilled, and semiskilled workers. What is interesting here is that these immigrants were not indentured labourers. They paid their own ship fares and immigration "taxes." It is arguable that many of the South Asian immigrants—those who were former members of the British Army or landowners from the Punjab region—were "underemployed" and that their skills were "underutilized" by those seeking the services of cheap, immigrant labour.

The "warm greeting" was short-lived, however. In contrast to the sentiments expressed by business owners and industrialists, many labour, political, and religious organizations raised objections to all forms of immigration whether the source of individuals was Asia, or South Asia, or northern or eastern Europe. On the Prairies, for example, suspicion and hatred of this kind were focused mainly on the first wave of Ukrainian (Roman Catholic and Orthodox) immigrants and the Doukhobors (Anabaptist) from Russia. For example, we have the following from the archives of Citizenship and Immigration Canada:

> Organized labour, of course, took a very jaundiced view of the hiring of unskilled immigrant labour by railways and manufacturing companies. One spokesman who did not hesitate to speak bluntly on the subject was James Wilks, a vice-president of the Trades and Labour Congress. In 1900, he wrote to Prime Minister Wilfrid Laurier about the impact that an influx of Scandinavians and Finns from Minnesota had on the Canadian labour market. Wilks beseeched the Laurier government to enforce the *Alien Labour Act*, a piece of legislation designed to prevent the importation of contract labour. Only rigorous enforcement of this law, claimed Wilks, would prevent Canada from being inundated with "ignorant, unfortunate . . . non-English-speaking aliens" who would do irreparable damage to the community. There was also widespread opposition to western pioneers from central and southeastern Europe. Excellent farmers they might have been, but in the eyes of many westerners this did not qualify them as desirable settlers. Only those who assimilated readily into the dominant Anglo-Saxon society were welcome.[23]

Despite these strenuous objections, Sir Clifford Sifton, who became Minister of the Interior in Sir Wilfrid Laurier's Liberal government in 1896, instituted a relatively tolerant immigration policy and simplified regulations to meet the economic needs of business, trade, manufacturing, and farming. Sifton actively persuaded immigrants, particularly from both eastern and western Europe, to come to Canada. He did this by appointing immigration agents in many countries, and advertising aggressively though handbills, newspapers, and periodicals. According to the archives of Statistics Canada, 2 500 000 immigrants from around the globe came to Canada between 1904 and 1913. The numbers of immigrants from India (almost exclusively male) were as follows:

1904–04	45
1905–06	387
1906–07	2124
1907–08	2623

Opposition to Asian and South Asian Immigration Rises: Legislation and the Riot

By 1907 the public outcry associated with perceptions of rising unemployment contributed to the problems of racial hatred, particularly against the Japanese and Chinese immigrants in British Columbia and the eastern Europeans in the Prairie provinces.[24] Headlines in the newspapers of the day demanded politicians act against the "rising tide" of immigrants, and the papers in British Columbia focused on the "crisis" created by the Chinese and South Asian immigrants "flooding the country" and working for "half the wages of a white worker." To keep these settlers out of Canada, the federal government increased their "head tax" on all Chinese immigrants from $50 in 1885 to $100 in 1900 and to $500 in 1903.[25] Immigration continued, however, and racial tensions came to a boil when white rioters in Vancouver (British Columbia), fuelled by racial propaganda supplied by hate groups such as the San Francisco–based Asiatic Exclusion League, attacked members of the Asian immigrant community in 1907 and destroyed property worth thousands of dollars (again, animosity focused largely on the Japanese and Chinese community).

Fortunately, members of the South Asian community were located in a different section of Vancouver and spared the brunt of the rioters' actions; however, they did receive the full impact of the British Columbia legislature. In 1907 the right to vote in provincial elections was denied to all Hindus (S.B.C. 1907 c.6) and persons from India, and in 1908 the Municipal Elections Act (S.B.C. 1908 c.14 s.13 [1]) denied any "Chinese, Japanese or other Asian or Indian person the right to vote in any municipal election."[26]

Federally, with Sifton's resignation in 1905, Laurier's government began to bow to pressure from the majority "white population"—one that was steadily growing with a new influx of immigrants from the British Isles—fellow politicians, various labour organizations, and members of the Christian clergy, and by 1907–08 introduced a Federal Order-in-Council demanding all Indian and Asian immigrants have at least $200 in their possession before entering Canada.[27]

Feeling the Federal Order-in-Council was not sufficient to stop immigrants of South Asian origin who could readily afford the entry requirements, the Laurier government under the direction of Sifton's successor, Frank Oliver, amended the 1906 *Immigration Act* and introduced a "continuous journey" regulation in 1908, which more than any other legislation was specifically directed to stem the flow of South Asian immigration into Canada. From the archives of Citizenship and Immigration Canada we have the following:

> Would-be immigrants, who all traveled by ship unless they were coming overland from the United States, were now required to arrive in Canada from the country of which they were natives or citizens, and on a ticket purchased in that country. The government, wanting to stop the flow of immigrants from Asia, had signed an agreement with Japan in 1907 limiting the number of male immigrants from that country to 400 per year, but no such agreement was in place with the government of British India. Since no shipping company provided direct passage from India to Canada, the new 1908 continuous-journey regulation effectively banned immigrants from India.[28]

What was particularly draconian about this legislation was that it meant that neither wives nor family members could immigrate overseas to join their husbands or fathers who had established themselves in Canada. The new legislation had, therefore, achieved its desired objective of limiting the establishment of a religio-cultural community. This legislation was certainly challenged by members of the South Asian community who were resolved to claiming their right to equal treatment as "white" citizens of the British Empire. For the most part these challenges failed. One of the more glaring examples of mistreatment was the incident starting 21 May 1914 and involving the S.S. *Komogata Maru* and 376 South Asian immigrants (330 Sikhs, 24 Muslims, 12 Hindus). In the "spirit" of Gandhi, the voyage was meant to challenge the "continuous journey" legislation, but they were still refused entry to Canada and forced to stay anchored off Vancouver for over two months. Eventually, a Navy gunboat escorted the ship from the port of Vancouver.[29]

By 1908, the South Asian population in Canada (primarily in British Columbia) was over 5000 individuals, but in the years 1909–10 the total number of South Asian immigrants allowed to enter Canada had fallen to ten people, to five individuals in the following year,[30] and to one in the years between 1914 and 1917. In large part this was due to the passage of the *Immigration Act of 1910*, which, unlike the 1906 *Act*, conferred on the Cabinet the authority to exclude "immigrants belonging to any race deemed unsuited to the climate or requirements of Canada." The *Act* also strengthened the government's power to deport individuals, such as anarchists, on the grounds of political and moral instability. Compounding the situation, an Order-in-Council required that all immigrants arriving in any season other than winter have $25 on them. This last measure ignited a storm of protest in Great Britain, because it meant that prospective immigrants would need to have $25 in addition to their ocean and inland transportation fares.

As mentioned above, by the year 1917 the total population of South Asian immigrants in Canada had dwindled through out-migration (primarily to the United States or back to India) to under 2000 individuals, and as Sampat-Mehta notes, "between 1914 and 1917 only one Indian successfully entered Canada as an immigrant."[31] Although these were the years of World War I, it is little excuse for such blatant inaction. The only positive legislation regarding South Asian immigration in the early part of the 20th century was initiated in 1919. As the result of political pressure from the remaining members of the Sikh community, a Federal Order-in-Council was passed to allow "British Hindus residing in Canada" to bring to Canada their immediate family members (wives and children). Between 1904 and 1920, however, only nine South Asian women were allowed to immigrate to Canada (a similar pattern of oppression was applied to the Chinese community as well). The "continuous-journey regulation" stayed in effect until 1947 and was finally repealed in 1977.

The Struggle and the Legislation

The year 1947 was important for both Canadian nationals and those who had immigrated to Canada. The war had confirmed Canada as a sovereign nation, and this led, thanks to the *Canadian Citizenship Act* passed in June 1946 (effective 1 January 1947), to the legal recognition of Canadian citizenship to those living in Canada (Order-in-Council, P.C. 4849), eliminating the official designation of Canadians as solely British subjects. This meant that a few immigrants of South Asian descent who met the criteria outlined in the *Act* (those who had "lived here for many years") were given the right to vote in Canadian elections; however, the reality is that for most (particularly those of Asian or South Asian origin or persons of colour) the qualifications continued to be both restrictive and discriminatory.

Although there had been significant revisions to the *Immigration Act of 1910*, it was not until 1952 that a new *Act* was finally introduced to Parliament. Unfortunately for both the South Asian community and other potential immigrants of colour, the wording meant that "Cabinet could prohibit or limit the admission of persons by reason of such factors as nationality, ethnic group, occupation, lifestyle, unsuitability with regard to Canada's climate, and perceived inability to become readily assimilated into Canadian society"[32]—provisions that were clearly designed to exclude non-white and non-Christian immigrants.

Although slow in coming, significant changes began to occur within the prevailing Canadian psyche by the late 1950s. For example, Prime Minister John Diefenbaker introduced a Bill of Rights in 1960 that rejected any discrimination by reason of race, colour, national origin, religion, or sex. The passage of the Bill meant that pressures were exerted on members of the federal government who attempted to justify the selection of immigrants on the basis of race, national origin, or "types perceived to be radicals." Ten years after the passing of the *Immigration Act of 1952*, Ellen Fairclough tabled new regulations to amend it, which prohibited the use of race, religion, colour, and national origin as criteria for the selection of new immigrants to Canada providing "(1) they had a specific job waiting for them in Canada or were able to support themselves until they found employment, (2) they were not criminals or terrorists, and (3) they did not suffer from a disease that endangered public health."[33]

The introduction of the "point system" in 1967—100 years after Canadian confederation—allowed all potential immigrants to be "given entry points" based on merit, language, and professional skills that eliminated the sole discriminatory provision in the *Act* and allowed European immigrants and immigrants from the Americas to sponsor a wider range of relatives. Former Prime Minister Pierre Trudeau's adoption of a policy of multiculturalism on 8 October 1971 institutionalized the idea that to be Canadian is to share a proud heritage of diverse cultural and religious backgrounds—the idea of the "Canadian cultural mosaic" had finally been articulated. In 1988, both Houses of Parliament unanimously passed the *Multiculturalism Act* first introduced in 1985 (see above) and Canada became the first country in the world to legislate specific goals for cultural and religious harmony.

CONCLUSION: WHERE DID WE COME FROM, AND WHERE DO WE GO FROM HERE?

Although our snapshot of Canadian history has shown fleeting moments of religious tolerance, for the most part the first three hundred years of her history has been marked by religious, legislative, and social intolerance, starting with the treatment of First Nations peoples and the destruction of their religion and culture through the process of missionary conversions (and later, the infamous residential schools run by various Christian denominations). The religious landscape of Canada had been dominated first by the Franco-Europeans, then by the Anglo-Europeans. With Britain's acquisition of the Canadian territories in 1763, we see a shift from economically based military aggression to cultural domination by the British over the French. This conflict was largely based upon the differences between the Protestant religious culture and that of the Roman Catholics, and continued with the Roman Catholic Irish immigrants of the late 1840s. Compounding the societal tension during this period, all three Christian groups discriminated against the growing Jewish communities in both Upper and Lower Canada.

As the "new Canada" developed during the late 1800s, and her populations pushed westward across the prairies to meet with the communities of British Columbia, the diversity and numbers of her ethnic populations also grew—and so did the cultural and religious conflict. During the 1870s, for example, there is the Métis Rebellion, and during the late 1880s and early 1900s we see specific social movements, often supported by legislation, directing their wrath against the Asian, South Asian, and Eastern European communities as they tried to establish themselves in the new Dominion of Canada (1867).

And these hostilities were not confined to the 19th century. Along with the legislation covered in the section above, we also observe cases of incarceration of her citizens along with the confiscation of their private property—one of the more infamous examples of this injustice being that visited upon members of the Japanese community during World War II. And though many arguments were made for this action, it might be said that part of the reason for this tragic miscarriage of justice was religious discrimination. That is, it was based on the pretext that these Canadians might somehow be loyal to the Japanese Emperor—a rationalization that appears to be fuelled by a limited understanding of the Japanese Shinto tradition, in which it is held that the Emperors are direct descendants of Japan's first Emperor Jimmu, a descendant of the Sun Goddess, Amaterasu.

Another group that was imprisoned solely for their religious beliefs were the Doukhobors of western Canada. This group of Anabaptists, originally from Russia, practise a strict form of religious pacifism and were imprisoned because of their protests against war. In fact, their pacifism and non-compliance with militarism led the Canadian authorities to confiscate Doukhobor farmland twice: once in Saskatchewan and once in the Kootenay region of British Columbia.

In other cases where the state disagrees with a particular religious practice or religio-political position, rather than confiscating property or imprisoning members of the community we find the state either removing a child or children from the homes of a particular individual or removing several children from the religious community at large. A more recent example occurred when a parent of a particular religious community refused to allow a child to be given a lifesaving blood transfusion.

Despite this mottled past, Canada has certainly matured during the past fifty years. Starting with Prime Minister John Diefenbaker's Bill of Rights (1960), and continuing with Prime Minister Pierre Trudeau's push for a multiculturalism policy (1971) and the establishment of the Canadian Constitution (1982), Canada has gained the status of a world leader in the multiculturalism project—a project that enshrined the concept that no one ethnic or religious group would dominate the cultural or political landscape.

For the most part these changes in political policy have been positive. From a religious perspective the recognition of multiculturalism has meant a rapid growth of the ecumenical movement, by increasing dialogue not only between the many Christian denominations but also between the diverse faith communities now represented in Canada. Organizations such as the Multifaith Council of Canada have significantly opened the lines of communication and social cooperation between many religious groups and denominations that seek mutual understanding and the sense of a common social cause.

On the other hand, many religious groups in Canada are beginning to view the modernist multicultural experiment as impotent, and consequently we have also seen the rise of the more conservative forms of fundamentalism in recent years. The question must be asked: Does this newly revived religious conservatism unite or divide the larger multicultural community? In some circumstances—the fundamentalist opposition to the gay marriage legislation being the most recent example—there appears to be unity among the conservative religious groups, but it certainly was not a rallying point for communities across this nation, or one that sought political or social consensus. And this brings us to a final question: If liberal religious views allowed for the rise of the social and religious ecumenical movement in this country, a movement that asks religious communities to put aside their own interests for the greater good of all communities—then will the rise of conservative fundamentalism (of all religions) return this country to a point where one religious viewpoint will dominate? Only time will tell.

CHAPTER SUMMARY

With the arrival of the new millennium, and the subsequent events surrounding 11 September 2001, it becomes apparent that religion can no longer be thought of as some antiquated cultural fossil made extinct by such commodities of modernism as science, economics, and modern statecraft. Instead, religion has come forth from the shadow of modernism to show that it is still very much a vital and significant cultural entity, an entity that continues to operate as part of the complex human condition.

The first part of this chapter illustrates that religion not only provides us with a lens into the great heritage of human civilization, but also provides Canadians with both a new discourse with which to understand the meaning and values of others within our multicultural mosaic and a distinctive opportunity to individually reflect, shape, and articulate our own unique vision of reality. But religion goes beyond the experience of the individual and individual identity formation. It also plays an important part in the human "cultural project"—that social enterprise or dialectic process that projects meaning into the empty vastness of the universe by creating a material, institutional, and, more importantly, religious culture reflecting that meaning.

The second part of this chapter shifts focus to the issues of religion as present within the rich cultural tapestry of the Canadian context. Whether the peoples be

First Nations, French, British, American, Irish, Scottish, Asiatic, South Asian, Ukrainian, Italian, African, Arab, or Russian, religion has always provided them with both a sense of cultural meaning within their Canadian experience and an acute awareness of order and justice. However, it has also been the basis of socio-religious intolerance and political bias. To outline the experiences of each immigrant group would be too vast an undertaking for this chapter, so the remainder of the section pays particular attention to both the Chinese and the South Asian immigration patterns, and the consequent political legislation. In the light of this historical content, the chapter concludes not so much with answers as questions concerning some of the future challenges for religion in this country.

KEY TERMS

anomie, p. 171
"big R" religion, p. 168
cults, p. 171
dialectical process, p. 168
essentialize, p. 160
externalization, p. 168
failed states, p. 175
fundamentalism, p. 174
internalization, p. 168
jiva, p. 165
karma, p. 161

"little r" religion, p. 168
moksha, p. 165
nirvana, p. 167
nomos, p. 169
objectivation, p. 168
products of the human cultural project, p. 169
sacred canopy, p. 170
samsara, p. 165
transcendent God, p. 161
Ultimate Reality, p. 161

DISCUSSION QUESTIONS

1. The *Multiculturalism Act* states not only that we should condone various religious practices, but also that we should enhance these practices. How *actively* should government agencies *enhance* religious practice within Canadian society?
2. The recent events surrounding the "cartoon issue" in which a Danish newspaper produced cartoons depicting the Islamic prophet, Mohammad, has raised some very serious issues (and violence) around the world. One particular issue concerns the right of free speech versus the right of a religious community to not to have its religious belief debased. The balance between these two premises is perhaps one of the greatest challenges of a multicultural society. In small groups, discuss the following: (a) Should the right of religious belief and practice supersede the right of free speech

(e.g., should publication of the Danish cartoons be forbidden because all depictions of the Prophet are considered sacrilegious by Muslims)? (b) Are there responsibilities associated with the right of free speech? What are they? (c) Do individuals or members of a particular group have the right to say anything they wish no matter how offensive it may be to another group?

3. Discuss the advantages and disadvantages of having one religious tradition dominate a particular culture. What are the advantages and disadvantages of granting all religions equal status within a particular culture?
4. Many religions, such as Roman Catholicism (Canon), Judaism (Halacha), and Islam (Shariah), have their own religious codes, traditions, and sacred laws—some of which, such as those governing marriage and the rights of inheritance, might conflict with Canadian law. Discuss when the application of these codes, traditions, and laws would be appropriate or inappropriate in the light of the Canadian context.
5. In both Canada and the United States, the rise of religious fundamentalism has created some concerns over this form of religious philosophy entering the political arena and influencing legislation. Though Canada recognizes and supports religious freedom of expression, should its legislative process involving such issues as abortion, gay marriage, evolution, etc. be influenced by religious doctrines?
6. As you understand religious fundamentalism, can you appreciate the relationship between the social theory of *anomie* and the concept of the sacred canopy? Is this a proficient response? As a group can you envisage other, alternative responses to *anomie*? Are they as successful and effective as the religious response?

NOTES

1. Ninian Smart, *The World's Religions* (Cambridge: Cambridge University Press, 1998), p. 10.
2. Max Weber, *The Protestant Ethic and the Spirit of Capitalism* (New York: Charles Scribner's Sons, 1958).
3. Emile Durkheim, *The Elementary Forms of Religious Life* (New York: Free Press, 1995), p. 44.
4. Max Weber, *The Sociology of Religion* (Boston: Beacon Press, 1964), p. 1.
5. Clifford Geertz, *The Interpretation of Cultures* (New York: Basic Books, 1973), p. 90.
6. Peter L. Berger, *The Sacred Canopy* (Toronto: Doubleday, 1967), p. 3.
7. Berger, p. 4.
8. Berger, pp. 14–15.
9. Berger, p. 100.
10. Lorne L. Dawson, *Comprehending Cults* (Toronto: Oxford University Press, 1998), pp. 52–53.
11. *Canadian Multiculturalism Act*, R.S., 1985, c. 24 (4th Supp.), available Canadian Heritage site <www.canadianheritage.gc.ca/progs/multi/policy/act_e.cfm>, accessed 4 April 2006.

12. Robert Choquette, *Canada's Religions* (Ottawa: University of Ottawa Press, 2004), pp. 433–434.
13. Choquette, p. 145.
14. Denise Kupferschmid-Moy, *Across the Generations: A History of the Chinese in Canada*, 2005, <http://collections.ic.gc.ca/generations/index2.html>, accessed 4 April 2006.
15. "A Tale of Perseverance: Chinese Immigration to Canada," *Life and Society*, CBC Archives, 2006 <http://archives.cbc.ca/IDD-1-69-1433/life_society/chinese_immigration>, accessed 4 April 2006.
16. Norman Buchignani, Dorren Indra, and Ram Srivastiva, *Continuous Journey: A Social History of South Asians in Canada* (Toronto: McClelland and Stewart, 1985), p. 113.
17. Roger Daniels, "The History of Indian Immigration to the United States: An Interpretive Essay," in Jagat Motwani, Mahin Gosine, and Jyoti Barot-Motwani, eds., *Global Indian Diaspora: Yesterday, Today and Tomorrow* (New York: Global Organization of People of Indian Origin, 1993), p. 440.
18. C.I. Petros, "Indo-Canadians" in J. Motwani, M. Gosine, and J. Barot-Motwani, eds., *Global Indian Diaspora: Yesterday, Today and Tomorrow* (pp. 475–484) (New York: Global Organization of People of Indian Origin, 1993), p. 475.
19. Petros, p. 475.
20. Government sources include: Citizenship and Immigration Canada (www.cic.gc.ca/english); Statistics Canada, *The Daily*, February 17, 1998, available (www.statcan.ca/Daily/English/980217/d980217.htm); Department of Justice of Canada (http://canada.justice.gc.ca/en); Heritage Canada (www.canadianheritage.gc.ca).
21. Buchignani et al., p. 8.
22. R. Sampat-Mehta, "First Fifty Years of South Asian Immigration: A Historical Perspective," in R.N. Kanungo, ed., *South Asians in a Canadian Mosaic* (Montreal: Kala Bharati, 1984), p. 13; Subash Ramcharan, "South Asian Immigration: Current Status and Adaptation Modes," in Kanungo, p. 33.
23. "The Arrival of the Europeans," *Forging Our Legacy*, October 2000, available Citizenship and Immigration Canada site <www.cic.gc.ca/english>, accessed 4 April 2006.
24. Sampat-Mehta, p. 24.
25. *Chinese Immigration Act*, S.C. 1900, c.32 S.6; *Chinese Immigration Act*, S.C. 1903, c.8 S.6.
26. Department of Justice of Canada site <www.chrc-ccdp.ca/en/timePortals/milestones/9mile.asp>, accessed 9 April 2006.
27. C.I. Petros, "Indo-Canadians," in J. Motwani, M. Gosine, and J. Barot-Motwani, eds., *Global Indian Diaspora: Yesterday, Today and Tomorrow* (New York: Global Organization of People of Indian Origin, 1993), pp. 475–484. Available <www.chrc-ccdp.ca/en/timePortals/milestones/8mile.asp>, accessed 9 April 2006.
28. Citizenship and Immigration Canada, "Chapter 1: Across by Boat, Overland by Train," *The Role of Transportation in Canadian Immigration 1900–2000*, March 2001, available <www.cic.gc.ca/english/department/transport/chap-1a.html>, accessed 10 May 2006.
29. For more details on this incident see Hugh Johnston, *The Voyage of the Komagata Maru: The Sikh Challenge to Canada's Colour Bar* (Vancouver: UBC Press, 1989).
30. Petros, p. 476.

31. Sampat-Mehta, p. 27.
32. "Towards the Canadian Citizenship Act," *Forging Our Legacy*, November 2000, available Citizenship and Immigration Canada site <www.cic.gc.ca/english/department/legacy/chap-5b.html>, accessed 4 April 2006.
33. "Trail-Blazing Initiatives," *Forging Our Legacy*, October 2000, available Citizenship and Immigration Canada site <www.cic.gc.ca/english/department/legacy/chap-6.html>, accessed 4 April 2006.

CHAPTER 7

Disability as Difference

Nancy Nicholls

[We make disability] in impersonal encounters, in school, on the job (and long before a job is gotten on), in the media, in human service agencies, within self-help groups, through governmental policy, within cultural belief (nurtured and displayed through all of the above and more), through technology, through sweeping societal changes, and even through the writings of observers of disability.
— Michelle Fine and Adrienne Asch, quoted in Higgins, 1992

Objectives

After reading this chapter, you should be able to

- clarify the relationship of historical ideas and the history of disability in Canada

- appreciate the role institutions, especially government, church, education, and family, have on issues affecting people with disabilities

- know and analyze the different models and theories for studying disability

- explore the importance of the theory of "normalization" in the treatment of individuals with disabilities

- recognize critical social policy issues for people with disabilities in Canada

INTRODUCTION

When people used to talk about **inclusive societies**, it was assumed that they were referring to issues of race and ethnicity. Increasingly, however, the notion of inclusivity also incorporates **disability**.

Disability is not a new phenomenon in Canada. To appreciate the issue of disability in the 21st century and beyond, we will begin with some background discussion about how people with disabilities were treated in the past. This historical perspective will (1) help shed light on the definition of disability, (2) outline the role institutions have played in the lives of those who have disabilities and in shaping the view others have of people with disabilities, (3) show how public interest has shaped the development of policies in the area, and (4) show how government has responded.

Not all disabilities affect all people the same way. We will look at the types of disabilities—physical, intellectual, and mental—and attempt to identify how differently they have been treated over time and how the various disabilities raise different issues for equality. We will highlight the major theoretical perspectives on the study of disability, with particular emphasis on the theory of "normalization."

HISTORICAL CONTEXT

The Early Years

Historically the view of those with disabilities was that they were being punished for their sins or those of their ancestors. Many felt that people with disabilities were possessed of "evil spirits." Often they were ridiculed and certainly they were ostracized by society—they were seen as outcasts and were unwelcome. In time, the view changed from a concern about evil spirits to one of seeing people with disabilities as weak, as less worthy, as persons in need of charity. Originally, the person's family was expected to provide this charity. The family was seen as the major institution to provide health, educational, and social care for all its members. In the early 16th century, people with disabilities were thought of as noncontributing members of society and as individuals who needed to be provided with only basic care (minimal food and shelter). If their families were unable to provide this care, then it would sometimes be provided by more fortunate members of society.

European settlers brought with them to Canada the views of individuals with disabilities that they had held in Europe. Individuals with disabilities were seen as weaker members of society in need of the protection of stronger members, and their family and friends were expected to care for them. For these settlers, life in the new land often involved physical hardship and physical strength; individuals who could not be adequately protected in the new country were sent back to Europe for care.

In Europe, the major provider of care outside the family had been the church, so it was natural, in terms of both culture and tradition, for the church to assume the same responsibilities in the new land. From about the mid-18th century, as the population grew in New France (now Quebec), the Roman Catholic Church began to provide care for individuals requiring special assistance. Even after Britain assumed control of New France, the Roman Catholic Church maintained this role as the major institutional provider of care in Quebec. This lasted until the 1950s, when the provincial government assumed more responsibility and as the Quiet Revolution began to take root.

The Church provided care for individuals with mental disabilities. One of the earliest known institutions for the care of those with disabilities was a hospital founded by the Sisters of Charity (Grey Nuns) in Montreal. In the late 18th century, Quebec's Legislative Assembly authorized expenditures for the care of "insane" persons in Quebec, and a per diem rate was given to the Grey Nuns to provide care for the "insane" in their existing institutions in Trois-Rivières, Quebec City, and Montreal. This is the first example in Canadian history of government expenditure for people with disabilities.

Services were expanded in the 19th century, largely because of increased population growth. An example of this expansion is the establishment of a hospital for the "insane" in Saint John, New Brunswick.

It was very difficult for those with severe physical disabilities to survive the hardships and challenges of the new country, and those that did not return to Europe often died young. In the 19th century, in response to population growth but also because of two major epidemics of cholera and smallpox, services for the physically ill were also expanded. Similar medical crises, polio and the effects of thalidomide, had a major impact on the treatment of people with disabilities in the 1950s. Events such as these touched many lives and brought the issues of people with disabilities to the forefront.

Confederation and the *British North America (BNA) Act*

Pivotal to an understanding of the Canadian government's role in caring for people with disabilities is the *British North America Act*, now called the *Constitution Act*, of 1867. This act established that health care and education, as well as what we now term "social services," are the responsibility of the provincial governments. It is important to note that the federal government retained the responsibility to provide health care to the Inuit populations and to veterans of the armed forces. Policy issues about the provision of services to people with disabilities often are entrenched in the division of powers as outlined in this act.

New Services in the 19th Century

An important player in the development of most services is the public. Without public pressure and interest, governments tend not to expand services. For people with mental disabilities, the public was satisfied with the government providing institutional care—citizens were not only concerned about the care given people with mental disabilities, but they were also equally, and some would argue more, concerned about the safety of the public. Institutional care provided in isolated areas, far from the public view, was considered best.

Public concern for people with physical disabilities, however, was growing, primarily for those who were deaf or blind. Sending such individuals back to Europe was no longer feasible or practical, as many of them were now a few generations removed from the "old country." In response to the pressure to provide special education to the deaf and the blind, asylums for the blind were funded in Halifax and Montreal, and the Ontario School for the Deaf was established in 1872 in Toronto. Alexander Graham Bell, who is well known for his invention of the telephone, was a major contributor both financially and in terms of public pressure to developing and influencing the provision of special education services to the deaf. This notion of education for individuals with physical disabilities influenced those concerned about other kinds of disabilities. In 1888 an institution was built in Orillia, Ontario, to provide education to the "**mentally deficient**."

It is sad to note that what started out as a movement for special education in many instances resulted in institutions that provided merely housing, often in less than adequate conditions.

Along with public pressure for the government to provide services came an impetus from the public to offer volunteer services. Volunteer organizations were established, such as the Canadian Red Cross in 1896 and the Victorian Order of Nurses in 1898. Volunteer services both increased the provision of care to people with disabilities and had a significant impact on the quality of care provided. As more people worked with individuals with disabilities through these volunteer efforts, public awareness grew about the issues facing people with disabilities, as did public support of increased financial and moral support.

The epidemic outbreak of serious diseases has often been the impetus for demanding services; another major impetus is civilian disaster. In 1917 an ammunition ship, the *Mont Blanc*, exploded in the port of Halifax. More than 1900 people were killed instantly, and within a year more than 2000 had died. Around 9000 more were injured, many permanently, 1000 with eye injuries alone. Virtually all of north-end Halifax was destroyed. The impact of this incident focused the public's attention on the needs of those with physical disabilities.

The voluntary sector saw children as "innocent victims" and thus the only ones deserving of help. Public pressure to care for children led to the establishment of institutions such as the Toronto Hospital for Sick Children and, later, the Ontario Society for Crippled Children.

The World Wars

World War I had an enormous impact on the delivery of services to people with physical disabilities in Canada. For the first time in Canadian history, many Canadians were faced with major physical problems, as soldiers returned from the front with a variety of serious illnesses and "**handicaps**," among them tuberculosis, blindness, mental illnesses, paralysis, and amputated limbs. The notion of who "deserved" help shifted. People with disabilities were no longer the unfortunate and the weak, but some of our "brightest and strongest"—our country's "hope for the future."

Because these disabilities had been incurred by our men in fighting for our country, the country felt an obligation to help them. And because, according to the *BNA Act*, the federal government was responsible for veterans, the federal government was obliged to take a leading role. Thus, help for people with disabilities took on a federal perspective with regard to government policy. Help also took on a federal perspective with regard to the voluntary sector: in 1918 both the Canadian Mental Health Association and the Canadian National Institute for the Blind were established.

Care for people with disabilities shifted from being merely the provision of food, shelter, and some educational services to the provision of rehabilitation services. The policy was to help veterans get into the labour market, and thus vocational training was seen as critical.

In 1915 the Military Hospitals Commission was created. On the basis of experience gained from World War I, a subcommittee of government workers and civilians was established shortly after war broke out in 1939 to develop a comprehensive plan for assisting veterans.

Another important result of the wars was the vast medical experience gained and the resulting innovations in **rehabilitation medicine**, including the development of "wonder drugs." After World War II, a group approach, now known as the "**rehabilitation team**" approach, was developed in Montreal to combat the many issues facing returning veterans. Team conferences and team planning are now standard practice in working with individuals who have disabilities, whether physical, mental, or intellectual. This approach recognizes that treating the effects of disabilities also includes the social aspects such as education, vocational training, housing, and employment and that professional expertise and access to a variety of services are needed.

The services developed in the war years for veterans were extended to include all Canadians. In 1948 national health grants were established by the federal government to assist all the provinces in providing rehabilitation services. These grants were also to be used for the training of medical personnel and for rehabilitation equipment. This initiative led to many services and facilities that we still have today, including Toronto's Variety

---

CRITICAL THINKING BOX 7.1

In what way did World War I affect the public's view of people with disabilities?

Village, which was originally a school for children with disabilities and today offers a wide range of services including recreational activities for people with disabilities.

Rehabilitation Planning and the Labour Market

In the early 20th century, the new **social movement** of organized labour pushed for protection of injured workers. This resulted in the Ontario *Workmen's Compensation Act* of 1914. At first the *Act* dealt with giving injured workers lost wages, but later it provided medical assistance and rehabilitation benefits as well. Today all provinces have such legislation, and in response to the position of women in the labour market, most such acts have changed their wording from "workmen" to "workers."

A major breakthrough in the rehabilitation field was the 1951 Touchstone Conference. This national conference saw the provinces reassert their role in the field of rehabilitation, at the same time encouraging a continuing presence of the federal government in planning and in coordinating a national rehabilitation program. The most significant outcome of the conference was the Federal–Provincial Vocational Rehabilitation of Disabled Persons (VRDP) initiative, which led to the *Vocational Rehabilitation of Disabled Persons Act* in 1961. The major tenet of this *Act* is that the federal government makes cost-sharing arrangements with the provinces for the provision of services to help people with disabilities enter or reenter the workforce. The *Act* also requires that there be a special section in the federal and provincial employment services for placement in the workforce of (primarily) people with physical disabilities.

Medicare

The federal *Hospital Insurance and Diagnostic Act* was passed in 1957. It provided people who have disabilities with free treatment and rehabilitation for insured services. In 1961 Saskatchewan established what has been called the first Medicare program in North America. This was followed in 1966 by the federal *Medical Care Act*, and with this legislation, a 50 percent cost-sharing arrangement for medical services was made between the federal government and the provinces. By 1971 all Canadians had medical coverage under this act.

The *Canada Pension Plan Act* (1965) was another major breakthrough in this area. It stipulates that for those who have contributed to the plan and who are later deemed unlikely to be able to work because of a "severe" disability, a pension will be paid.

In 1964 a conference, similar in importance to the Touchstone Conference, was held for the "mentally retarded." Since the early 1960s there have been major breakthroughs for people with what are now called **developmental disabilities**. In 1963 a National Institute on Mental Retardation was founded together with York University in Toronto. Today, other such foundations exist at universities throughout Canada, which has greatly enhanced research and consultation in the field.

Parents of children diagnosed as having developmental disabilities have arguably had the greatest impact on care for any disabled group in Canadian history. The first such parents' council was formed in 1951; similar associations are in evidence throughout the country today, and there is a national association. These groups have successfully lobbied for services and for **antidiscriminatory** legislation. It is they who are largely responsible for the present-day inclusion, or mainstreaming, in many schools of children with special needs. Another major facilitator of changes in the treatment of those with developmental disabilities is the theory of "normalization" (to be discussed later in this chapter).

Human Rights Codes

Rehabilitation is not only an issue in Canada. Over the years the international community has put great emphasis on the prevention of various disabilities and on the integration of people with disabilities into society. This has manifested itself in a variety of ways, including international conferences such as the one dedicated to people with disabilities sponsored by the United Nations in 1981, the International Year of Disabled Persons. In part this was in reaction to lobbying by people with disabilities themselves, through a variety of self-help groups, in particular by Vietnam War veterans in the United States. Canada's response to the dedicated year was to table a special committee report entitled *Obstacles*.

In 1980, in Canada, a special committee was created to deal with issues affecting people with disabilities, including the effect of programs and services in the government and in the voluntary sector and the overlap and interlocking of such services. Perhaps most important, however, this committee addressed the issue of human rights. It strongly recommended that these people should have full and equal protection under the law. Today, individuals with mental or physical disabilities are protected in human rights documents in Canada.

Care for the people with mental illnesses has developed along similar lines to those outlined for people with physical disabilities and for those with developmental disabilities. However, many would argue that the stigma is often greater for mental illness. The development of psychoactive drugs in the 1970s led to the discharge of vast numbers of patients from psychiatric institutions and allowed for care on an outpatient basis. These drugs did much to decrease the number of individuals getting custodial rather than rehabilitative care.

Advocacy

A major shift in the treatment of all individuals with disabilities has been "the consumer movement." No longer are people with disabilities unable to speak for themselves, and the 1980s trend toward self-help groups was critical in bringing about this change. This increased involvement is a result of several things, including better medical treatment and thus improvement in the mental and physical well-being of consumers; numerous improvements in assistive devices brought about by advanced technology; changes in expectations on the part of people with disabilities themselves and society as a whole; and increased access to and involvement in education and the workforce. To appreciate this change it is important to consider not only the historical changes that have occurred, but also the changes in theoretical perspectives.

THEORETICAL PERSPECTIVE

Three distinct models of disability have been developed, and each produces its own expertise and experts: (1) the biomedical model, (2) the economic model, and (3) the sociopolitical model. The operation of these models is reflected in the historical perspective, that is, throughout certain periods of Canadian history, implementation of one or the other model appears to have dominated.

Biomedical Model

The biomedical model emphasizes **impairment**, which the World Health Organization (WHO) in 1980 defined as "any abnormality of physiological or anatomical structure or function." Disablement is divided into respective varieties, and, according to Canadian lawyer and philosopher Jerome Bickenbach, social action concerns itself with issues of "prevention, cure, containment, pain management, rehabilitation, amelioration, and palliation."[1] This view largely resembles the medical model of treatment, and the individual is seen as a patient who is "sick" or hurt or, in some cases, a victim of bad luck. This model lends itself to social goals that focus on accommodation, that is, society is obligated to provide a basic level of medical care and health services. This model also had a role to play in concerns about eligibility criteria and the emphasis on assessment in determining need. Much of the early history of disability in Canada and society's reaction to disability before World War I can be understood by using this model.

> ### BOX 7.1
>
> *Canadian Human Rights Act, 1985*
>
> **Section 2. Purpose**
> The purpose of this Act is to extend the laws in Canada to give effect, within the purview of matters coming within the legislative authority of Parliament, to the principle that every individual should have an equal opportunity with other individuals to make for himself or herself the life that he or she is able and wishes to have, consistent with his or her duties and obligations as a member of society, without being hindered in or prevented from doing so by discriminatory practices based on race, national or ethnic origin, colour, religion, age, sex, marital status, family status, disability or conviction for an offence for which a pardon has been granted. . . .
>
> **(g) Discrimination Based on Handicap**
> Handicap was added as a ground of discrimination in 1981 (S.O. 1981, c. 53, ss. 1, 2, 4: now ss. 1, 2, 5). While boards of inquiry had recognized an employer's duty to make reasonable accommodation of an employee's characteristics as a requirement of the Code, the 1986 amendments (S.O. 1986, c. 64, s. 18(15) introduced into section 23(2) (now s. 24(2)) an express reference to reasonable accommodation and a provision (s. 16(1)(a); now s. 17(2)) dealing specifically with the duty of accommodation in relation to the handicapped (S.O., 1986, c. 64, s. 18(10)).
>
> *Source:* Human Rights Legislation: An Office Consolidation (Toronto: Butterworths, 1991).

Economic Model

The economic model emphasizes the economics of disablement. How does one's impairment affect one's capabilities? This model is concerned with the effects of a disability not only on the individual, but also on society as a whole and on the individual's ability, or inability, to contribute to society. Concerns are framed as labour market issues. Any disability is significant in that it limits a worker's productive potential and skills. Using a cost–benefit analysis, people with disabilities are seen as an economic cost. Policies must take into account that people with disabilities may not be able to work to their full potential and thus not contribute fully to the labour market. These people are also a cost in terms of the provision of health and other services such as housing. This view of people with disabilities emphasizes the need to address any barriers that prohibit full participation of people with disabilities in the workforce, and it leads to the development of social policy that strives to fully integrate people with disabilities into the economic market.

The economic model can serve to explain the shift to rehabilitative services that occurred after World War I. Also, from this perspective many of the policies toward people with disabilities adopted by Canada in this century can be explained—for example, the establishment of vocational services, social insurance programs, and social assistance programs.

Sociopolitical Model

The third model is the one most representative of current trends in the area of disability policy. From this perspective, disability is viewed as a type of social injustice. This social injustice stems from the way people with disabilities are stigmatized in society and from the way they are treated, that is, from discriminatory practices. This sociopolitical model blends sociological theories and psychosocial theories about disability with an advocacy and civil rights perspective. It is society that determines disability; it is discriminatory social attitudes that become part of the social fabric, part of how institutions are organized and operated, and it is these institutions and their practices that often disadvantage and marginalize people with disabilities. In this model society disables people.

The sociopolitical model holds that disability needs to be destigmatized and that respect and rights must go hand in hand in shaping policy for people with disabilities. Thus, disability issues resemble other equality issues, such as racial discrimination and women's issues.

The sociopolitical model is reflective of the rise of self-help groups and the increased participation of people with disabilities in the sociopolitical realm. The independent living movement stands as a clear example of a shift toward **empowerment**. This model is the one most concerned with reform.

As with most theories and models, social policy is often explained by a combination of models. The policies developed for people with disabilities are extremely diverse. Jerome Bickenbach cites 14 policy areas for people with disabilities. These include

---

CRITICAL THINKING BOX 7.2

As a blind child today, how would your life be different from what it would have been in the 1980s?

biomedical services (research, therapy, chronic care), independent living (group homes, the deinstitutionalization movement), employment issues, housing concerns, communication (media) issues, and human rights. In part because of the diversity of issues and concerns and because of the variety of disabilities (physical, mental, and intellectual), much of the policy development has been fragmented and is ad hoc in nature.[2]

Obstacles (1981), mentioned above, discussed three goals sought by disabled persons, and these have marked policy development in Canada for people with disabilities ever since. Individuals and groups representing disabled persons have differed on which goal they want emphasized. Some goals have been emphasized more than others at particular times. Nevertheless, the three goals are as applicable today as they were in 1981. People with disabilities want

1. *to be treated with respect.* This includes the idea that people with disabilities should participate in decision making and in developing their own services and agencies.
2. *to have the same opportunities as other Canadians to participate socially, economically, educationally, recreationally, and in all other ways in the social life of the country.* Participation should not be limited by discriminatory acts or practices, and the means necessary to achieve this participation should be provided.
3. *services and assistance to ensure that their needs are met.* Environments, including the workplace, should accommodate people with disabilities so that full social participation can be achieved.[3]

One issue that has complicated the development of consistent social policy in the area of disability is the conflict between programs that discourage people with disabilities from working with programs that encourage them to work. For example, is a person with a disability actually unable to work and thus in need and deserving of financial assistance? Or could he or she work if certain accommodations were made in the workplace? The needs of people with disabilities are as varied as their disabilities, their capabilities, and their skills. Much of the confusion around policy stems from the various definitions of disability.

CRITICAL THINKING BOX 7.3

The sociopolitical model argues that society disables people. In what ways might society disable an individual?

Normalization Theory

A milestone in the treatment of people with disabilities is the "theory of **normalization**," developed in the 1970s. This theory led to the belief that all individuals are entitled to lives that are "as normal as possible." In many ways this theory explains the massive **deinstitutionalization** of those with developmental disabilities and mental illnesses. With the move of many individuals with disabilities from institutions and asylums to group homes and other forms of independent living, people with disabilities are better able to participate in the daily social life of society. Deinstitutionalization has allowed many people with disabilities to have more participatory lives, but some argue that the real impetus behind this normalization was to save public money. Governments often moved people with disabilities out of institutions without providing them with adequate community programs and financial resources. Especially in urban areas, this policy has led to many people, particularly those with mental illnesses, becoming homeless.

BOX 7.2

Case Scenario

Judith Snow, a woman with disabilities who lives in Ontario, describes how society's view of people with disabilities and of their potential—and not her physical disability itself—sustained her disadvantages. Snow describes how after completing university, her world became limited by the practices of government and their institutions, who seemed insensitive to her needs as an individual.

Over a three-and-a-half-year period, Snow was bounced from one chronic care hospital to another because each facility was unable to provide her with the minimum five hours of daily attendant care that she required.

In an effort to preserve her mental health and to keep active and connected with life outside the institution, she continued to work four days a week at York University. However, Snow states, "The money I earned was used up in paying for a semi-private room, a private nurse in the mornings and my transportation . . . my health began to break down because of the institution, its policies, atmosphere and staff. No matter how hard I worked to explain how I needed to be active, I was always pushing against the life of the institution."

Source: Judith Snow, "Imprinting Our Image on the World," in Diane Driedger and Susan Gray, eds., *An International Anthology of Women's Disabilities* (Charlottetown: Gynergy, 1992), pp. 81–82.

Inherent in the theory of normalization is that "normal" is a desired outcome and that it is possible to determine what normal is. If what is normal can be determined, it follows that we can also determine what is not normal. Sociologists have written at great length about how a dominant group regards its experience as "natural," as "normal," and experiences that differ from these are considered less than normal. Therefore, when it comes to disability, it is society that socially constructs the difference. Paul Higgins points out[4] that **learning disability** is a new term for a condition that in earlier times was called "word blindness." This example illustrates how views of people with disabilities are ever-changing and that the issue of disability is an ongoing one. The example also helps to remind us that throughout history, "normal" has been a narrow concept and because of that, society has been able to shut out many from the mainstream, be it an ethnic group, women, or people with disabilities. Individuals with disabilities have often said that it is not the physical, mental, or intellectual disability that represents the major obstacle in their lives, but the barriers that society creates.

WHAT ABOUT THE FUTURE?

Individuals with a variety of disabilities now take part in the day-to-day life of Canadian society. Much remains to be done, and some accomplishments are being threatened. The 1990s saw the cancellation of many programs that provided assistance for people with disabilities, as major cuts to government spending occurred in all areas of health care. Unemployment rates rose, and there was greater competition for fewer jobs. For example, employment equity legislation was curtailed by the Conservative government in Ontario. It is difficult to determine what the ultimate consequences of such actions will be on the lives of people with disabilities in Ontario and in other provinces, as more and more groups continue to compete for limited government resources.

CRITICAL THINKING BOX 7.4

Think of a particular disability (physical, mental, or intellectual) and describe two obstacles that the person who has this disability faces because of the disability itself. Now think of three obstacles the person faces that represent barriers imposed by society.

Trends for the 21st Century

Great strides have been made in the way disabled individuals are viewed and in terms of increasing the participation of people with disabilities in decision making that affects their own lives (**self-determination**). The inclusion of individuals with disabilities in advertisement, although still minimal, would not have been thought of five years ago.

In the early years of Canadian history, as our pioneers fought for their very existence, no one thought much about any individual's—let alone any individual with a disability's—recreational needs. The history of sport and of the Olympics is a relatively new one, yet today, wheelchair athletes take part in the Paralympics. We have witnessed an individual climb the CN Tower in a wheelchair.

Although we have seen the first quadriplegic member of Parliament, many individuals still experience the stigma of a disability and are afraid to acknowledge learning disabilities and mental illnesses.

As with many developments, change takes place when people are aware of and become involved in issues. As our population ages, more and more Canadians are going to face physical disability and this, in turn, will increase awareness of the needs of people with disabilities. Since women have a longer life expectancy than men do, the needs of disabled women likely will increase in disproportion to those of men (see Figure 7.1).

Figure 7.1 Characteristics of Ontarians with Disabilities

Proportion of men and women with disabilities by age group
- 15 to 64
- 65 and over

(Men: 45% / 55%; Women: 49% / 51%)

Percentage of men and women with disabilities living in households (all ages)

(Men: 34%; Women: 46%)

Percentage of men and women with disabilities aged 65+ living in households and institutions
- Living in households
- Living in institutions

(Men: 59% / 41%; Women: 73% / 26%)

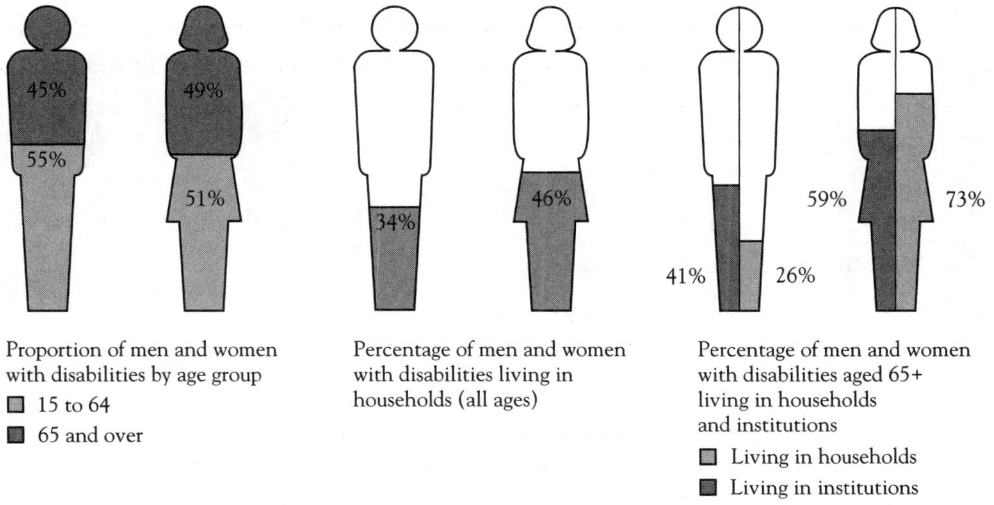

Source: "Characteristics of Ontarians with Disabilities," Adapted from the Statistics Canada publication "Adults with disabilities, their employment and education characteristics," Catalogue 82-554, Released July 28, 1993.

The Many Faces of Diversity

Table 7.1   People with Disabilities: Most Difficult Goal to Achieve

		Type of Client Disability				
	Responding Organizations (65)	Mobility/ Agility (39)	Emotional/ Mental (32)	Hearing/ Speech (32)	Vision (27)	Multiple Disabilities (40)
Ensure equality of opportunity and access	23%	26%	19%	25%	22%	25%
Help people with disabilities to integrate into communities	15	13	19	13	15	8
Ensure full range of services for people with disabilities	11	5	9	6	0	8
Help disabled persons enter and succeed in workforce	9	13	13	13	22	15
Help disabled persons to be independent	9	10	13	9	11	8
Make sure that people with disabilities receive a fair share of government services and funding	9	8	16	13	7	13
Promote acceptance of people with disabilities among the general public	5	8	3	9	7	8
Develop sense of community among those with similar disabilities	3	3	3	3	4	3
Reach as many as possible who have same disabling conditions	3	5	0	0	7	3
Obtain parallel or separate services	2	3	0	3	4	3
Don't know or no answer	11	8	6	6	0	10

Note: Totals may not add up to 100% because of rounding.

Source: "Most Difficult Goal to Achieve," Adapted from the Statistics Canada publication "Adults with disabilities, their employment and education characteristics," Catalogue 82-554, Released July 28, 1993.

In 1989 in Ontario, the then Office for Disabled Persons engaged Environics Research Group Limited to conduct a survey of the needs and attitudes of disabled Ontarians. The major findings are representative of a significant number of individuals with disabilities and organizations serving them. Most of all, the survey provided people with disabilities with an opportunity to share their views. It remains one of the most comprehensive studies of the views of the disabled, and thus, although it is over a decade old, it remains relevant and in many ways has influenced new legislation in Ontario. For most "clients" the issue of "equality of opportunity and access" remains paramount (see Table 7.1). They cite income as the area most in need of improvement. (See Table 7.2; note that the data refer to a period before the recession of the last decade.) Table 7.3 gives further evidence that, although gains have been made, people with disabilities in Ontario are still disadvantaged in comparison with Ontario's "general population." It is also interesting to note that how a disability limits a person is related to the severity and type of disability (see Figure 7.2).

The census reports that as of 2003, 17.5% of the population aged 12 and older experience moderate to severe "functional health problems."[5] In other words, physical and or

Table 7.2 People with Disabilities: Priorities for Improvement

Area of Priority	% of Respondents Who Chose Area
Income	29
Education	17
Accessibility	11
Recreation/leisure opportunities	9
Housing	7
Health care	7
Assistive devices	7
Transportation	6
Employment	3
Rehabilitation	2
Other	2

Note: Figures may not add up to 100% because of rounding.

Source: "Priorities for Improvement," Adapted from the Statistics Canada publication "Adults with disabilities, their employment and education characteristics," Catalogue 82-554, Released July 28, 1993.

Table 7.3 People with Disabilities: Comparisons with the General Population

Characteristics	Respondents with Disabilities (%)	General Population (%)
Gender		
Males	39	49
Females	61	51
Ages (years)		
15–34	21	43
35–54	26	31
55–64	19	12
65+	32	14
Educational Attainment		
Elementary school	18	12
Secondary school	44	45
Some postsecondary	34	25
University	9	11
Employment Status		
Working full time	23	56
Working part time	9	10
Unemployment/looking for work	4	5
Students	5	11
Not in labour force	60	18
Annual Household Income		
<$10 000	14	12
$10 000–$20 000	22	16
$20 000–$30 000	17	16
$30 000–$40 000	10	17
$40 000–$50 000	8	14
>$50 000	10	25

Note: Totals may not add up to 100% because of rounding.

Source: "Comparisons with the General Population," Adapted from the Statistics Canada publication "Adults with disabilities, their employment and education characteristics," Catalogue 82-554, Released July 28, 1993.

Figure 7.2 Type and Severity of Most Limiting Disabilities

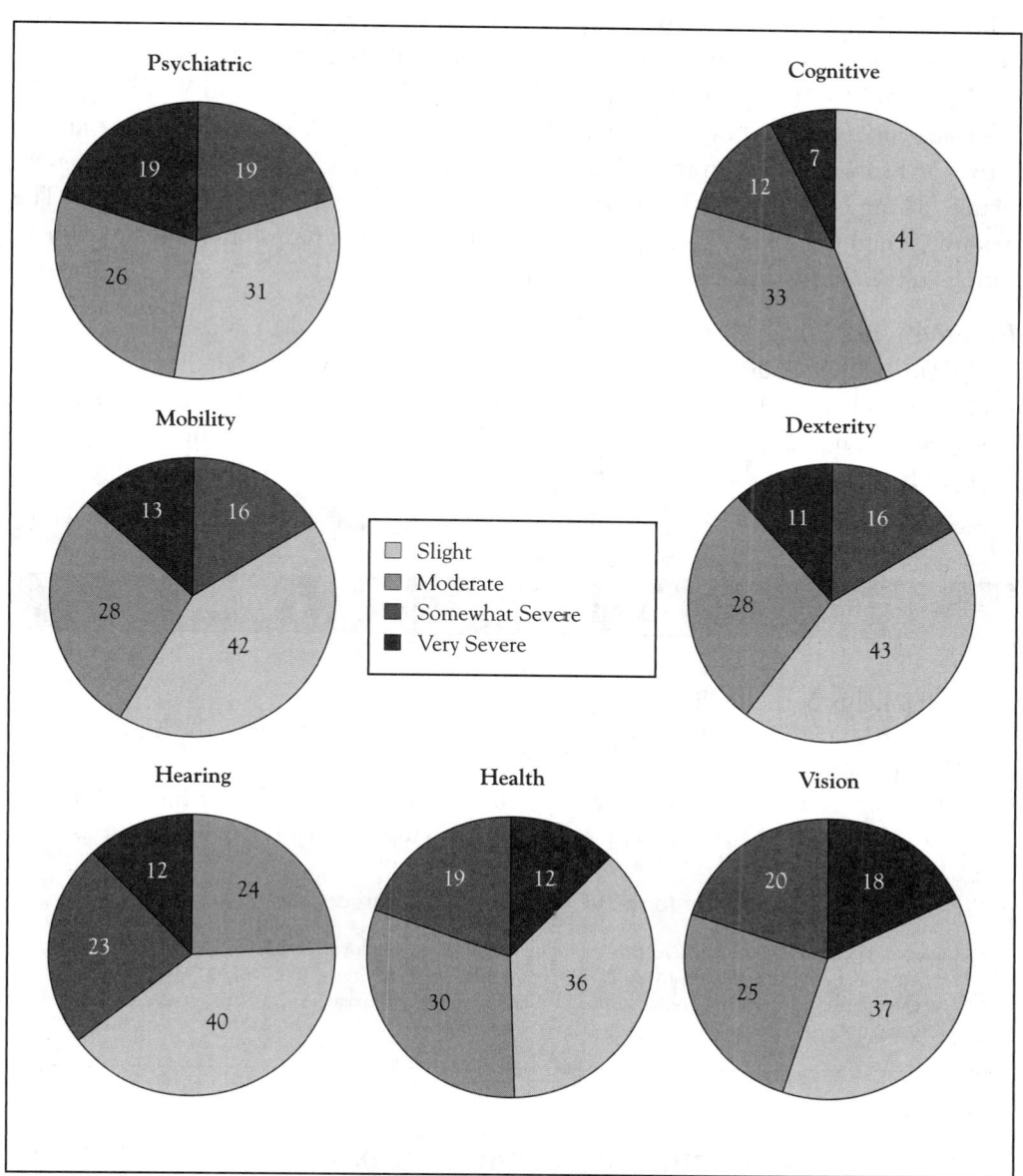

Source: "Type and Severity of Most Limiting Disabilities," Adapted from the Statistics Canada publication "Adults with disabilities, their employment and education characteristics," Catalogue 82-554, Released July 28, 1993, page 17.

medical conditions restrict or limit activities of a vast number of Canadians in work, school, and home areas. Yet little analysis of the data and the issues affecting the disabled is available, and this indicates that the disabled are still not a public policy priority.

A research project of the Canadian Council on Rehabilitation and Work, Diversity Planning for Inclusive Employment (DPIE), looked at trends in employer attitudes in terms of hiring and retaining individuals with disabilities. Over 700 responses were obtained from identified employers across Canada by e-mail surveys and focus groups. The research found that "little or no progress" had been made in terms of employer attitudes hiring and retaining individuals with disabilities (see Box 7.3).

Positive Trends

In addition to major advances in technology and assistive devices and in the overall inclusion of individuals with all forms of disability in society as a whole, there have been some significant changes and shifts in social policy. One of the most significant is that in 2005 Ontario finally passed the *Accessibility for Ontarians with Disabilities Act*.

BOX 7.3

Highlights of the findings of the DPIE project were:
- that attitudes and cultural biases continued to be a barrier to the hiring of persons with disabilities
- that awareness of statutory obligations under the *Employment Equity Act* (EEA) of 2003/2004 to recruit, hire, and employ persons with disabilities was low
- that accountability for achievement of EEA targets was mostly poor

Source: Halifax Global Inc., "Diversity Planning for Inclusive Employment (DPIE) Employer Survey," 7 September 2005, available Paths to Equal Opportunity site (Government of Ontario) <www.equalopportunity.on.ca/eng_t/subject/index.asp?action=search_7&page_id=788&file_id=26146>, accessed 6 April 2006.

CRITICAL THINKING BOX 7.5

What do you think accounts for the fact that the disabled are still experiencing barriers in obtaining and retaining employment?

The *Americans with Disabilities Act* came into effect in 1993, and while it was in advance of much work in Canada, the new Ontario act, although staged over 20 years, promises to accomplish much and may turn out to be a model for other provinces. The Act announces the need to develop, implement, and enforce accessibility standards in all areas of social life, including accommodation, services, and employment, by the year 2025, and ensures the provision of involvement of disabled individuals in meeting the standards. It also deals with enforcement issues.

Perhaps the recognition of the importance of the disabled was best expressed in a speech by the President of the Treasury Board of Canada to the Disabled People's International Conference in 2004:

> We want a Canada in which citizens with disabilities have the opportunity to contribute to and benefit from Canada's prosperity—as learners, workers, volunteers, and family members.[6]

CHAPTER SUMMARY

The issue of disability is not simple, and the interplay of both historical events and theoretical perspectives affects the day-to-day lives of people with disabilities. In Canada generally there is a move to greater empowerment of individuals with disabilities. Yet, in many ways, people with disabilities are still not accorded full participation in our society.

Wars, the labour movement, public disasters, epidemics, and the resultant changes in social policy and social programs have had an impact on policies that affect people with disabilities.

The initial institutional provider of services to people with disabilities was the Church. This responsibility shifted to the government, especially because of World War I and II. The rise of the voluntary sector and increased participation and awareness of the needs of people with disabilities among the Canadian public at large led to a sharing of responsibility between the government and the private sector. Now, with reduced government funding, the role of the voluntary sector is likely to increase.

A major trend in the past two decades has been for greater involvement in all aspects of social policy and programs for people with disabilities by people with disabilities.

Three models help to explain the direction of policy initiatives: (1) the biomedical model, (2) the economic model, and (3) the sociopolitical model. The theory of normalization has led to fuller participation of people with disabilities in all areas of everyday life (work, recreation, and housing), and yet this same movement, many would argue, accounts for there being vast numbers of individuals who need care and have no place to go for help.

As with other issues of equality, acceptance and attitudinal change must occur before practice changes. The needs, issues, and demands of people with disabilities resemble those of other groups in Canadian society who are disadvantaged or marginalized. There are parallels with the fight of people with disabilities for empowerment and equality with various Aboriginal and ethnic groups, with immigrants and refugees, and with women. The experience of people with disabilities, however, is unique in this country, as is reflected in the history and in the development of theories related to disability.

KEY TERMS

antidiscriminatory, p. 200
deinstitutionalization, p. 205
developmental disability, p. 200
disability, p. 195
empowerment, p. 203
"handicaps," p. 198
impairment, p. 201
inclusive societies, p. 195
learning disability, p. 206
"mentally deficient," p. 197
normalization, p. 205
rehabilitation medicine, p. 198
rehabilitation team, p. 198
self-determination, p. 207
social movement, p. 199

DISCUSSION QUESTIONS

1. Compare and contrast three major theoretical perspectives about people with disabilities.
2. What accounts for the great changes in social policy toward people with disabilities that were brought about by the two world wars? In other words, why did the wars have such a great impact on the lives of people with disabilities?
3. From the perspective of the sociopolitical model, what indicators are there for increased empowerment of people with disabilities in the 21st century? Consider the fields of health, education, and welfare in your answer.
4. What might be some negative consequences to the "theory of normalization"?

NOTES

1. Jerome Bickenbach, *Physical Disability and Social Policy* (Toronto: University of Toronto Press, 1993), p. 12.
2. Bickenbach, p. 12.

3. Special Committee on the Disabled and the Handicapped, *Obstacles*, First Session, Thirty-Second Parliament, 1980–81, The Third Report, February 1981, available Department of Human Resources and Social Development site <www.sdc.gc.ca/asp/gateway.asp?hr=/en/hip/odi/documents/obstacles/00_toc.shtml&hs=vxi>, accessed 6 April 2006.
4. Paul C. Higgins, *Making Disability: Exploring the Social Transformation of Human Variation* (Springfield, IL: Charles C. Thomas, 1992).
5. Statistics Canada, "Functional Health States," 2003.
6. Reg Alcock (President, Treasury Board of Canada), "The World Needs More Canada: Strengthening Support for People with Disabilities," speech to Disabled People's International Summit, Winnipeg, MB, 10 September 2004, available President of Treasury Board of Canada Secretariat site <www.tbs-sct.gc.ca/media/ps-dp/2004/0913_e.asp>, accessed 6 April 2006.

CHAPTER 8

Diversity and Conformity: The Role of Gender

Leslie Butler

I now see the women's movement for equality as simply the necessary first stage of a much larger sex role revolution. . . . What had to be changed was the obsolete feminine and masculine sex roles. . . . It seemed to me men weren't really the enemy—they were fellow victims, suffering from an outmoded masculine mystique that made them feel unnecessarily inadequate when there were no more bears to kill.
— Betty Friedan, *The Feminine Mystique*

God created men and women different—then let them remain each in their own position.
— Queen Victoria

Objectives

After reading this chapter, you should be able to

- define gender identity and gender spheres

- describe the different spheres men and women occupy

- give social and biological explanations for why men and women choose or are channelled in different life directions

- discuss the ways in which gender divisions are positive or negative for individuals and society

- form opinions about what, if anything, to do about the gender issues raised in this chapter

INTRODUCTION

The differences between men and women are perhaps the most celebrated of all human diversities. Poets and philosophers have probed the mysteries of love and the wonders of sexual difference for centuries, but only recently have sex and gender become the subject of scientific study. These studies confirm what many might already have known intuitively: your sex and gender influence every aspect of your life. Whether you are born a boy or a girl will affect everything from what kinds of jobs you will have and how frequently you want sex with your partner, to whether you are likely to be poor and how long you might live. Our masculinity or femininity is a physical and psychological lens that filters our very perception of reality. Knowing something about how sex and gender influence our personal and social lives can help us to better understand our experiences. But first, we need to understand what these terms mean.

To begin with, sex and gender are two different things, even though the terms are sometimes used interchangeably. Your **sex** is determined at conception and refers to the reproductive organs you are born with and your hormonal makeup. Your **gender** is a social role governed by how your culture defines masculinity and femininity. Sex is therefore something unchangeable, whereas gender changes as cultures evolve. (Twenty-first-century Canadian women wearing suits or playing sports would seem terribly unfeminine to their 19th-century counterparts.) This distinction between sex and gender is important, because if we want to build a fair and equal society we need to have the wisdom to know what we can change and what we cannot.

WHAT IS GENDER?

Gender Identity

Despite the many changes of the 20th century, there are still clusters of traits we designate masculine and feminine. These clusters of traits make up our conception of masculine and feminine **gender identity**, or what it means to be a man or a woman. In North American culture, a truly masculine person is, above all, competent. He is physically strong and sexually virile, aggressive, logical, unemotional, decisive, and protective. A feminine person is, above all, warm and nurturing. She is physically "soft," intuitive, emotional, indecisive, and in need of protection. Women talk, men act; women conciliate, men confront. These associations are powerful determinants of how we feel and how we act as men and women. Some feminists have pointed out that masculine and feminine traits are not valued equally, that society pays lip service to the value of nurturing and warmth but actually rewards masculine traits more concretely than feminine ones.

These clusters of traits are **gender stereotypes**, or generalizations, about the way most men and women are expected to behave. Individuals certainly deviate from these stereotypes, and modern Canadian society tolerates a much wider range of opposite gender behaviour than in the past. However, despite the many social changes of this century, statistical evidence shows there is a surprisingly high degree of conformity to traditional roles.

It is tempting to believe we are unaffected by stereotypes, that we are free to look and behave as we choose. But think of the penalty a man pays for being effeminate and a woman pays for being "butch." Penalties such as ridicule, ostracism, and workplace discrimination are called **social controls**, and they help enforce rigid gender behaviour codes. When men are feeling emotional or vulnerable, they may feel conflicted about feeling "feminine" and may suppress those feelings for fear of being ridiculed. Women may similarly suppress their masculine side. Enforced by social controls, the gender stereotype becomes a self-fulfilling prophecy. Biology, too, may limit our freedom to choose our gender identity. We will discuss the social and biological influences on gender in more detail later in this chapter.

Diversity or Conformity?

It is interesting to note here that because gender is a social role and controls are in place to make sure most people play the right role, gender may actually enforce conformity rather than encourage diversity. In other words, gender may limit rather than liberate us.

Consider for a moment the expected attitude of men and women toward sex. Men are expected to (and do) initiate sex, to want it frequently, and to value sex over other kinds of intimacy; women are expected to value love over sex, to desire sex less frequently, and to see love as a prerequisite for sex. Because our very identity as men and women includes these messages about sex, these role definitions may profoundly affect and perhaps limit our sexual freedom. Women may feel unfeminine, and thus uncomfortable, if they are sexually aggressive; men may feel unmasculine, and thus uncomfortable, if they prefer submissive roles in sexual interplay. Women may feel guilty or conflicted about having purely sexual relationships that do not have emotional commitments. It is finally okay for women to want sex, but they must not want it too much for fear of being labelled "easy" or a slut. There is simply no male equivalent in our language for slut. Apparently, only women can be too easy and are far more likely than men to feel the guilt or shame associated with sexual promiscuity. Alternatively, men who are anything less than obsessed with sex may find themselves ridiculed or questioning their own masculinity. These powerful messages about appropriate sexual feelings and behaviour are one example of how gender identity may be restrictive rather than liberating.

Challenges to Traditional Gender Identity

It has become more commonplace to see people openly challenging these behaviour codes. **Transvestites** are those men and women who adopt the dress and behaviour of the opposite sex. Their refusal to behave in a gender-appropriate way challenges cultural definitions of masculinity and femininity. The often-harsh reaction to transvestites is a measure of how deeply ingrained is the idea that women should behave like women and men like men. **Transsexuals** are genetically of one sex but have a psychological urge to belong to the opposite sex. Some transsexuals seek surgery to modify their sexual organs to bring their biological self in line with their psychological self.

Gay sexuality has become so openly expressed in North American society that early tentative attempts to portray homosexuality in a sensitive rather than denigrating ways (movies such as *The Kiss of the Spider Woman* and *The Crying Game*) have been followed by mainstream sitcoms such as *Will and Grace*. Gay marriage was a defining issue in the 2004 and 2006 federal elections in Canada, and *People* magazine announces gay celebrity breakups as casually as straight ones. For the first time in 2001, the Canadian census counted the number of same-sex common-law couples in Canada and reported there are 32 000 same sex couples representing .5 per cent of all couples.[1]

Although the challenges to rigid gender identity continue and even move into the mainstream of our culture, they are still the exception rather than the rule. The strict definitions of masculinity and femininity persist, along with the expectation that most people will adopt the "right" kind of behaviour. And, in fact, the vast majority do.

Gender Patterns

Both your sex and gender have a powerful influence on the kinds of experiences you will have in your life. When parents gaze at their children, for example, they may not know that compared with her brother, their daughter is more likely to

- live longer and be widowed
- earn less
- live in poverty
- marry younger
- be a single parent

Their son, on the other hand, is more likely than his sister to

- commit suicide
- be a victim of violence
- remarry if divorced
- lose custody of his children in divorce
- get cancer or AIDS

> **BOX 8.1**
>
> *Feminist and Feminine?*
>
> In her groundbreaking book of the early 1960s *The Feminine Mystique*, Betty Friedan chronicled women's struggle to break out of the confines of their domestic straightjacket. But Friedan also argued that the new role of women didn't mean they had to hate men or give up fulfilling romantic and sexual relationships with men. Much of the backlash against feminism in the past 30 years has centred on feminists as man-haters. Some believe being a feminist means women have to hate men or give up on marriage. Others say simply that the idea of romantic love is so entwined with traditional gender roles that it dies when those roles change and overlap. Still others say men react to the feminist challenge with a powerful backlash that includes violence. Friedan herself believed that feminism was not incompatible with romantic love and that the liberation of women would lead to the liberation of men.

Canadian statistics continue to bear out these patterns. A woman can expect to live to age 82, whereas a man can only expect to live to age 77. At all ages Canadian women are more likely to be widows than men, and when women reach age 80, two-thirds will be widows in contrast with just less than half of men aged 80. Nearly 40 percent of elderly women live alone.[2] Women who work full time continue to earn only 71 percent of what full-time male workers earn,[3] and when women have children, the wage gap widens further. Women have part-time jobs much more often than men, and they continue to shoulder a greater share of the childcare and domestic work.[4] In Canada, women head four out of every five single-parent households,[5] and 38 percent of those households headed by single women have incomes that fall below the poverty line.[6]

The patterns for men are equally compelling. It may surprise some readers to know that North American men are more likely to experience violence than women are, especially in the light of our heightened awareness of the frequency with which women are abused. In the United States, men are three times more likely to be killed than women.[7] Men of all ages commit suicide more frequently than their female counterparts, and men over 65 are five times more likely to commit suicide than women.[8] Men are still much more likely than women to be employed in senior managerial positions. This tendency to be shut out of high-level jobs is sometimes called the **glass ceiling**. At all ages men as a group are more likely than women to be employed and they make more money.[9]

Individual men and women will not always follow these patterns. Nevertheless, there is striking consistency in the kinds of experiences men and women can expect to have. Let's look at some in more detail.

WHAT ARE GENDER SPHERES?

Because people conform to fairly well-defined notions of masculinity and femininity, patterns of male and female behaviour emerge. These patterns create what we might call **gender spheres**. By this, we simply mean there are areas of work, school, and play in which men dominate and areas in which women dominate. Although there has been a lot of social change in Canada, these traditional divisions between what men and women do are surprisingly persistent.

Occupational Spheres

In the area of work, men still dominate in administration, technology, the professional fields, and jobs requiring physical strength. Women still find themselves dominating in the "caring" professions such as nursing, social work, and teaching. Women also predominate in the clerical and service sectors. "In 2004, two thirds of all employed women were working in teaching, nursing and related health occupations, clerical or administrative positions and sales and service occupations."[10] Table 8.1, which is based on data from the year 2005, compares the number of men and women employed in selected industry sectors. The data show clearly that men and women still find themselves in predictable male and female occupations.

Educational Spheres

Like the workplace, education also has its gender spheres. We might expect this, because educational institutions feed the labour market and as such are a kind of gate that young men and women pass through on their way to jobs. If men and women self-select or are selected out (through biased hiring standards) of certain educational programs, the gate to jobs requiring that training will close. Data from the 2001 Census show that Canadian men are far more likely than women to study architecture, engineering, and applied sciences. Women predominate in secretarial, community and social services, nursing, and the humanities. Women significantly outnumber men in education, health professions, social sciences, and the humanities; men are far more likely than women to enter engineering, mathematics, and physical science.[11]

Leisure Spheres

Men and women have different amounts of free time, and they often choose to spend it differently. Although men spend more time than women doing paid work, they have

Table 8.1   Employment by Industry and Sex, 2005

Industry/Sector	Men (n)	Women (n)
Manufacturing	1 586 300	628 900
Construction	913 200	110 000
Transportation and warehousing	606 600	194 000
Forestry, fishing, mining, oil	257 000	50 000
Agriculture	241 100	103 000
Health care, social assistance	308 600	1 419 500
Education services	378 400	724 700
Accommodation and food services	400 000	599 500

Source: "Employment by Industry and Sex, 2005," Adapted from the Statistics Canada CANSIM database <http://cansim2.statcan.ca>, Table 282.

slightly more free time than women because women do more unpaid work such as childcare and housework.

As the gender stereotype would predict, men are more physically active at all ages than women. In 1998, 44 percent of Canadian men reported that they regularly participate in sports, in contrast to only 26 percent of women.[12] Men are far more likely than women to choose competitive and physically rough sports such as football and hockey; women are more likely to choose swimming, cross-country skiing, and bowling.[13]

To summarize, men and women continue to choose or be forced into separate work, school, and leisure activities. Figure 8.1, compiled from 1998 data, gives a clear illustration of where masculine and feminine spheres remain separate and where they have begun to overlap.

The Issue of Equality

Once we recognize that men and women occupy different occupational and educational spheres, it is natural to wonder whether these spheres are separate but equal, or separate but unequal. In fact, the issue of equality is central to most questions surrounding sex and gender, and we will return to it several times in this chapter.

It is hard to ignore the economic inequality between the sexes in Canada: women simply do not get an equal share of the pay rewards, whether they work outside the home or inside. The work of raising children and maintaining a household goes unpaid. Women's occupations—childcare, clerical work, nursing, service jobs—for the most part

Figure 8.1 Gender Spheres

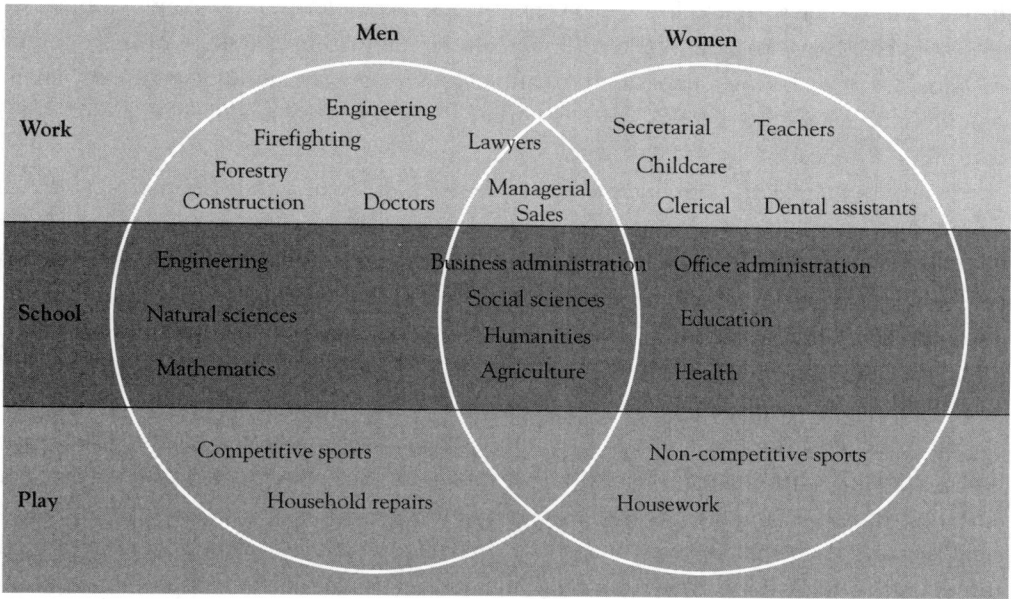

Sources: "Gender Spheres," Adapted from the Statistics Canada CANSIM database <http://cansim2.statcan.ca>, Table 282-0008, and also adapted from the following Statistics Canada publications "The Daily", Catalogue 11-001, Women in Canada, Tuesday, March 7, 2006, and from "Education in Canada: Major Field of Study, 2001 Census," Catalogue 97F0018XCB2001043, July 3, 2003.

CRITICAL THINKING BOX 8.1

Superiority in Sports

Is male superiority in athletics a result of nature or nurture? Consider how women have closed the gap in record time for the Boston Marathon. In 1964 the men's marathon record time was 1 hour and 13 minutes faster than the women's. In 2004, only 10 minutes and 30 seconds separated the men's record from the women's. With more women participating, better training for women, and larger monetary rewards, will women one day close the gap altogether?

pay less than men's—business, manufacturing, and professional jobs. Is it just a coincidence that women's work is less valued than men's? Is childcare, for example, inherently less valuable than janitorial work? Did men simply take an unfair share of the material rewards? Or does having a pool of cheap and even unpaid (female) labour simply benefit employers? (Wages are kept down when there is a reserve army of labour ready and waiting to fill jobs.)

A stark example of economic inequality between the sexes can be found in Table 8.2, which shows what men and women can expect to earn after graduating in different fields of study. The survey compared men and women who achieved the highest degree, certificate, or diploma in each field, and found that men make more than their female counterparts in every field, including teaching and nursing, which have traditionally been female-dominated. So even when men venture into nontraditional roles they can still expect to earn more than their female colleagues.

If women are the losers in the economic sphere, there is strong evidence that men are losers in the interpersonal sphere. As breadwinners forced to be absent from the home, they are more likely to forgo close relationships with their children. Confined by rigid definitions of masculinity, many men may be unable to achieve emotional and intellectual intimacy with their partners. Men rarely get sole custody of their children in divorces, and often play a secondary role in raising children after divorce (see Box 8.2).

Table 8.2 Wage Gap by Selected Field of Study

Major Field of Study	Average Employment Income ($)		Difference for Men
	Men	Women	
Medicine	121 010	75 003	+46 007
Dentistry	112 512	60 932	+51 580
Law and jurisprudence	93 458	52 196	+41 362
Business and commerce	64 447	35 275	+29 172
Civil engineering	57 564	34 090	+23 474
Primary education	32 742	23 770	+ 8 972
Nursing	42 458	35 524	+ 6 934
Hairdressing and other applied arts	27 632	17 043	+10 589

Source: Adapted from Statistics Canada, "Education in Canada: Major Fields of Study," 2001 Census, 97F0018XCB2001043, available <www12.statcan.ca/english/census01/products/standard/themes>, accessed 23 April 2006.

> **BOX 8.2**
>
> *Child Custody Awards*
>
> A 1998 survey of 5000 Canadian divorce cases showed women get sole custody of children in 80 percent of Canadian divorces. Men get sole custody only 8.4 percent of the time. Some would say this is concrete evidence of gender discrimination against men. Intentionally or unintentionally, the courts may assume that men are unsuited for nurturing and raising children, leaving men with the pain of losing their children. Men's rights groups and websites devoted to promoting men's interests in custody awards have proliferated in recent years.
>
> Source: Lorne D. Bertrand and Joseph P. Hornick, "The Survey of Child Support Awards: Preliminary Analysis of Phase 2 Data (October 1998–May 1999)," 25 August 2000, available <www.justice.gc.ca/en/ps/sup/pub/rap/phase2.html>, accessed 23 April 2006.

For many men, the stress of having the family's material well-being thrust on their shoulders results in mental and physical health problems. The bare fact is, men die younger than women, and more men than women suffer from chronic stress-related diseases such as heart disease and alcoholism. Being valued as a meal ticket is probably every bit as confining for men as being valued as sex object or mother figure is for women. A picture of two solitudes continues to emerge, with men cast in the role of provider and women in the role of caregiver.

Sexual Politics

Women who have historically found themselves on the short end of the economic stick have nonetheless achieved some degree of equality through what might be called *sexual politics*. This is the delicate, usually unacknowledged, power struggle between men who desire, or appear to desire, sex more than their female partners, who have the power to give or withhold sex. The fact that men have a greater sexual appetite than women, or at least a more openly acknowledged sexual appetite than women, is evident in sex trade statistics. The job market for female prostitutes, strippers, exotic dancers, and pornographic subjects far exceeds the demand for their male counterparts. This apparent difference in the sex drives of each gender has caused many feminists to observe that the only two professions in which women are able to out-earn men are prostitution and fashion modelling.

> ### CRITICAL THINKING BOX 8.2
>
> *Sexual Politics*
>
> How real is the issue of sexual politics in most relationships between men and women? Do women use men's more openly acknowledged desire for sex to get some of the things they want out of a relationship (e.g., financial security, help around the home)?

Whether men hire prostitutes, exchange their paycheques for regular sex in marriage, or merely use money (along with flattery and favours) when courting and wooing, men's apparently greater need for sex gives women some leverage in negotiating with their partners. Before women could advance themselves in paid work outside the home, they could improve their socioeconomic standing by being sexually desirable to a wealthy man. Today, the process of sexual negotiation is much more subtle, and may even be invisible to those involved in it. But in the day-to-day give and take of male–female relationships, sexual politics continues to play an often unacknowledged role in determining who gets what out of the partnership.

The Myth of Progress

It may surprise some of you to see such stark evidence of persisting gender spheres. After all, aren't the bad old days gone?

Much *has* changed. The days when women were prevented by law from entering politics, and when no man would choose to be a secretary or nurse for fear of total ostracism, are gone. Economic imperatives, such as the need for two-income families, have resulted in real changes in the workplace and the home. Attitudes about what is acceptable for men and women have also changed. Women now have professional hockey and basketball leagues, and some men choose to be nurses and secretaries.

But the majority of men and women are still channelled into "acceptable" roles by powerful social controls. It may be that our liberated attitudes toward gender have changed much faster than social reality has.

We may have trouble accepting that gender spheres persist because deeply ingrained in our consciousness is the notion of progress. Our worldview is dominated by the idea that things are getting better and that this inexorable force will march ever forward and never backward. But the number of women in politics, for example, goes up and down.

There were significantly more women in the Ontario legislature in the Bob Rae government of the early 1990s than there were in the Mike Harris government. It is clear that gains made by women can be lost. Even so, some people think that very little should actively be done to eliminate the inequalities between the sexes because things will just get better in the not-too-distant future.

WHY GENDER SPHERES EXIST

Nature and Nurture

Once we recognize that gender spheres exist, and that men and women do find themselves playing fairly predictable roles, it is natural to wonder why. Do we behave as men and women because we were born that way or because we learned how to behave? When a man is masculine, is he strong, aggressive, and unemotional because of biological factors or social factors? Does testosterone make him so? Or is he merely conforming to rigidly maintained social definitions of masculinity? Does a woman's biological makeup lead her to be nurturing, emotional, and conciliatory? Or does she learn there will be penalties if she behaves in an unfeminine way? You will recognize this as the **nature/nurture** debate central to most questions about human behaviour, a debate that rages on. Although it is not likely to be resolved any time soon, the debate is worth looking at, because it reveals some interesting ideological alignments and helps us understand how politically charged is the issue of sex and gender.

Progressivism and Conservatism

Biological explanations for why men and women seem to behave differently and occupy different social roles have a **conservative ideology** because nothing can be done to change the status quo. Taken to its extreme, the biological argument would say there is little point in trying to make men more nurturing, compassionate, and conciliatory because men are driven by their biology to adopt exactly the opposite behaviour. Similarly pointless are attempts to make women more aggressive, logical, and decisive. Going a step further in the conservative position, most men will be less effective than women if they move into occupations such as childrearing and nursing that require traditionally feminine nurturing qualities. Women who move into politics and business will be hampered by their emotional nature. It is therefore sensible, when we are building our social world, to exploit the natural abilities of men and women rather than trying to buck the natural trend.

Social explanations for why men and women seem to behave differently demonstrate a **progressive ideology** because they include the possibility that the status quo can be changed. In other words, if behaviour is learned and not forced on us by unchanging laws of biology, then we can simply learn differently and change will happen. If we want to broaden the range of acceptable masculine and feminine behaviours, we can change the social environment in which we raise boys and girls: dolls for boys and trucks for girls.

It is, however, unlikely that very many people side wholly on the nature or the nurture side of the question (and are therefore entirely progressive or entirely conservative). This is probably because in reality, we know people with both masculine and feminine traits. There is a widely held notion, probably based on the work of psychologists such as Carl Jung, that all men and women have both a masculine and a feminine side to their psyche. This mixing of masculine and feminine traits is what probably leads many people to conclude that a combination of biological and social factors shape human behaviour. Therefore, most people probably lean toward one side or the other rather than believing entirely in either nature or nurture.

It is worth noting here that whether you lean toward the nature or the nurture side may depend more on self-interest than on how logical you find the arguments on that side. For example, women who are dissatisfied with rigid role definitions that they perceive have relegated them to low-status, low-paying jobs may align themselves with the progressive, nurture position. At least hope for change exists there. Similarly, men who like their positions of relative prestige and power might align themselves with the conservative, nature position and believe that they are merely living out an inevitable role thrust on them by laws of nature. Alternatively, men who find themselves limited by the social dictate that they should be strong, unemotional breadwinners may also find themselves looking to break down rigid gender roles through social change.

TWO THEORIES ABOUT GENDER SPHERES

Sociologists also have pondered why gender spheres exist. They tend to examine the behaviour of men and women in groups, rather than as individuals, and they look primarily to the social world, not the biological world, for explanations. The two theories that follow reflect two common divisions in the way people explain why men and women occupy distinct spheres: (1) gender spheres either result naturally and provide a useful way to organize our social world or (2) they result from one group trying to maintain advantage over the other.

Structural–Functionalist Theory

The theory that says that gender spheres evolved naturally because of the fact that men and women had different biological strengths and weaknesses is called **structural–functionalism**. Men hunted because of their physical strength and women could not hunt because they were tied to the children by their capacity to bear and breastfeed them. As societies became more complex, male and female roles grew out of this fundamental division of labour: women at home, men out hunting and protecting the family. This perspective stresses that the roles played by men and women are natural and serve to help society function more or less smoothly. Men and women have clear roles and those roles stay the same over time. The idea is that societies, like organisms, evolve to survive. Social structures evolve to serve different purposes in the functioning of society. Structural–functionalists would point to the fact that it makes sense, from a group survival standpoint, to have men do dangerous work such as hunting, firefighting, or mining because they are reproductively expendable. (We only need one man to produce 20 babies in a year if he impregnates 20 women, but we need all 20 women to produce 20 babies.)

For structural–functionalists, the exclusion of women from paid work or the devaluing of women's work also serves a useful function in distributing wealth fairly. At first glance this may seem absurd, but consider that the family is the basic unit and each family has a male breadwinner. Paying only the men makes it more likely that each family will have access to at least one of the scarce good jobs. When both men and women can take the good jobs, it is easier for families to "hog" wealth by taking two good jobs. This phenomenon is evident today when a male doctor marries a female lawyer.

Conflict Theory

In contrast, another sociological theory takes a different view of gender spheres. **Conflict theory** sees society as consisting of many different groups that have competing, or conflicting, interests. Each group is busy pursuing its self-interest even if it means disadvantaging or even exploiting another group. Conflict theorists would say that having clear male and female spheres and excluding women from the paid work sphere effectively cuts the competition for good jobs and monetary rewards in half. This was no doubt good for men, who used every means to maintain this advantage. But as economic necessities brought women into the workforce in large numbers, the competition for good jobs intensified. Much of the backlash against feminism in North

---

CRITICAL THINKING BOX 8.3

Reverse Discrimination?

Do employment equity laws now give women an unfair advantage over men in the employment sphere? Or do these laws merely level the playing field for men and women?

America results from the fact that males today face much greater competition from females than their fathers faced. If you are male, you may face nearly twice the competition for jobs that your father did.

Because men benefited from limited competition, they would naturally use any means at their disposal to maintain the status quo. Conflict theorists might look at the means by which men maintained their advantage over women historically. These means are called instruments of social control and are seen not as natural but as "man"-made phenomena. Laws that prevented women from owning property, for example, were an important means by which men kept women out of the competition. Because owning property was a necessary criterion for just about any political or economic activity, women were effectively excluded from anything that shaped society. For conflict theory then, the law becomes a social institution that does not benefit all equally, but rather an instrument used by one group to gain advantage over another.

It is vital to point out here that conflict theorists do not see gender as the only, or even the most important, distinction in understanding who dominates whom. Equally important are differences in social class. When women's work is undervalued, it is not necessarily men controlling women, but rather people with power and money (the ruling class) controlling those without power and money (the working class). An underemployed man earning a fraction of a professional woman's salary can more directly relate to class difference than gender difference. Employers can be, and are increasingly, women. These women may even benefit from sexual discrimination against their sisters.

Conflict theorists might similarly identify ways women benefited from gender division of labour and used their biological childbearing roles to exert power over men. Some feminist scholars have unearthed evidence of ancient matriarchal societies in which women's central role as procreator put them in a position of power over men. A concrete

---

CRITICAL THINKING BOX 8.4

The Disappearance of Men's Work

Since men have occupied the sphere of hard physical work, what will happen to men as the physical work in society increasingly is done by machines or disappears altogether? Will men lose their traditional territory as blue-collar jobs disappear and white-collar jobs—which women are equally or better qualified for—make up the majority of work?

modern example of how men are excluded from childrearing is seen in the disproportionate number of child custody suits that are won by women (see Box 8.2).

So gender spheres may simply serve a function in the smooth operation of society by ensuring that jobs get done by those best equipped to do them, or gender spheres may be evidence of an ongoing battle of the sexes and social classes.

Merit Versus Social Connections

Whether gender spheres evolved in order to help society function or because of conflicting gender interests, why do they persist, especially in a society that prides itself on providing equal opportunity for all? If people are free to choose their destiny, if they can truly be whatever they want to be, why do men and women mysteriously want to choose gender-appropriate activities? Why do they behave so predictably? Or could it be that people are not as free as they think, that strong social, economic, legal, and psychological controls steer men and women into predictable destinies?

If you believe that people are free to choose their path in life, that individual character and hard work can help women overcome barriers preventing them from entering a male sphere and men from entering female spheres, then you believe in a **meritocracy**. This simply refers to the idea that people get their rewards in life on the basis of their individual merits (say talent, skill, or intelligence). In a meritocracy, skills and talents are rewarded fairly and equally regardless of factors such as gender, race, or social class. Therefore, women who manage to acquire the talents and skills necessary to do a job will ultimately get that job. Along this line of reasoning, it would be foolish and ultimately self-defeating for employers to turn away well-qualified women just because they are women. Therefore, if women find themselves ghettoized in low-paying, low-status jobs, the way out is to get the skills and education necessary for better jobs. Those skills will be rewarded. Similarly, men who are barred from female spheres also need to get the right qualifications to compete equally with women. Therefore, those who buy into the idea that we have a meritocracy would say gender spheres persist because individual men and women have not applied themselves to getting the right kinds of qualifications.

Others, however, doubt that a meritocracy can or does exist. They believe that the competition for jobs, money, and status is not a fair one at all and that the rewards are not given out to those with the most merit but instead to those with social connections to the group that controls hiring. Therefore, if women control the selection of candidates for early childhood education in schools, men will be excluded by virtue of their sex. Women simply will not value the kinds of qualities men bring to the job. Those who doubt we have a meritocracy tend to believe that women and men are prevented from entering some fields of education and occupation because of often-invisible **systemic barriers**, not because they lack the skills or talents. These barriers are built right into the system in the form of laws, discriminatory hiring practices, and psychological barriers.

Laws

An example of how law works as a built-in barrier preventing women from entering the male sphere can be seen in the fact that women are still prohibited from combat roles in the military. History is crammed with examples of laws that prevented women from entering male spheres. For example, in the early 20th century, women were not defined as persons under the law in Canada and therefore could not hold political office.

Discriminatory Hiring

Discriminatory hiring practices might include a host of subtle or not-so-subtle ways to screen out the wrong gender. An experienced male primary-school teacher in Ontario was privately told by the principal to stop applying for the kindergarten job at his school because parents were simply not ready to have a man take their daughters to the washroom.

Gender-biased hiring standards can be another form of discrimination in hiring. Recently, women have challenged the high value placed on height or physical strength in screening applicants to police academies, arguing that the vast majority of police work requires skills such as mediation, conciliation, and communication (feminine strengths). To the men who created the tests, the gender bias would probably not even be visible, and they may see any attempt to change the standards as "lowering" the standards for women.

CRITICAL THINKING BOX 8.5

British professor Alison Wolf argues that workplace equality for women has led to an exodus of women from caring roles such as motherhood, teaching, health care, and elder care. Whether the work was paid or volunteer, women are abandoning these helping roles as they access the greater rewards of money and status in the workplace. Birth rates are plunging in industrialized countries. And despite ever-greater expenditures on education and health care, there is still a huge vacuum left by the loss of female altruism. Has women's liberation come at the cost of a healthy, caring society? What can be done to replace the female altruism that was "profoundly important to modern industrial societies?" [14]

Psychological Barriers

Psychological barriers are less tangible but may be no less powerful in preventing men and women from being free to choose their path. Girls and boys may never even want to do non-traditional jobs because of powerful messages given to them about what is an appropriate activity for males and females. Education officials may wonder why social work still attracts very few males, and they may even institute a program to attract more male applicants to social work programs. But if boys sense that social work, with its emphasis on interpersonal skills and its relatively low pay reward, is an inappropriate activity for them, no amount of "affirmative action" will bring more men into the programs.

WHY IT ALL MATTERS AND WHAT TO DO ABOUT IT

Individual Choice

We know that men and women are different, that they are expected to and do behave differently, and that as a result they occupy different spheres in our social world. We might even have some ideas about why gender divisions exist and persist. The logical next question is: Why does it matter?

It matters to individuals because men and women suffer when they are forced into rigid masculine and feminine roles that limit their potential or simply do not fit them. Stereotypes about how people should behave restrict people's freedom to act the way they want to and to pursue their real interests and strengths. Men who want to take a secretarial job and who have no desire for promotion or further responsibility are deterred from doing so by a disapproving society. (Does he have enough testosterone?)

It is difficult to overstate the profound infringement this is on a person's individual liberty. If the state were to pass a law prohibiting men from staying home to raise children, for example, Canadians would be outraged and challenges to the *Charter of Rights and Freedoms* would abound. But we quietly tolerate a powerful set of social controls that virtually guarantee men and women will "self-censor" their activities and conform to society's expectations. Although challenges to accepted gender roles have always existed and may over time change our conceptions of masculinity and femininity, they are still peripheral and cause the challengers much pain.

But conforming to society's expectations might cause as much pain as challenging them. Feminist scholars are now looking at the high rate of mental illness among women historically as evidence that some women literally lost their minds when they were prevented from using their talents in the world outside the home. Some were brilliant women who were forced into mundane domestic chores for which they were wholly unsuited.

Men, too, have suffered terribly, because the masculine stereotype left them unable to communicate their emotions or achieve intimacy. Literature has many examples of emotionally crippled men who are isolated from their loved ones because they cannot communicate their feelings.

Constructing a Fair Society

On a societal level, our construction of society along rigid gender lines may not be as functional as we think it is, nor may it be very fair. What a waste of talent and potential we have created if we exclude talented men and women from certain jobs simply because they are men and women! It is an even greater waste if they exclude themselves by never even wanting to do other than what men and women have always done. What should we do to try to open up opportunities for individuals and to ensure that society uses its human resources to the best potential? How do we stop gender from limiting our life choices?

Education

One popular notion about how to break down gender barriers is to educate people. The assumption is that education will erase many gender stereotypes that restrict human potential. If we teach people that men and women have a very broad range of skills and talents that often do not conform to traditional male and female stereotypes, we may liberate the potential of men and women alike. An enlightened population is presumably more open to change than an unenlightened one.

Equity Laws

Another way to remove gender barriers is to simply cut to the chase and engineer the result we want through laws and regulations (see Critical Thinking Box 8.6). If we want more men in secretarial fields or childcare, simply establish hiring quotas. If we want women to enter firefighting or engineering, make rules forcing educational institutions and employers to bring in more women. This is, of course, very controversial. In Canada, we have not gone as far as establishing quotas, but we have had employment equity legislation at both the federal and the provincial level. These laws promote the hiring of not just women but also minorities and people with disabilities by requiring employers to set their own hiring targets and to meet them. The debate about equity laws can be inflammatory at times. Some believe laws are necessary to balance out existing systemic barriers and overt discrimination against these groups; others believe these laws cause

> **CRITICAL THINKING BOX 8.6**
>
> *Are Employment Equity Laws Necessary?*
>
> Employment equity laws are designed to engineer changes in our workplaces to ensure that certain groups (women, people with disabilities, racial minorities, and Native people) are fairly represented in Canada's workplaces. But some people believe hiring quotas will cause reverse discrimination against qualified white males and could lead to unqualified women and minorities getting jobs. Advocates of equity laws counter by saying there are plenty of well-qualified women and minorities and there is no need to lower hiring standards. What do you think?

reverse discrimination by excluding qualified males and lowering hiring standards. Those who favour employment equity also argue that gender stereotypes and the psychological barriers that prevent people from choosing nontraditional roles will be broken down most effectively when young children see male daycare workers and female engineers as role models.

Proportional Representation

Along these same lines, another way to eliminate gender spheres is to have proportional representation of women (and other disadvantaged groups) on hiring committees and educational selection committees. This approach admits that consciously or unconsciously, we will value the attributes of those most like ourselves. Proportional representation (e.g., the number of women on any committee should mirror the percentage of women in that community) would then lead to more qualified women being selected for employment or postsecondary programs. If women help define hiring standards, it might also lead to less-gender-biased selection processes. In the case of police academies, this might lead to a higher valuing of communication and mediation skills and a deemphasis on physical strength. If women made up 51 percent of the House of Commons, women's concerns would be more urgently addressed by government. Proportional representation would also help open the doors for men in fields in which they have suffered systemic discrimination.

Chickens and Eggs

Employment equity and proportional representation raise an interesting point that divides many people about how change happens. Those who want to cut to the chase with hiring quotas and proportional representation believe that you have to change reality first and then changes in people's attitudes and behaviour will follow. In other words, only *after* people see that women can be aggressive decision makers and men can be excellent nurturers will gender stereotypes break down. This view stresses that our attitudes are shaped in response to what we see around us. The opposite view stresses that we must change people's attitudes first, and then social change will follow. This view stresses the role of education in social change. If we teach boys and girls to be open-minded about gender and to question traditional stereotypes, those attitudinal changes will lead to the kinds of real changes contemplated by employment equity.

The Wisdom to Know the Difference

There may be some facts about gender that we simply cannot change, or we cannot change any time soon. Perhaps we need to find the wisdom to know what they are. For example, we know women must bear children, and they suffer economic penalties as a result. We know men pay dearly in health and longevity for their role as primary breadwinners. Whether gender divisions are caused by biological or social forces, we may want to build a society that accommodates these facts about gender rather than a society that denies them. For example, society could accommodate women's childbearing imperative by requiring all employers to have decent maternity leave programs, job-sharing provisions, and fewer pay and promotion penalties for women who have interrupted careers. Similarly, measures for men such as more frequent vacations, voluntary unpaid leaves, and flextime working hours might help reduce the incidence of stress-related diseases such as alcoholism and heart disease that men suffer and die of.

CONCLUSION

Do gender divisions help society function smoothly or do they lead to conflict between the sexes? How you answer this question probably depends on the kinds of experiences you have had as a man or woman. The existence of a strong feminist movement throughout most of the past century in Canada may be evidence that some women find gender restrictive. Men, too, may want to free themselves from restrictive male stereotypes.

If we want to make changes, we must begin to understand how gender affects men and women individually and in groups. Only then can we make the kinds of changes that will enable men and women to live out their destinies as freely and as fully as they can.

CHAPTER SUMMARY

Despite many changes in Canadian society, men and women are still encouraged to adopt a fairly rigidly defined masculine or feminine gender identity. Gender identity is a socially defined role for men and women. When men and women consistently adopt traditional male and female roles, gender spheres are created. Gender spheres are simply areas of work, school, and play that tend to be dominated by either men or women.

Whether gender identity and the resulting gender spheres are created mostly by unchangeable biological factors or by changeable social factors is still hotly debated. Also at issue is whether the creation of male and female territories helps society to function smoothly overall or whether it creates conflict between the sexes. Some people believe men and women are free to choose their gender identity. Others believe it is forced on them by powerful social controls that reward proper masculine and feminine behaviour and penalize nonconformity.

Issues of freedom of choice and social equality are central to the gender question. Is individual freedom severely limited by rigid gender stereotypes and the social controls that enforce them? Do women lose out in the economic sphere when women's work is undervalued? Do men lose out in the interpersonal sphere when they are barred from nurturing roles? As you further your education and life experiences, you will form your own views on these important gender issues.

KEY TERMS

conflict theory, p. 229
conservative ideology, p. 227
gender, p. 217
gender identity, p. 217
gender spheres, p. 221
gender stereotypes, p. 218
glass ceiling, p. 220
meritocracy, p. 231

nature/nurture, p. 227
progressive ideology, p. 227
sex, p. 217
social controls, p. 218
structural–functionalism, p. 228
systemic barriers, p. 232
transsexuals, p. 219
transvestites, p. 218

DISCUSSION QUESTIONS

1. In small groups, discuss the gender roles of men and women in the culture your family originated from. How do the roles differ from culture to culture? How are they the same? Consider some other cultures, such as Afghanistan and the Taliban's laws regarding women.
2. Discuss the question of how rigid our gender roles are. Would the men you know be turned off by a sexually aggressive, dominant woman who works in a construction job? Would the women you know be attracted to or turned off by a submissive man who loves his secretarial job and has no desire to move up the ladder?
3. Do you agree with employment equity laws? Why or why not?
4. Parents frequently remark that their male and female children were different from the moment they were born. Despite their attempts to treat their boys and girls the same, the children seem to follow predictable patterns of male and female behaviour. Which is more significant in forming our gender identity, nature or nurture?
5. Discuss some of the penalties people pay for violating society's expectations of them as men or women. Consider minor deviations such as men wearing long hair and major ones such as transsexualism.
6. Do you think the ability of men and women to form happy, lasting relationships is made more difficult by the gender inequalities in our society?

NOTES

1. Statistics Canada, "Same Sex Common-Law Couples," 2001 Census, available <www12.statcan.ca/English/Census01/Products/Analytic/companion/fam/canada.cfm#same_sex_common_law>, accessed 23 April 2006.
2. Statistics Canada, "Women in Canada," *The Daily*, 7 March 2006, available <www.statcan.ca/Daily/English/060307/d060307a.htm>, accessed 23 April 2006.
3. Statistics Canada, "Women in Canada."
4. Statistics Canada, "Women in Canada."
5. Vanier Institute of the Family, "Family Facts," *Virtual Library*, available <www.vifamily.ca/library/facts/facts.html>, accessed 23 April 2006.
6. Statistics Canada, "Women in Canada."
7. Bureau of Justice Statistics, U.S. Department of Justice, "Trends by Gender," *Homicide Trends in the U.S.*, available <www.ojp.usdoj.gov/bjs/homicide/gender.htm>, accessed 23 April 2006.
8. Statistics Canada, "Suicides, and Suicide Rate, by Sex and by Age Group," available <www40.statcan.ca/l01/cst01/health01.htm?sdi=suicide%20rate%20sex>, accessed 23 April 2006.
9. Statistics Canada, "Women in Canada."
10. Statistics Canada, "Women in Canada."

11. Statistics Canada, "Education in Canada: Major Fields of Study," 2001 Census, 97F0018XCB2001043, available <www12.statcan.ca/english/census01/products/standard/themes>, accessed 23 April 2006.
12. Statistics Canada, "Sports Involvement by Sex," *General Social Survey, 1998*, available <www40.statcan.ca/l01/cst01/arts18.htm>, accessed 23 April 2006.
13. Statistics Canada, "Most Popular Sports," *General Social Survey, 1998*, available <www40.statcan.ca/l01/cst01/arts16.htm>, accessed 23 April 2006. (Table is noted as discontinued from 25 February 2005.)
14. Alison Wolf, "Working Girls, Broken Society," *Toronto Star*, 2 April 2006, p. D-1.

CHAPTER 9

Sexuality: Emergent Understanding in Diversity

Brigitte Guetter

Sex. What a powerful three-letter word. What a joy. What a burden. After millions of years of living with it, we are as awed and as baffled by its power as we are by the meaning of life itself.
— Fritz Klein, M.D.

Traditional history's search for origins in great moral truths is entirely misguided; everything is subject to history's disintegrating gaze. There are no absolutes.
— Lydia Alix Fillingham (on Michel Foucault)

Objectives

After reading this chapter, you should be able to

- discuss the relationship between human rights and sexual rights

- evaluate Bill C-250 and its effect on the protection of sexual orientation

- consider Bill C-38 and assess the effect of same-sex marriage on its participants and on the country

- examine same-sex parenting in the light of your own opinion on this kind of family arrangement

- define institutional completeness and give an account of de-assimilation by the homosexual community
- analyze both Kinsey's and Klein's models of sexuality and experiment by locating yourself on one or both grids
- appreciate the complexity of factors involved in the production of the diversity of sexualities

LESBIAN, GAY, BISEXUAL, TRANSGENDER/ TRANSSEXUAL (LGBT) EMERGENCE

Introduction

To talk about emergent trends in diversity of human **sexuality**, we need to begin with some consensus on how to categorize sexuality, and we quickly see that for every rule that we might find about sex in the past or the future, there is usually a counter-rule. For every concept that shows how much we've changed, there seems to be a competing truth that says we're run by the exact same angels and demons that have been wiring our genes since our paths first diverged from the Missing Link. Basic human sexuality really hasn't changed much over the past few millennia. We all still have pretty much the same body parts, the same physical capabilities, and the same feelings of attraction, desire, arousal, affection, disgust, obsession, heartbreak, and love. It is our philosophy of sexuality that is constantly changing—our approach to sex, our sexual attitudes, interpretations, values, laws, customs, turn-ons, and turn-offs. And the single thing that has most changed how we deal with sex over the course of the 20th century and into the 21st has been technology. Technology, along with its academic partner, science, "has given and continues to give us a less superstitious and irrational, more liberated and realistic understanding of our sexuality. Of course, sex drives technology. And technology takes sex for a ride. It's a marriage made on Earth," says forward-thinking sex therapist Dr. Susan Block.[1] And like any marriage, it has the ecstasies and the heartbreak to show for it. In this chapter, we will attempt to decipher some of those aspects in the currently compelling area of sexual diversity.

HUMAN AND SEXUAL RIGHTS

"Whereas recognition of the inherent dignity and of the equal and inalienable rights of all members of the human family is the foundation of freedom, justice and peace in the world. . . ." So begins the preamble to the Universal Declaration of Human Rights adopted and proclaimed on 10 December 1948 by the General Assembly of the United Nations and currently in force.[2] Three of its 30 articles read as follows:

- All human beings are born free and equal in dignity and rights. They are endowed with reason and conscience and should act towards one another in a spirit of brotherhood. (#1)
- Everyone is entitled to all the rights and freedoms set forth in this Declaration, without distinction of any kind, such as race, colour, sex, language, religion, political or other opinion, national or social origin, property, birth or other status.... (#2)
- No one shall be subjected to arbitrary interference with his privacy, family, home or correspondence, nor to attacks upon his honour and reputation.... (#12)

Then on 26 August 1999, during the 14th World Congress of **Sexology** held in Hong Kong, sexual rights were affirmed to be fundamental and universal human rights, and a set of 11 sexual rights were revised and approved. Some of the currently more urgent and relevant ones read:

- *The right to sexual freedom:* Sexual freedom encompasses the possibility for individuals to express their full sexual potential. However, this excludes all forms of sexual coercion, exploitation, and abuse at any time and situations in life. (#1)
- *The right to sexual equity:* This refers to freedom from all forms of discrimination regardless of sex, gender, sexual orientation, age, race, social class, religion, or physical and emotional disability. (#4)
- *The right to sexually associate freely:* This means the possibility to marry or not, to divorce, and to establish other types of responsible sexual associations. (#7)
- *The right to make free and responsible reproductive choices:* This encompasses the right to decide whether or not to have children, the number and spacing of children, and the right to full access to the means of fertility regulation. (#8)

Progress Around the World

Notwithstanding the declaration of these rights, progress in implementation has been slow and uneven in different countries around the world. The legal rights of **lesbian, gay, bisexual**, and **transgender/transsexual** (LGBT) people vary widely, but nowhere do they yet have true equality with **heterosexuals**. For example, while South Africa, Fiji, and Ecuador include **sexual orientation** as a category protected from discrimination in their constitutions, many other countries maintain laws that prohibit or regulate sexual activity between consenting adults of the same sex. Some statutes, often called "**sodomy** laws," still regulate specific sexual acts, for example anal sex, regardless of gender or sexual orientation, while others prohibit a range of same-sex sexual activities. Many laws are quite broad in scope, prohibiting any "unnatural" or "indecent" sexual act without providing any clear definitions. In some countries, in particular Muslim nations where conservative forms of Shariah law are followed such as Saudi Arabia and Iran, homosexual acts are illegal and subject to a maximum penalty of death. In other countries, most often in the case of lesbian relationships, national laws do not mention either criminalization or legality.

Progress in Canada: Bills C-150, C-242, and C-55

By comparison, Canada has advanced dramatically in the fight for same-sex rights, exceeding many other liberal nations in both laws and benefits for same-sex couples and members of the transgender/transsexual communities. Between the 1960s to the 1980s, things began to change. It was a time of protests, legal fights, and backlash. With a growing sense of solidarity, gays and lesbians became more visible in Canadian society. Homosexuality gradually became more familiar and accepted as more Canadians decided to **come out (of the closet)** to demand equality under the law.

In 1965, the court case of Everett Klippert caused much discussion of homosexuality among Canadians. As part of an arson investigation in the Northwest Territories, Klippert was interrogated by the police, arrested, and sentenced to three years in prison. While in prison, Klippert admitted to psychiatrists that he was gay and had had sex with other men over a 24-year period. It was determined that he was unlikely to stop having sex with men and he was declared a dangerous offender. He was then sentenced to life in prison.

In 1967, then–Justice Minister Pierre Elliott Trudeau proposed amendments to the *Criminal Code* calling for massive changes, including revision to abortion laws, legalization of lotteries, new gun ownership restrictions, and breathalyzer tests. Contrary to popular belief, however, Bill C-150 did not attempt to decriminalize homosexuality in general, but instead established a distinction between public and private sexual acts. Discussing the amendments, Trudeau said, "It's certainly the most extensive revision of the Criminal Code since the 1950s and, in terms of the subject matter it deals with, I feel that it has knocked down a lot of totems and overridden a lot of taboos and I feel that in that sense it is new. It's bringing the laws of the land up to contemporary society I think. Take this thing on homosexuality. I think the view we take here is that there's no place for the state in the bedrooms of the nation. I think that what's done in private between adults [age 21 at the time] doesn't concern the Criminal Code. When it becomes public this is a different matter, or when it relates to minors this is a different matter." Bill C-150 became law on 28 August 1969. Everett Klippert was released on 20 July 1971,[3] just in time for the first gay march on 28 August 1971, timed to coincide with the second anniversary of the **Omnibus Bill** C-150.

In the pouring rain, a group of about 100 men and women, in the first large-scale public protest for gay rights in Canada, walked to Parliament Hill in Ottawa, carrying signs and chanting "Two-four-six-eight! Gay is just as good as straight!" A week earlier, they had submitted a brief called "We Demand" to the federal government, which among other requests, asked that gays be permitted to serve in the military, and that the terms "gross indecency" and "indecent act" be removed from the Criminal Code and replaced with specific offences. It also asked that all references to homosexuals be removed from the Immigration Act.[4]

On 2 May 1980, Bill C-242, an act to prohibit discrimination on grounds of sexual orientation, which aimed to insert "sexual orientation" into the *Canadian Human Rights Act*, got its first reading in the House of Commons by MP Pat Carney but didn't pass. MP Svend Robinson introduced similar bills in 1983, 1985, 1986, 1989, and 1991. In 1991, Robinson tried to get the definition of "spouse" in the *Income Tax Act* and the *Canada Pension Plan Act* to include "or of the same sex." In 1992, he tried to get the "opposite sex" definition of "spouse" removed from Bill C-55, which would add the definition to survivor benefits provisions of the Canada Pension Plan. All the proposed bills were defeated.

Addition of Sexual Orientation to Protected Identifiable Groups: Bills C-33 and C-250

In 1996, the federal government passed Bill C-33, which added "sexual orientation" to the *Canadian Human Rights Act*. However, it took eight more years, to 3 February 2004, for Bill C-250 (originally called Bill C-415), a private members' bill introduced by B.C. NDP MP Svend Robinson, to get first reading in the House of Commons and royal assent. It was passed into law on 29 April 2004. *An Act to Amend the Criminal Code* expanded the definition "identifiable group" relating to the area of **hate propaganda** (sections 318 and 319) in the *Criminal Code* to include any section of the public distinguished by "sexual orientation."[5] The bill would protect gays, lesbians, and bisexuals. However, transgender people are not explicitly covered, although they may have protection on the grounds of "sex." A major criticism of this bill, which is designed to protect sexual minority groups from hate propaganda, is that it will have a significant impact on free speech and religious freedoms in Canada. The Focus on the Family (Canada) Association, for example, charges that it is vague and poorly crafted.[6] The Catholic Legate further charges that sexual orientation is not defined. Is **pedophilia** a sexual orientation, they ask? Can a citizen be imprisoned for advocating prison sentences for **sexual predators**? Also, "hate" is not statutorily defined. The Supreme Court of Canada has defined it as connoting an emotion of intense and extreme nature clearly associated with vilification and detestation. But hatred is a highly speculative judgment, the dissenters charge. What is understood as hatred to some is considered criticism to others. Also, while subsection 319(2) provides an exemption from being convicted by referring to a religious text, there is no provision for a non-religious defence. What if a citizen appeals purely to anthropology, science, anatomy, natural law, or mere personal conviction?[7]

Perhaps you will find it a little amusing that Real Women of Canada have **outed** Svend Robinson, noting that before he went public as the first openly gay member of Parliament in 1988, he had been married to his high school sweetheart, Patricia, from 1972 until 1978. He had described his sexual relationship with her as good.[8] Of course, an attack on a person's credibility is often resorted to when other strategies aren't working well, but it must be noted that this can be true on both sides of any issue. Sometimes it might be relevant, but often it's impossible to tell, because we mostly don't have all the information and our emotions can more easily move us under those circumstances.

When an important bill is passed, there is usually an audible sigh of relief, but Amnesty International's brief on Bill C-250 struck a note of caution. The organization has noted "a rise in **homophobic** abuse related to a 'backlash' effect. In Canada and throughout the world, the claims of people to equality regardless of sexual orientation are assuming an unprecedented visibility. This struggle is the struggle for human rights—the right of people to be who they are, free from violence, harassment and discrimination. But AI has observed that as fast as change is happening, so is the reaction. Increased visibility often leads to increased hostility and ill-treatment." It is the task of governments, asserts AI, to ensure that the pursuit of safety and justice by sexual minorities does not result in increased repression calculated to frighten equality seekers into silence, submission, and invisibility.[9]

SAME-SEX MARRIAGE IN CANADA: BILL C-38

Developments in Canada

Over a number of years, there have been court challenges in Ontario, British Columbia, and Quebec, and several more later in other provinces and territories, seeking equality in marriage. Although the cases were initiated in the provinces, the federal government was involved in each case, because it is the federal government that under the Constitution has jurisdiction over who can marry.

In Ontario, a court application was filed on behalf of eight same-sex couples who were denied the right to marry. One couple separated before the hearing of the appeal and did not wish to continue to participate in the proceedings.[10] Also joining the challenge was Metropolitan Community Church of Toronto after the province refused to register the marriages of two couples in January 2001, using the tradition of publication of marriage banns, in accordance with Ontario's *Marriage Act*. The cases were heard in November 2001. On 12 July 2002, the Ontario Superior Court ruled 3-0 that denying same-sex couples the equal right to marry was unconstitutional and violates the *Charter of Rights and Freedoms*. It gave Parliament two years to fix the law. Canada's Justice Minister announced the government would appeal the judgment. He also sent the issue of **same-sex marriage** to the Commons Committee on Justice and Human Rights for public hearings across Canada. On 10 June 2003, the Ontario Court of Appeal unanimously affirmed that the prohibition on same-sex marriage is unconstitutional. However, the Court went further than previous courts and reformulated the definition of marriage to include same-sex couples, ruling that the new definition would have immediate effect. That same day, 10 June 2003, same-sex couples began to legally marry in Ontario.

On 17 June 2003, then–Prime Minister Jean Chrétien announced legislation to make same-sex marriages legal. The draft legislation would also recognize the rights of churches and other religious groups to "sanctify marriage as they see it." Ottawa would not appeal two provincial court rulings allowing same-sex unions. "There is an evolution in society," Chrétien recognized.

On 8 July 2003, British Columbia became the second province, after Ontario, to legalize same-sex marriages. The British Columbia Court of Appeal lifted its ban on same-sex marriages, giving couples in the province the right to marry immediately.

On 17 July 2003, Ottawa revealed the exact wording of historic legislation that would allow gay couples to marry. The *Act Respecting Certain Aspects of Legal Capacity for Marriage for Civil Purposes* was sent to the Supreme Court of Canada for review. According to the draft bill, "marriage for civil purposes is the lawful union of two persons to the exclusion of all others." Also, the Supreme Court is being asked whether Parliament has the exclusive legal authority to define marriage, whether the proposed act is compatible with the *Charter of Rights and Freedoms*, and whether the Constitution protects religious leaders who refuse to sanctify same-sex marriages.[11]

The issue caused an uproar among many church leaders and traditionalists, who argued that the government does not have the right to redefine marriage. Prime Minister Jean Chrétien, however, vowed not to let religious objections alter his stand on same-sex marriage, saying that members of Parliament will be allowed to vote freely on the bill when it is introduced in the House of Commons after his retirement in 2004. A "free" vote means that members of Parliament can vote according to their conscience and not along party lines. A significant number of Liberal MPs said that they did not support same-sex unions and would vote against the legislation.

On 14 August 2003, the United Church of Canada voted overwhelmingly to support same-sex marriages, a decision following an extensive and emotional debate. The majority of delegates at the church's general council meeting in Wolfville, Nova Scotia voted to ask Ottawa to recognize same-sex marriage in the same way as heterosexual ones. However, a few days later, on 18 August 2003, the Catholic archbishop of St. John's, Newfoundland defended his censure of a local parish priest, saying Father Paul Lundrigan's comments were unacceptable within the Catholic Church. In a sermon a week earlier, Lundrigan had challenged the Church's campaign against legalizing same-sex marriage. He called the Catholic

Church hypocritical, criticizing it for fighting same-sex marriage while remaining silent about sexual abuse by members of the clergy. The Anglican Church of Canada has put off its decision on same-sex marriage until 2007, saying that they are in the midst of a conversation about it.

On 27 November 2003, then–Canadian Alliance Leader Stephen Harper, in a somewhat unexpected move, fired MP Larry Spencer as the Alliance's family issues critic after Spencer told the *Vancouver Sun* that homosexuality is part of a "well orchestrated" conspiracy that should be outlawed.

On 19 March 2004, the Quebec Court of Appeal ruled that homosexuals have the right to marry, and that the traditional definition of marriage was discriminatory and unjustified. The ruling upholds a lower-court decision and follows similar decisions in Ontario and British Columbia.

On 1 February 2005, the federal government introduced its same-sex marriage Bill C-38 in the House of Commons. The bill aims to give married same-sex partners the same legal recognition as other married couples, but newly elected Catholic Prime Minister Paul Martin reaffirmed that no church, no temple, no synagogue, no mosque, no religious official would be asked or forced to perform a marriage contrary to his or her beliefs.

To keep a written promise made by the Liberals to the Bloc Québécois that the legislation would be voted on before the end of the parliamentary session, MPs voted on 23 June 2005 to extend the sitting of the House of Commons, giving the government more time to push Bill C-38 through.

On 28 June 2005, the Liberals' controversial Bill C-38 passed final reading in the House of Commons in a 158–133 vote, supported by most Liberals, the NDP, and the Bloc Québécois.

On 20 July 2005, Bill C-38 became law. After being passed by the Senate, the same-sex marriage legislation, referred to by its short title as *The Civil Marriage Act*, received royal assent as Chief Justice Beverley McLachlin, acting in her role as Deputy Governor General, signed it into law.[12] This made Canada the fourth country in the world to officially recognize same-sex marriage, after the Netherlands, Belgium, and Spain.

Same-Sex Marriage to Be Revisited

The vote, however, came at a price for Paul Martin's new minority government. The minister responsible for Northern Ontario, Joe Comuzzi, resigned from the Cabinet and voted against the bill—an open rebuke of the government legislation. Conservative Leader Stephen Harper said that if his party formed the next government, the law would be revisited. His party did in fact form the next government, and in a news conference on 5 April 2006, he reiterated his intention to revisit the same-sex marriage issue, but said it likely won't happen before the Commons break for summer vacation.

That has stirred protest among those who want them to drop the matter rather than make the well over 3000 already-married gay couples and those contemplating marriage endure months of uncertainty. "Live and let live," urged Laurie Arron of Canadians for Equal Marriage.[13]

Reasons for Wanting to Get Married

Why do people want to get married, whether same-sex or opposite-sex marriage? The Ontario Court of Appeal, in its introduction for its 10 June 2003 decision on behalf of the seven same-sex couples, states: "Marriage is, without dispute, one of the most significant forms of personal relationships. For centuries, marriage has been a basic element of social organization in societies around the world. Through the institution of marriage, individuals can publicly express their love and commitment to each other. Through this institution, society publicly recognizes expressions of love and commitment between individuals, granting them respect and legitimacy as a couple. . . . This can only enhance an individual's sense of self-worth and dignity." The Court acknowledges that to participate in this fundamental societal institution is something most Canadians take for granted, but same-sex couples have been denied this opportunity simply on the basis of their sexual orientation. It views sexual orientation as "a deeply personal characteristic that is either unchangeable or changeable only at unacceptable personal costs."[14]

Same-Sex Marriage Around the World

Same-sex partnership arrangements vary widely in the United States and the rest of the world. Some countries have legalized same-sex marriage and others have **domestic**

CRITICAL THINKING BOX 9.1

The Ethnocentrism of Sexuality

Ethnocentrism is the tendency to see your own ethnic group and its social and cultural standards as the basis for evaluative judgments about the practices of others—with the implication that your own standards are superior. Do you think ethnocentrism could interfere with students' willingness to learn about and accept cultural differences in sexual values and practices? Could such acceptance be desirable or is it preferable that each group think and do as they prefer?

partnerships or **civil partnerships** or **civil unions**, and changes happen almost daily. In the United States at time of this writing, only Massachusetts has legalized same-sex marriage, but the state has included a provision that only partners who live and will continue to live in Massachusetts can get married there. U.S. Vice-President Dick Cheney has an adult lesbian daughter and feels that decisions about same-sex marriage should be made by individual states; he does not support a federal ban by means of a constitutional amendment, which President George W. Bush continues to strive for despite having been rebuffed so far and saying that he will continue this fight.

In England, **gay marriage** became legal on 21 December 2005, just in time for Christmas and two days after they began in Northern Ireland. Officially called "civil partnerships," Sir Elton John, 58, and his longtime partner, David Furnish, 43, tied the knot on the first possible day. The two, called the highest-profile gay couple in the world, have been together since they met in 1993 at a dinner party, and "from day one have led a charmed life with lavish homes around the world, luxury cars, more than 20 dogs and an enormous staff."[15] It has been reported that John has bequeathed Furnish a vast majority of his $400 million fortune. And despite their enormous wealth, there has been little if any chatter about a prenup. They have said publicly that they probably won't adopt children, mostly because of John's age—59 by publication time. Like MP Svend Robinson, Elton John had previously been married to a woman, Renate Blauel, in 1984 but divorced four years later.

Married for Life?

So far the "happy" news. Now for the other side—and it is not a little ironic. Two women, who had been together almost 10 years, married on 18 June 2003, a week after the Ontario Court of Appeal legalized same-sex marriage. Five days later they separated. This was to be the first same-sex divorce not just in Canada, but also in the world, in history. At the request of one of their lawyers, a publication ban was imposed on their identities, and they became known only by the initials of their respective lawyers, M.M. and J.H. The problem was, no law was available for their divorce. In an adaptive and historic ruling, Madam Justice Ruth Mesbur approved the couple's application, ruling the definition of spouse in the *Divorce Act* to be unconstitutional. No clear reasons for the separation were reported, but there was no possibility of reconciliation. It is reminiscent of a "Pardon My Planet" cartoon in which a middle-aged, apparently long-married couple are sitting in their own easy chairs, reading. Looking up from his newspaper, the husband laments: "Gay marriage is just so wrong. Haven't they suffered enough?"

It seems that at the moment of getting married, no one, straight or gay, feels they have suffered enough. Equality at all costs!

FAMILIES WITH LESBIAN, GAY, BISEXUAL, AND TRANSGENDER/TRANSSEXUAL PARENTS

The White House Easter Egg Roll

Hundreds of lesbian and gay parents began lining up on the evening of Friday, 14 April 2006 to make sure they got tickets for the Monday event, the annual White House Easter Egg Roll to be held on the South Lawn of the White House. Thousands of time-specific tickets—an estimated 16 000 last year—were to be given away on a first-come, first-served basis beginning 7:30 a.m. on Saturday, 15 April. According to the White House, the Egg Roll has been an annual tradition since 1878, and is the White House's largest public event for children six years old and under. This time children would use spoons to push coloured eggs through the grass in a race. Last year, there was a petting zoo, and there are usually music and games, face painting, visits by celebrities, and storytelling.

First Lady Laura Bush's office had issued a statement saying all families were welcome. In a steady rain on that chilly Monday, about 100 lesbian and gay couples and their children participated, many wearing leis made of yellow, purple, red, green, blue, and orange silk and plastic flowers instead of T-shirts with colourful protest slogans. But this was not to be a protest. It was really about showing the country that gay and lesbian families exist and are part of the American fabric, and that they participate in American traditions, said Jennifer Chrisler, executive director of the Family Pride Coalition.[16]

Last year, President Bush said in a *New York Times* interview that while "children can receive love from gay couples," he believed that "studies have shown that the ideal is where a child is raised in a married family with a man and a woman."[17] But "ideal" is an elusive concept, which can be difficult, if not impossible, to strive for. So how does one deal with that?

Pope Benedict XVI, in his harshest criticism of same-sex couples to date, prayed for the "filth" that surrounds society to be cleansed and said the world is in the grip of "a diabolical pride aimed at eliminating the family." Although he did not specifically name same-sex relationships, the message was clear. Pope John Paul II, Benedict's predecessor, had condemned Spain and Canada's moves to legalize gay marriage and Britain's civil partnership.

San Francisco's Openly Gay Castro District: Infiltration by Families

The Castro, a quarter-mile-square area and one of San Francisco's most openly gay neighbourhoods, has been attracting heterosexual parents to raise their families for more than a decade now. People are drawn to its quaint Victorian homes and small-town atmosphere. Like any neighbourhood, the Castro has always been evolving. In the 1970s, the former Irish-Catholic enclave saw the arrival of gay-owned bars such as Toad Hall and the Missouri Mule. The neighbourhood soon became a haven for gays attracted by a defiantly buzzing new counterculture. In the 1980s, the Castro endured the AIDS epidemic while struggling to remain a bohemian stronghold for gay bartenders, artists, and musicians.

But rising rents in the 1990s drove out many with lower incomes, leaving only the wealthiest. Then came young families. But sometimes these new families didn't feel welcome. Gay waiters scowled while taking their order. Men sneered at them on the street. Concern is expressed by gay men who had originally moved to the Castro for its lively gay culture that family-friendly sensibilities will quash the neighbourhood's spirit. "What surprise is next? Are they going to outlaw the **Gay Pride Parade**?" asked one gay man. "This is the Castro, not the Vatican," he complained.[18] The racy storefront displays have pitted protective parents against equally militant gay residents. But many parents, both heterosexual and gay, say the suggestive ads are inappropriate for children. Gay activists want to preserve the sexually liberated atmosphere that embraces such gay-themed holidays as "Leather Day" that reflects the **sadomasochism** culture practised by many gays and "Bear Day," in celebration of hairy men. Last year, a lesbian mother of two children, who are now six and two years old, complained about a sadomasochistic exhibit in a clothing shop window that featured a male mannequin chained to a toilet. "As an adult I find this disgusting," she wrote to city officials. "As a parent I find it unconscionable."[19] Another parent complained when an antiques store displayed a kitschy life-size statue of an aroused naked man. The owner reluctantly covered the offending portion, but only after police intervened. Some complain that gay culture itself, which has long enjoyed free sexual expression, is under attack, not just by straights, but also by gays and lesbians.

On the other hand, Fred Kirkbride, who owns a store on Castro Street, points out that a community that has for years argued for tolerance by heterosexuals should be more accepting. "Isn't it amazing how long we fought to be accepted by straight society? Now we want to keep straights and their children out of here," he said.[20] He agreed that many store displays have gone too far. He used to keep his parents away from those windows, he said, because he didn't want them to think all gays were into animal **bestiality**. And, slowly, it seems that both sides are showing compromise. For example, because parents often take their kids to the Gay Lesbian Bisexual and Transgender Community Center, nudity in the hallways is now forbidden, and the centre's **bondage** classes are required to stay behind closed doors. Twenty years ago, said the Center director, people would have fought such rules.

The Annual Pride Parade

Compromise has also reached San Francisco's Gay Pride Parade, the nation's most prominent gay celebration. Now organizers provide a children's area with licensed daycare. Families are encouraged to attend the two-day event on Saturday, which will feature fewer risqué events. But organizers don't want to put too many parameters around that, feeling it important that the gay community have this place to express itself. There is, it seems, a ways to go. Many men remain annoyed whenever they spot another baby stroller invading their beloved gay mecca.

SAME-SEX PARENTING

The Challenges of Research

Let me begin by saying that **same-sex parenting** is extremely difficult to study. The current body of research has grown partly out of court cases in which lesbian and gay parents (or co-parents) sought to defend or get custody of children. Many researchers approached the matter with a sympathetic or protective attitude toward the children and families they studied. Critics have accused researchers of downplaying differences between children of gay and straight parents, especially if those differences might be interpreted unfavourably. The best defence against bias, of course, is to judge each study, whatever its author's motivation, critically and on its merits, which is not an easy or obvious task. Same-sex parenting researchers, from the very beginning, faced daunting methodological challenges. For example, difficulty finding representative samples; small sample sizes with meagre funding; who should be used as comparison groups; subject-group heterogeneity, meaning that same-sex families can be structurally very different from one another; measurement issues and knowing how and from whom (from parents or directly from children) to collect the data; statistical issues, which were a more significant problem prior to today's advanced software availability; and finally, putting the research challenges into perspective—that is, the current research is probably on a par with the standards done since the 1970s on other families, such as foster and adoptive families.[21]

The Positive Side

So what have the studies found? Summarizing the research, the American Psychological Association (APA) stated in its July 2004 "Resolution on Sexual Orientation, Parents, and Children" that

> There is no scientific basis for concluding that lesbian mothers or gay fathers are unfit parents on the basis of their sexual orientation. . . . On the contrary, results of research suggest that lesbian and gay parents are as likely as heterosexual parents to provide supportive and healthy environments for their children. . . . Overall, results of research suggest that the development, adjustment, and well-being of children with lesbian and gay parents do not differ markedly from that of children with heterosexual parents.[22]

A review of the evidence by the organization The Future of Children in their journal *Marriage and Child Wellbeing* shows it to be consistent with the APA's characterization. It supports, they say, four conclusions:[23]

1. Lesbian mothers and gay fathers (about whom less is known) are much like other parents. Where differences are found, they sometimes favour same-sex parents.

2. There is no evidence that children of lesbian and gay parents are confused about their gender identity, not in childhood or in adulthood, or that they are more likely to be homosexual. Evidence on gender behaviour (as opposed to identification) is mixed; some studies show no differences, but others show that girls raised by lesbians may be more "masculine" in play and aspirations and that boys raised by lesbians are less aggressive. Some studies say that children, especially daughters, of lesbian parents adopt more accepting and open attitudes toward various sexual identities and are more willing to question their own sexuality. Others report that young women raised in lesbian-headed families are more likely to have homosexual friends and to say that they have had or would consider having same-sex sexual relationships. (How to view such differences? Conservatives may see lax or immoral sexual standards, whereas liberals may see commendably open-minded attitudes.)

3. In general, children raised in same-sex environments show no difference in cognitive abilities, behaviours, general emotional development, or such specific areas of emotional development as self-esteem, depression, or anxiety. One study reports that preschool children of lesbian mothers tend to be less aggressive, bossy, and domineering than children of heterosexual mothers. The only negative suggestion to have been noted about the emotional development of children of same-sex parents is a fear on the part of the children that they might be homosexual, which seems to dissipate during adolescence when sexual orientation is first expressed.

4. Many gay and lesbian parents worry about their children being teased, and children often expend emotional energy hiding or in other ways controlling information about their parents, mainly to avoid ridicule. It is not clear, however, whether the children actually have more difficulty with peers, with more studies finding no particular problems.

Concerns of the Opposing Side

A very different view is held, however, by the American College of Pediatricians. The stated mission and philosophy of the College is 'to enable all children to reach their optimal, physical and emotional health and well-being ... [and] that children are the future of our nation and society. As such, they deserve to be reared in the best possible family environment and supported by physicians committed to ensuring their optimal health and well-being."[24] The "Parenting Issues" section on the College's website points out that the current studies that appear to indicate neutral-to-favourable results of same-sex parenting have critical flaws such as non-longitudinal design, inadequate sample size, biased sample selection, lack of proper controls, and failure to account for confounding variables. The risks of a homosexual lifestyle to children are specified as follows:

- Violence among homosexual partners is two to three times more common than among married heterosexual couples.
- Homosexual partnerships are significantly more prone to dissolution than heterosexual marriages; the average homosexual relationship lasts two to three years.
- Homosexual men and women are reported to be inordinately promiscuous, involving serial sex partners, even within what are loosely termed "committed relationships."
- Individuals who practise a homosexual lifestyle are more likely than heterosexuals to experience mental illness, substance abuse, suicidal tendencies, and shortened life spans.
- The same dysfunctions exist at inordinately high levels among homosexuals in cultures where the practice is more widely accepted, making societal pressure an unlikely recourse.
- Children reared in homosexual households are more likely to experience sexual confusion, practise homosexual behaviour, and engage in sexual experimentation.
- Adolescents and young adults who adopt the homosexual lifestyle, like their adult counterparts, are at increased risk of mental health problems, including major depression, anxiety disorder, conduct disorder, substance dependence, and especially suicidal ideation and suicide attempts.

The American College of Pediatricians believes, given the current body of research, that it is inappropriate, potentially hazardous to children, and dangerously irresponsible to change the age-old prohibition on same-sex parenting, whether by adoption, foster care, or by reproductive manipulation. This position, they contend, is rooted in the best available science, and the burden is on the proponents of homosexual parenting to prove, with unambiguous scientific studies, that moving farther away from the heterosexual parenting model is appropriate and safe for children.

How to Interpret the Interpretations

We are faced here with two diametrically opposite interpretations of the research. Can this dilemma be resolved? Judith Stacey and Timothy Biblarz, two sociology professors at the University of Southern California, believe that both sides of the argument are right, at least partially. So they started from the beginning by reexamining 21 studies conducted between 1981 and 1998 and found that the conventional wisdom is wrong, and

> **CRITICAL THINKING BOX 9.2**
>
> *Thinking Critically ... About Critical Thinking*
>
> Critical thinking involves a set of thinking skills that include thoughtful analysis and probing of claims and arguments—your own or those of others. It uses careful observation, asking questions, seeing connections among ideas, and evaluating the evidence on which arguments are based and the logic of conclusions. Now consider how critical thinking skills can be applied to the study of the diversity of human sexuality. Why not simply accept the findings of experts?

they challenge the predominant claim that the sexual orientation of parents does not matter at all. Contrary to earlier assertions, children of same-sex parents exhibit significant differences from children raised by heterosexual couples. They claim that it would be difficult to conceive of a credible theory of sexual development that would not expect the adult children of "lesbigay" (lesbian-bisexual-gay) parents to show somewhat higher incidence of **homoerotic** desire, behaviour, and identity than children of heterosexual parents. They conclude that earlier researchers downplayed those differences when they found them, which in turn stunted research that might further highlight and explain these differences. "The pervasiveness of social prejudice and **institutionalized discrimination** against lesbians and gay men ... exerts a powerful policing affect on the basic terms of psychological research and public discourse on the significance of parental sexual orientation," Stacey and Biblarz write in a report in the *American Sociological Review*.[25] Claiming that few respectable scholars today oppose same-sex parenting, they suggest that most scholars fear that highlighting the differences will be used by opponents of homosexual parenting and marriage to oppose gay adoption and gay marriage. Researchers, they say, ought to be honest about their personal convictions and let the political chips fall where they may. Stacey and Biblarz admit in their own review that they believe in a "diverse" and "pluralistic" family structure that does not discriminate against same-sex households. Any "differences" found in research on children do not necessarily constitute "deficits," they say, and ought to be acknowledged and studied more thoroughly.

David Murray of the Washington-based Statistical Assessment Service and co-author of *It Ain't Necessarily So: How Media Make and Unmake the Scientific Picture of Reality*, agrees that most of the research on homosexual parenting is politically contaminated.

"We have allowed the politicization of this issue to erode our capacity to see clearly and to effectively decide policy issues," Murray said.[26] Of course, flawed science is not new. Right now it's swirling around the controversial area of sexuality.

LGBT COMMUNITY: INSTITUTIONAL COMPLETENESS AND ITS DEVELOPMENT

Institutional Completeness

"After living in Vancouver for three years, returning to my home town of Toronto felt like coming back to a city gone lezzie. Dyke content is seemingly rampant, available across the city in solo and group shows, cabarets, exhibits, screenings and so on," wrote Jane Farrow in a critique in *Girrly Pictures* on her 1994 return to the theatre scene in Toronto.[27] Torontonians, she thought, were really a bit spoiled. Today, you can go to www.gayguidetoronto.com or www.gaytorontotourism.com and quickly and confidently find exhaustive directories on just about any LGBT specializing services and facilities in Toronto and elsewhere, such as bars and clubs, baths, bookstores, cafés, clothing, florists, gift stores, gyms, hair salons, health centres, health stores, job search, lawyers, music stores, real estate agents, restaurants, sex stores, skin care, theatres, party planning, hotels & B&Bs, hostels, travel agencies, and community groups, along with downtown and "village" maps to direct you to gay-friendly areas and people. You can find dozens of websites for local and around-the-world interests, as well as gay web search engines. You can also find information on same-sex marriage, where to get your licence, and much more.

A point of interest: while virtually the rest of the world has come to use the acronym "LGBT," Toronto's websites resolutely use "GLBT." The reason, quite possibly, is that the Toronto acronym was developed first and was never changed.

Montreal hoteliers expect to sell as many as 88 000 room nights when the first World Out Games kick off there on 29 July 2006, and are expected to attract about 250 000 visitors over a 10-day span, generating $170 million in economic spinoffs. This event will be the largest cultural and sporting event in the world for lesbian, gay, bisexual, and transgender athletes and their supporters.[28]

Then there is Pride Week with its culmination in the Pride Parade, which large cities and many smaller ones present every year. From the first official event in 1981, in which 1500 people gathered for an afternoon celebration, the Toronto event is now the largest of its kind in Canada, and one of the largest in the world.

Institutional completeness requires that a **community** have developed a fairly complete set of basic social services, so that within it a member might live virtually all his or her life and have all his or her important needs met. It is relatively easy to show that most large cities now contain an elaborate set of gay-specific institutions and that smaller cities are developing more of them, experiencing "institutional elaboration," as Stephen O. Murray calls it in *American Gay*.[29]

But it hasn't always been so.

Development and Organization of Modern Gay Communities

The term **homosexual** was not coined until 1869, although one cannot suppose that no one ever noticed a pattern of same-sex sexuality or that there were no congregations of those seeking such pleasures before. According to Murray, what is distinctive about the "modern" (gay/egalitarian) cultural conception of social organization of homosexuality has four salient features:

1. awareness of **group distinctiveness** and a willingness to assert the legitimacy of that distinction
2. de-assimilation from the mainstream culture and the development of separate institutions to serve the community
3. primacy of egalitarian same-sex relationships rather than ones that involve marked age differences and submission of the young (as in ancient Greece) or **gender-role-bound** ones that imitate male/female roles (as in many Third World cultures)
4. exclusive same-sex relationships rather than bisexuality (as in most cultures where homosexual behaviour is tolerated or institutionalized)

Lesbian communities, Murray points out, are less fully developed on all four characteristics. He attributes this to "lesser numbers of those committed to and defining themselves as 'lesbian' (in contrast to gay men), to discrimination against women (especially to women unaccompanied by men in 'public'), and to primary socialization (which remains primarily women's work, though eradicating 'sissy' traits is more shared than most other childrearing tasks are)."[30]

Important factors making the formation of gay communities possible were first of all economic changes, long-term trends from farming and manufacturing to service occupations providing work for men and women seeking autonomy from their families. Then the greater geographical mobility provided by the "car culture" that followed World War II, permitting single men and women to move away from home to take new jobs and to express their sexuality. Also, the rapid growth of the welfare state created a sort of social and economic safety net that previously only families had been able to provide, allowing would-be gays more autonomy from the watchful eyes of their families and neighbours. And perhaps the ideological changes provided by the greater openness about sex fostered by the Kinsey report and also the anti-orthodoxy political climate on 1960s college campuses may have played a role in laying the groundwork for gay liberation. Once gays began to cluster in large cities, not only with an awareness of their own numbers and moral legitimacy, but also with a publicly defensible set of arguments for that legitimacy, the gay community began to attract more like-minded people simply by existing, and began the process of selective de-assimilation from the mainstream that continues to this day.

The Process of De-assimilation

At first, there were the gathering places, the bars, though preceded by friendship networks and private parties. Drinking together seems to represent a kind of solidarity, which creates a sense of social equality, undermining **socially constructed roles**. One important factor in the process of **de-assimilation** was the challenge made by gay men to the cultural stereotype of gay men as effeminate. Feminine appearance came to be regarded with suspicion and increasingly stigmatized both in lesbian and in gay male circles in the late 1970s, although the male and female circles hardly overlapped during that time of pronounced separatism. Relations ideally were between men aspiring to masculinity and between women aspiring to be neither masculine nor feminine. Gay men now cultivated an aura of masculinity as a concomitant of gay pride. Gay gyms became a new community institution and men began working out in order to try to become the sort of man they knew they were attracted to, assuming that he, in turn, would be attracted to them. Many gay men took on the "clone" look with well-built bodies, checked shirts and/or tightly fitting T-shirts, denim jeans (most often blue), and the most invariable elements of the look being cropped hair and neatly trimmed mustaches.

With several decades of continued development of the gay community, this self-presentation has been somewhat moderated by younger gays now coming out. With the negative stereotypes less pressing, perhaps they do not feel the need to resist them quite so assertively. A new term, **lipstick lesbians**, first appearing in the 1990s, refers to some lesbians countering the older model of **sexual inversion**, in which the lesbian is assumed to be really a man trapped in a woman's body. Lipstick lesbians, by contrast, assert their feminine side by adopting the dress and behaviour of feminine heterosexual women, often in high finance and upper management, and choosing not necessarily a **butch** partner but another **femme**.

The change from gay "exogamy" (sexual involvement with **straight trade** outside the gay community) to "endogamy" (sexual involvement with other self-identified gays within the community) seems to have been a key component of the ability to exist to some degree separately from mainstream culture and mostly in the company of other gays. This made for fewer pressures from the outside culture, and whatever were to be the natural ways of being gay could develop and flourish.[31] For many young gays, joining the gay community at first involves not so much learning how to be homosexual but, in order to develop an authentic sense of self, unlearning the false notions of how to be homosexual (the "homosexual social role") they had absorbed. The gay community is thought of by many to be a male phenomenon. This may be partly correct, because there is little reason to think there are as many lesbians as gay men. Multiple sources of research lead one to conclude that there are probably three or four self-identified exclusively homosexual men for every self-identified exclusively homosexual woman.

What data there are suggest that partners stand a better chance of staying together if they have relatively equal success in the world. It may be a failure on this count that tends to undermine lesbian couples, whose relationships tend to be more unstable than gay male or heterosexual relationships. Gay men and women also differ in their approach to sex outside the relationship. Gay men are relatively casual about sex outside the relationship—provided it doesn't mean anything. By contrast, lesbians tend to view sex outside the relationship as indicating a lack of commitment to the relationship or even betrayal. The greater stability of gay male relationships may be due in large part to this willingness to mutually and openly allow outside sex, while lesbians may break up over such behaviour.[32] Lesbians' view of sex outside the relationship corresponds closely to heterosexual women's view and may be, at least partly, due to upbringing. On the other hand, social evolution theory would imply that it is to the female's advantage to prefer a mate who will stay for the long haul to help nurture the offspring, which in the case of lesbians may or may not be relevant.

A Word About AIDS

Despite widespread belief to the contrary, Murray says he is doubtful that AIDS has caused there to be more gay couples now than previously, at least not more durable gay couples. Even before AIDS, some gay men were already losing enthusiasm for the fast-lane lifestyle, and by the early 1980s, members of the first wave of gay liberation had grown older and were about to slow down somewhat anyway.[33] He points out that there was never any evidence that going to bathhouses was a risk factor for contracting AIDS, and some evidence to the contrary. Most of the sexual acts at bathhouses were without significant risk. Nor, says Murray, has "professional" safe-sex education had significant impact. Most gay community gay men had already changed their behaviour, and the professional AIDS education has turned out to have little impact even now on preventing new cohorts of gay men from becoming infected, particularly those from minorities.[34]

Public Morality: How Will We Decide?

"What you don't know can't hurt you," right? Well, at least, "What you don't know, you don't know." So we haven't heard much commotion about sex clubs with group sex having recently been legalized. Except for the odd police raid, not many people are crying out about massage parlours, bawdy houses, or prostitution in general, unless of course it's visible on your street.

The Many Faces of Diversity

And whatever homosexuals did wasn't of huge concern to most people as long as they did it in the privacy of their bedrooms. But when it bursts into the streets, onto the public scene, it becomes fair game to be assessed and regulated by our combined moral code. And there's the rub. Morality is a fluid commodity. It changes—that is, we change it as we change. Once a fleeting glimpse of a maiden's ankle was cause for great excitement and embarrassed blushing. Today, women have won equality with bare-chested men by also being permitted to be topless on the beach and not to be arrested—so long, of course, as they weren't a nuisance or caused a disturbance. What significant criticism about moral issues there is comes mostly from churches or groups with religious affiliations, which prompts many people to discount it. But to discount anything without having evaluated it is no wiser and no more scientific than automatically accepting what is "politically correct."

The National Association for Research & Therapy of Homosexuality (NARTH), founded in 1992, is composed of psychiatrists, psychoanalytically informed psychologists, certified social workers, and other behavioural scientists, as well as laymen in fields such as law, religion, and education. They are a nonprofit educational organization dedicated to affirming a complementary, male-female model of gender and sexuality.[35] They offer, among other things, therapy to homosexuals wishing to end that lifestyle. NARTH recently reported on a street fair sponsored in part by the Human Rights Campaign and by National Gay and Lesbian Task Force, two very prominent groups committed to mainstreaming and normalizing homosexuality. The street fair "featured public whippings, body piercing, public sex, sado-masochism, and public nakedness by parade marchers. Fair booths sold bumper stickers that said, 'God masturbates,' and 'I Worship Satan,' and merchants peddled studded dog collars and leather whips (not for their dogs). On the sidelines of the public fair, a man dressed as a Catholic nun was strapped to a cross with his buttocks exposed, and onlookers were invited to whip him for a two-dollar donation."[36]

Now consider this: There have been instances of lesbians and gay men trying to reincorporate themselves into conventional morality by joining in denigrating various conceptions of unruly types who must be controlled and/or rejected by right-thinking "respectable" homosexuals. There is a long-running or recurrent split between those who demand that gay people be accepted as they are, in all their diversity, and those who believe that what makes straight people uncomfortable can and should be suppressed so that only homosexuals who are otherwise entirely "respectable" should exist—or at least that only conventional ones should ever appear in public.

I don't know the answer. But an unnamed philosophy lecturer at the University of Maryland writes, "I do know this: From my vantage point **Queer Subcultures** are a far healthier foundation from which to rebuild a just, improved society than is white, racist, sexist, homophobic Heterosoc [his term for heterosexual society]. To create the utopian just society I want would require the queering of Heterosoc."[37] He doesn't think enough heterosexuals have the proper "sensibilities" to function in such a society and so doubts it will happen. But for the present he counts himself fortunate that there is a rich queer subculture that is his home.

He does have a final confession, saying that he is too much of a subversive, transgressive, deviant individual to ever be happy being ordinary. "Thank God I am queer," he says. "Indeed, it is getting uncomfortably respectable to be queer, and I find that constricting and suffocating. Thank God I am a kinky **leather-fag** queer and that there is a sub-sub culture for my kind. That way I can avoid being homogenized into a socially acceptable, non-deviant, vanilla gay."[38] His biggest fear is that, with the strides queer liberation has made, being queer, too, will become "just common."

DETERMINING SEXUAL ORIENTATION

Determining a person's sexual orientation would at first glance seem to be a simple matter. We could just ask people about their inclinations or observe them in their choices of sex partners. For those who have always been and still are exclusively heterosexual or homosexual, that method might suffice. But what about people who have had a few same-sex experiences, perhaps experimenting in childhood or when they were teenagers? What about those who have had mostly same-sex attractions, but have attempted and even sometimes enjoyed relationships with the other gender? What about long-term prisoners who find themselves in situations where human warmth and contact can only come from a same-sex partner? What about those who reject such relationships and prefer loveless isolation? What about people who marry, have children, and perhaps many years later, much to their surprise and even dismay, find themselves in love with a same-sex partner? What about those who "knew all along"? What about gays and lesbians who may feel occasional **heteroerotic** attractions, and heterosexuals who experience sporadic homoerotic interests? What about the transgendered, who through dress, behaviour, or body modification, display attributes which are different from the social and cultural expectations of their gender? And what about transsexuals, who experience an incompatibility between their anatomical and psychological sex, many of whom, in order to try to resolve the conflict, undergo hormone therapy followed by **gender reassignment surgery**? These are some of the factors that have plagued and continue to plague conscientious researchers in sexuality.

Kinsey's Seven-Point Continuum of Sexual Orientation

Researching in the 1940s, it certainly appeared to Kinsey that people might have varying degrees of heterosexual and homosexual feelings and experiences. Kinsey and his colleagues recognized the blurry boundaries and proposed a continuum of sexual orientation rather than the previously accepted mutually exclusive and opposite poles (see Figure 9.1).

Figure 9.1   Two Models of Sexual Orientation

1. Original Two-Point Typology

Heterosexual ◄─────────── or ───────────► Homosexual

The typology model places heterosexuality and homosexuality on polar opposites and mutually exclusive points.

2. Kinsey's Seven-Point Continuum

0	1	2	3	4	5	6
Exclusively heterosexual experience	Mostly heterosexual, with incidents of homosexual experience	Primarily heterosexual, with substantial homosexual experience	Equal amounts of heterosexual and homosexual experience	Primarily homosexual, with substantial heterosexual experience	Mostly homosexual, with incidents of heterosexual experience	Exclusively homosexual experience

Kinsey's model proposes a continuum between heterosexuality and homosexuality. People in category 0, who accounted for the majority of Kinsey's subjects, were considered exclusively heterosexual. People in category 6 were considered exclusively homosexual.

Source: A.C. Kinsey, W.B. Pomeroy, and C.E. Martin, *Sexual Behavior in the Human Male* (Philadelphia: Saunders, 1948). Reproduced with the permission of the Kinsey Institute for Research in Sex, Gender, and Reproduction, Inc.

This seven-point continuum ranges from zero, exclusive heterosexuality, to six, exclusive homosexuality, and includes five graded levels in between, the midpoint being bisexuality. In spite of many known, and at the time unresolvable, shortcomings in Kinsey's research methods, his continuum view of sexual orientation continues to influence today's understanding of sexuality.

The Klein Sexual Orientation Grid

Fritz Klein points out that Kinsey did not separate psychological reactions from overt experiences. And, Klein thought, there is a large difference between thought and action, between fantasy and experience. He wanted to test his idea that sexual orientation was a dynamic, multi-variable process, so he developed the Klein Sexual Orientation Grid (KSOG).[39] He saw an individual's sexual orientation as composed of sexual and nonsexual social variables, all of which can change over time. For him, the definition of sexual orientation must take into account seven distinct variables he labelled A, B, C, D, E, F, and G:

A. *Sexual attraction*, which is not synonymous with sexual behaviour. You can be attracted to one gender and yet have sex with the other. Reference is made to Anna Freud's view that the sex of one's masturbatory fantasies is the ultimate criterion in homo- or heterosexual preference.

B. *Sexual behaviour* looks at actual behaviour as opposed to sexual attraction. With whom do you have sex?
C. *Sexual fantasies* can occur while daydreaming, during masturbation, as part of our real lives, or purely in our imagination. Over a period of time, changes occur, sometimes radically during the course of one's adult life.
D. *Emotional preference* differentiates this aspect of sexual orientation from the previous three sexual variables. Some people prefer to have sex with one gender but are emotionally involved with the other. Our emotions directly influence the actual physical act of love.
E. *Social preference* refers to the degree to which you like to socialize with members of each sex. You may love only women, but spend most of your social life with men, or vice versa.
F. *Heterosexual/homosexual lifestyle* refers to the degree to which you like to socialize with members of your own sex and to what degree with the other sex.
G. *Self-identification* is one's own view of one's orientation. It is a strong variable, since self-image strongly affects one's thought patterns and behaviour.

See Figure 9.2. In this grid, one chooses three numbers for each variable (A, B, etc.), one for each of three aspects of your life: your past (up to a year ago), your present (preceding year up to now), and your ideal (as if it were a matter of choice or will). For some, these three numbers on each variable will be different, for others they may be the same. Taking all of the scales or grid locations as a whole gives a picture of one's sexual orientation over time. If one wants a single averaging number for all seven variables for each of the three time spans, total each column and divide each total by seven.

Sexual identity—how people think of themselves—sometimes has little to do with their sexual behaviour. For example, three different people may have the same distribution of sexual behaviour in the past and/or present, but have three different sexual identities: heterosexual, bisexual, or homosexual. Also, people's identity, behaviour, or fantasies may change over time. There seems to be significant fluidity in self-identification. As you reflect on any fluidity in your own ratings, consider how your particular self-identification and self-understanding might have been valid for you at each particular time in your life.

The Laumann Study

There have been many sex surveys over the past few decades, but the problem is that until very recently virtually all were methodologically flawed, making their data unreliable, uninterpretable, and impossible to use to understand sexual behaviour. The Laumann study, however, was based on a survey of a **statistically representative sample**, 3432 completed

Figure 9.2   The Klein Sexual Orientation Grid

	Variable	PAST	PRESENT	IDEAL
A	Sexual Attraction			
B	Sexual Behaviour			
C	Sexual Fantasies			
D	Emotional Preference			
E	Social Preference			
F	Heterosexual/Homosexual Lifestyle			
G	Self-Identification			

For variables A to E:

1 = Other sex only
2 = Other sex mostly
3 = Other sex somewhat more
4 = Both sexes
5 = Same sex somewhat more
6 = Same sex mostly
7 = Same sex only

For variables F and G:

1 = Heterosexual only
2 = Heterosexual mostly
3 = Heterosexual somewhat more
4 = Hetero/Gay-Lesbian equally
5 = Gay-Lesbian somewhat more
6 = Gay-Lesbian mostly
7 = Gay-Lesbian only

Source: Fritz Klein, *The Bisexual Option*, 2nd ed. (Binghamton, NY: Haworth/Harrington Park Press, 1993), p. 19. Reproduced with the permission of the publisher and author, Fritz Klein, who can be reached at fritzklein@cox.net.

interviews of American adults between the ages of 18 and 59, and is universally recognized as definitive. Since its publication in 1994, numerous large-scale epidemiological surveys conducted in all the English-speaking and many other industrialized nations have repeatedly confirmed and strengthened its findings.

One of the major points of the Laumann study, which the authors themselves did not expect, is that "homosexuality" as a fixed trait scarcely even seems to exist.[40] Most Americans believe that the factors that determine their sex lives lie mostly or solely within themselves. Their sexual drives, their hormones, their individual desires, are all that matter. This is in large part due to the long history of trying to study and control sexuality, dating back to studies in the past century that focused on "deviants" and sex criminals. So this long history of attempts to study sexuality had as a dominant theme this idea that sexuality comes from within, that it is a feature of the individual, and that to understand sexual behaviour we have to understand the individual's sex drives and hormonal surges and even genetic predisposition. This shaped the popular explanations of sexual behaviour and the belief that the individual is the sole actor on the sexual stage.[41]

However, the findings present a very different viewpoint. The authors of the *Sex in America* study say that they are convinced, and the data bear them out, that "sexual behaviour is shaped by our social surroundings. We behave the way we do, we even desire what we do, under the strong influence of the particular social groups we belong to. We do not ... have all the choices in the world when we decide what to do in bed. The choices we make about our sex lives are dramatically affected by our social circumstances."[42]

Only 2.8 percent of men who had a sex partner in the past 12 months had exclusive male partners; 0.6 percent had both male and female sex partners. That leaves 96.6 percent of men having exclusively female sex partners. Of the women who had a sex partner in the past 12 months, only 0.4 percent had exclusively female partners; 0.2 percent had both male and female sex partners. That leaves 99.4 percent of women having exclusively male sex partners. Also, the study points out three significant reasons why one cannot so simply say that a person is or is not gay. First, people often change their sexual behaviour during their lifetime, making it impossible to state that a particular set of behaviours defines a person as gay. Second, there is no one set of sexual desires or self-identification that uniquely defines homosexuality. And third, homosexual behaviour is not easily measured. Even though the recent struggles of gay men and lesbians to gain acceptance have had an effect on the public's mood, the history of persecution has a lasting effect both on what people are willing to say about their sexual behaviour and on what they actually do.[43]

FACTORS INVOLVED IN PRODUCING SEXUALITIES

Identifying a person's gender is far more complex than most people imagine. There are no absolutes in nature, only statistical probabilities. We all begin life with a common anatomy, which begins to differentiate if there is a Y chromosome present. This activates the production of testosterone, appropriate receptors in the brain, and the formation of the testes. Genetic, physical, and hormonal gender complexities occur in an estimated 1 in every 60 persons.[44] Now, the question often asked is "What causes homosexuality?" The question should really be "What causes sexual differentiation of any kind?" Is it biological or does it result from some sort of exceptional childhood experience? Much research has been done, but so far no single biological factor has been found to be clearly and irrefutably associated with variations in human sexual orientation. Theories proposed are neurohormonal differentiation, differences in portions of the brain, genes, finger length, inner ear differences, eye blinking, and more. And even though the Laumann study has found social factors to be key, specific theories such as distant or absent fathers, dominant mothers, childhood sexual trauma, or dysfunctional families have proved inconclusive as blanket explanations for variations in human sexual orientation.

No matter which new theories are expounded at any given time, there will be relief and disappointment or both on the part of heterosexuals, homosexuals, bisexuals, and others. In fact, what has been driving the interpretation and focus on various theories is precisely what reactions were expected by which side. Would there be relief or disappointment? Would it lead to greater harmony and peace or to more confrontations and protest marches? If, however, you are interested in pursuing the bewildering complexity of the various studies that have been described, and are fully prepared not to emerge with any satisfying and clear conclusion, see the regularly updated website of Simon LeVay, a British-born neuroanatomist, at http://members.aol.com/slevay/page22.html. LeVay is an extremely knowledgeable, remarkably unbiased gay man, searching for the truth, whether it turns out to be what he would prefer or not.

In the final analysis what is needed is a deeper and more widespread societal understanding of the struggles of homosexuals, increased compassion for the hurts they have encountered, and decreased hostility. Tolerance and respect reduce fear and anxiety, making it more possible to do good research and to find, share, and accept the truth whatever it may be.

CHAPTER SUMMARY

Beginning with the 1962 introduction and free availability of "The Pill" in Canada—the first reliable birth control in history (aside from abstinence) and one over which women had control—the entire social-sexual landscape erupted and began to inexorably change. The iron grip that reproduction had on sexuality was forcibly loosened, and sex for pleasure presented itself for consideration. This began to level the playing field by the reconsideration of the worth and value of the various activities of the various sexualities. The claim of superiority of heterosexual sex was left as, at best, a partial claim.

Human rights, not just in Canada but also around the world, took on greater importance and stricter and more consistent enforcement. Inevitably, sexual rights came to be seen as human rights, and the crusade for not only tolerance and acceptance but also recognition and respect reached critical mass. Sexual orientation was added to the list of identifiable groups to be accorded special protection under the law. Same-sex marriage followed, and Canada was one of the first countries in the world to enshrine it in law. Same-sex parenting came more into focus. It was not new because lesbians and gay men had long been producing babies from previous heterosexual relationships or from other arrangements. But concern by religious and conservative groups raised the profile of such arrangements. The development and organization of modern gay communities and their self-sufficiency with institutional completeness, clustering more in cities and larger towns, was discussed.

Current views of diversity in sexualities was looked at through the instruments of Kinsey's seven-point continuum and the Klein Sexual Orientation Grid. The KSOG used a similar seven-point analysis to Kinsey's but added social aspects and change-over-time processes. The extreme complexity of factors involved in producing the various sexual orientations and gender issues were considered. At this time, there are no rock-solid theories about what is responsible for the variations. What seems to be emerging, however, especially as hostilities among the factions seem to be easing, is that probably a number of factors, both biological and social-environmental, will be seen to play a role in how sexuality develops in any given individual. Better research with carefully designed methodologies and complete and unbiased interpretation, together with the willingness to accept the findings, will go a long way toward a better understanding of all of our sexualities. Compassion and respect are crucial in the positive resolution for all people.

KEY TERMS

bestiality, p. 251
bisexual, p. 242
bondage, p. 251
butch, p. 258
civil partnerships, p. 249
civil unions, p. 249
come out (of the closet), p. 243

community, p. 256
de-assimilation, p. 258
domestic partnerships, p. 248
femme, p. 258
gay, p. 242
gay marriage, p. 249
Gay Pride Parade, p. 251

gender reassignment surgery, p. 261
gender-role-bound, p. 257
group distinctiveness, p. 257
hate propaganda, p. 244
heteroerotic, p. 261
heterosexual, p. 242
homoerotic, p. 255
homophobic, p. 245
homosexual, p. 257
institutional completeness, p. 256
institutionalized discrimination, p. 255
leather-fag, p. 261
lesbian, p. 242
lipstick lesbians, p. 258
Omnibus Bill, p. 243
outed, p. 245
pedophilia, p. 244
queer subcultures, p. 260
sadomasochism, p. 251
same-sex marriage, p. 245
same-sex parenting, p. 252
sexology, p. 242
sexual inversion, p. 258
sexual orientation, p. 242
sexual predator, p. 244
sexuality, p. 241
socially constructed roles, p. 258
sodomy, p. 242
statistically representative sample, p. 263
straight trade, p. 258
transgender, p. 242
transsexual, p. 242

DISCUSSION QUESTIONS

1. With a partner, discuss whether sexual rights qualify as human rights. What about the rights of priests or the Pope? Can one voluntarily give up a human right?
2. If you were the judge in a "hate propaganda" case and had to decide the fate of a member of the clergy who spoke out against homosexual practices, how would you decide?
3. Gay men who are in committed same-sex relationships or who are married generally agree to allow outside sexual encounters. Do you think men in heterosexual marriages should negotiate for the same arrangement with their wives? Discuss this with your current partner and report your conclusions to the class.
4. If you and your partner felt forced to give up an unplanned newborn baby because you are not in a position to care for it, would you want a same-sex (male or female) couple to adopt it? What would be your preference? Why?
5. Homosexual communities, especially in larger towns or cities, tend to de-assimilate in order to conduct most of their lives within their own community. How do you think this is similar to or different from other communities, for example, ethnic communities, wealthy communities, etc.?

6. Assuming that you at this moment consider yourself to be exclusively heterosexual, whether you have tested yourself by having sex or not, try to imagine having sex with a member of your own sex. Could you imagine it more easily with one rather than another of your same-sex friends? Not to say that you ever would, but if the gun were to your head, which one would it be? Note your reaction—physically, emotionally. Report to your small group or whole class.

NOTES

1. Susan Block, "Sex & Technology," *Dr. Susan Block's Journal*, available <www.drsusanblock.com/editorial/sextech.htm>, accessed 14 May 2006.
2. "Universal Declaration of Human Rights," United Nations site, 2004, available <www.un.org/Overview/rights.html>, accessed 18 April 2006.
3. "Indepth: Same Sex Rights," CBC News Online, 29 June 2005, available <www.cbc.ca/printablestory.jsp>, accessed 18 April 2006.
4. Arthur Lewis, "Clip: The First Gay March," TV broadcast, 28 August 1971, CBC Archives, available <http://archives.cbc.ca/400i.asp?IDCat=69&IDDos=599&IDCli=3227&IDLan=1&No>, accessed 18 April 2006.
5. "Chapter 14, An Act to Amend the Criminal Code (Hate Propaganda)," *Canada Gazette*, available <http://canadagazette.gc.ca/partIII/2004/index-e.html>, accessed 20 February 2006.
6. "Bill C-250 Sparks Protest Rally in Ottawa," Focus on the Family (Canada) Association site, 20 April 2004, available <www.fotf.ca/familyfacts/tfn/2004/042004.html>, accessed 7 October 2005.
7. John Pacheo, "The Problems with Bill C-250," The Catholic Legate site, 11 April 2004, available <www.catholic-legate.com/articles/billc250.html>, accessed 14 May 2006.
8. "NDP MP Svend Robinson," *REALity* 23(1) (January/February 2004), available Real Women of Canada site <www.realwomenca.com/newsletter/2004_jan_feb/article_3.html>, accessed 14 May 2006.
9. "Amnesty International Brief on Bill C-250," Amnesty International Canada site, available <www.amnesty.ca>, accessed 7 October 2005.
10. "Court of Appeal for Ontario—*Halpern et al. v. Attorney General of Canada et al.*," June 10, 2003, Guide to Ontario Courts site, available <www.ontariocourts.on.ca/decisions/2003/june/halpernC39172.htm>, accessed 11 February 2006.
11. "Indepth: Same Sex Rights."
12. House of Commons of Canada, "Bill C-38: An Act Respecting Certain Aspects of Legal Capacity for Marriage for Civil Purposes," Parliament of Canada site, available <www.parl.gc.ca/PDF/38/1/parlbus/chambus/house/bills/government/C-38_1.PDF>, accessed 14 May 2006.
13. "Tories Talk Same-Sex in Fall," *Windsor Star*, 6 April 2006, p. B1.
14. "Court of Appeal for Ontario—*Halpern et al. v. Attorney General of Canada et al.*"
15. "Mum Staying Mum about Gay Wedding," *Windsor Star*, 20 December 2006, p. B9.

16. James Gerstenzang, "On a Gray Day, Same-Sex Couples Make Colorful Point at Egg Roll," *Los Angeles Times*, available <www.latimes.com/news/printedition/asection/la-na-eggroll18apr18,1,2548213.story>, accessed 14 May 2006.
17. "White House Easter: Gay Friendly?" CBS News site, 13 April 2006, available <www.cbsnews.com/stories/2006/04/13/national/main1496408.shtml>, accessed 14 May 2006.
18. John M. Glionna, "A Haven's Sex and Sensibility," Column One, *Los Angeles Times*, 21 April 2006, available <www.latimes.com/news/local/la-me-castro21apr21,0,4320735.story?coll=la-home-headlines>, accessed 14 May 2006.
19. Ibid.
20. Ibid.
21. William Meezan and Jonathan Rauch, "Gay Marriage, Same-Sex Parenting, and America's Children," *Marriage and Child Wellbeing* 15(2): 97–115, available <www.futureofchildren.org/usr_doc/Marriage_vol15_no2__fall05.pdf>, accessed 14 May 2006.
22. APA Council of Representatives, "APA Policy Statement: Sexual Orientation, Parents, & Children," American Psychological Association, APA [American Psychological Association] Online, 28 & 30 July 2004, available <www.apa.org/pi/lgbc/policy/parents.html>, accessed 14 May 2006.
23. William Meezan and Jonathan Rauch, "Gay Marriage, Same-Sex Parenting, and America's Children," *Marriage and Child Wellbeing* 15(2): 97–115, available <www.futureofchildren.org/usr_doc/Marriage_vol15_no2__fall05.pdf>, accessed 14 May 2006.
24. American College of Pediatricians, "Homosexual Parenting: Is It Time for Change?" American College of Pediatricians site, 22 January 2004, available <www.acpeds.org/?CONTEXT=art&cat=10005&art=50&BISKIT=2325235577>, accessed 14 May 2006.
25. Judith Stacey and Timothy J. Biblarz, "(How) Does the Sexual Orientation of Parents Matter?" *American Sociological Review* 66(2) (April 2001): 159–83.
26. Kelley O. Beaucar, "Homosexual Parenting Studies Are Flawed, Report Says," FOXNews.com, 18 July 2001, available <www.foxnews.com/printer_friendly_story/0,3566,29901,00.html>, accessed 14 May 2006.
27. Jane Farrow, "Tits, Tats and Testes," *Girrly Pictures: A Series of Five Exhibitions in the Project Room*, Canada's Digital Collections site, available <http://collections.ic.gc.ca/mercer/353.html>, accessed 14 May 2006.
28. "LGBT Event Good News for Montreal Hoteliers," *Windsor Star*, 20 March 2006, p. B1.
29. Stephen O. Murray, *American Gay* (Chicago: University of Chicago Press, 1996), p. 191.
30. Murray, p. 2.
31. Murray, p. 192.
32. Murray, p. 174.
33. Murray, p. 177.
34. Murray, pp. 120–123.
35. National Association for Research and Therapy of Homosexuality (NARTH) site, available <www.narth.com/>, accessed 4 January 2006.
36. Joseph Nicolosi, "Why Reveal the Dark Side of the Gay Movement?" National Association for Research and Therapy of Homosexuality (NARTH) site, 21 September 2004, available <www.narth.com/docs/whyreveal.html>, accessed 14 May 2006.

37. "Gay & Lesbian Philosophy," course description, University of Maryland, Fall 1995, available <http://carnap.umd.edu/queer/1995Lectures.html>, accessed 14 May 2006.
38. Ibid.
39. Fritz Klein, *The Bisexual Option*, 2nd ed. (New York: The Harrington Press, 1993), pp. 16–19.
40. Edward O. Laumann, John H. Gagnon, Robert T. Michael, and Stuart Michaels, *The Social Organization of Sexuality: Sexual Practices in the United States* (Chicago: University of Chicago, 1994).
41. Robert T. Michael, John H. Gagnon, Edward O. Laumann, and Gina Kolata, *Sex in America: A Definitive Survey* (Boston: Little, Brown, 1994), p. 16.
42. Ibid.
43. Op. cit., p. 172.
44. Judith Mackay, *The Penguin Atlas of Human Sexual Behavior* (Penguin, 2000).

CHAPTER 10

Diversity in Canadian Families: Traditional Values and Beyond

Geoff Ondercin-Bourne

There is no golden age of family life to long for, no past pattern that, if we only had the moral will to return to, would guarantee us happiness and security. Family life is always bound up with the economic, demographic, and cultural predicaments of specific times and places.
— Arlene Skolnick, *Embattled Paradise*

Family: *from the Latin, familia, "the slaves of a household"*
— *Random House Dictionary of the English Language*

Objectives

After reading this chapter, you should be able to

- define family in a way that accounts for the many variations of families in Canada today
- critically assess the underlying assumptions of the "traditional family"
- compare several theoretical approaches used to understand the evolution of families and relations among family members
- describe recent changes in Canadian families
- link the changes in familial structures and relationships to broader changes in society

INTRODUCTION

The family, we are told, is an institution in crisis. News headlines announce, "New Megatrends Reflect Family Decline."[1] Social service agencies such as Children's Aid with support from the courts restrict the use of corporal punishment by traditional parents. So-called "special interest groups" lobby governments on behalf of nontraditional families such as single parents or gays and lesbians. In June 2005, the federal government passed Bill C-38, which acknowledged the constitutionality of same-sex marriage. As the conservative lobby group, Focus Canada, laments, "With all the attacks now aimed at the natural family, is it any wonder that the majority of Canadians feel that 'the state of the family is a national crisis?'"[2]

If the family is now "under siege," then presumably there was a golden age when the traditional family thrived. This golden age was represented on television in the 1950s and early 1960s by such programs as *Leave It to Beaver*, *Father Knows Best*, and *Ozzie and Harriet*. Arlene Skolnick opened her book on the family with the tongue-in-cheek title "Who Killed Ozzie and Harriet?" This title suggests the apparent deterioration of modern family life, a "Paradise Lost," if you will. More recent television shows, like the top-rated *Everybody Loves Raymond* and *Gilmore Girls*, have provided more up-to-date versions of family life that nonetheless retain many of the traditional values associated with that same "golden age."

As the title of this book implies, society can best be understood through its diversity. In this diversity the very notion of the family is misleading at best. Families are evolving, as they always have, in response to social, political, and economic change. Consequently, to answer Skolnick's tongue-in-cheek question, no one killed Ozzie and Harriet, because in fact they never existed. The traditional family is an idea or belief, not a fact.

In this chapter we will (1) provide some definitions of families and outline the difficulties of doing so, (2) explain some theoretical approaches to the study of families, making a key distinction between what we describe as unification theories and liberation theories, and (3) look at how families in Canada have changed, as well as the challenges facing Canadian families in the twenty-first century.

CRITICAL THINKING BOX 10.1

In today's multi-channel universe, you can still see many of the TV shows your parents (and grandparents!) watched. Compare the family life portrayed in some of those shows with that portrayed in your favourite, current family TV show. How have family relationships on TV changed since the early days of television? In what ways have they remained the same?

DEFINING FAMILY

Traditional Definitions

Definitions of family have generally been based on the inclusion or exclusion of individuals from the group; either you are in, or you are out. An example is George Murdock's definition, quoted in many sociological studies of family: "a social group characterized by common residence, economic cooperation, and reproduction. It includes adults of both sexes, at least two of whom maintain a socially approved sexual relationship, and one or more children, owned or adopted, of the cohabiting adults."[3] A more recent definition, by Rose Laub Coser, is clearly influenced by Murdock: "[The family] finds its origin in marriage; it consists of husband, wife, and children born in their wedlock, though other relatives may find their place close to this nuclear group, and the group is united by moral, legal, economic, religious and social rights and obligations."[4]

The traditional definitions highlight the importance of the biological function of families. Families that consist of a mother, a father, and their children are called **nuclear families**. However, does such a definition adequately describe the modern family? For example, can there be families that are not based on producing children?

Some critics see many limitations with the traditional concept of the nuclear family. For example, by defining family in biological terms, we seem to legitimize the notion of family as a "natural" unit. The implication here is that any relationship or grouping that falls outside this exclusive definition is necessarily "unnatural," whether or not it is generally accepted in society as a family. Among those excluded are childfree couples, same-sex couples with or without children, single-parent families, commuting families, and remarried families.

Finally, let's look at the definition used in Canada's census, which, although more inclusive than the traditional definitions, is still heavily influenced by them. According to Statistics Canada, "census families include married couples and common-law couples with or without never-married children living at home, as well as lone-parent families."[5] However, as we shall see, even this definition omits many living arrangements that are considered "families" in today's society.

Two further issues raise questions about traditional definitions of family. For one, cultural differences are not factored into them. To what extent, for example, do North American Aboriginal families meet the criteria of the nuclear family? Does the structure of families immigrating from all over the world resemble that in the definitions of Murdock and Coser?

If we accept a definition of family based on the nuclear family, we may be marginalizing the impact of the **extended family** in many cultures, including our own. An extended family comprises two or more nuclear families joined together through blood ties. The classic example of an extended family is a husband and wife, their unmarried children, their married children, and the spouses and children of their married children. By focusing solely on the nuclear components that make up an extended family, the analysis ignores the unique character of that particular set of relationships. On the basis of these examples, it is hard to imagine a realistic "one size fits all cultures" definition of family.

The second issue is the changing nature of family. How applicable are the definitions of nuclear family to the families of the past? Consider the assumption of parents and children living under one roof. In Canada, this has not always been the case for everyone. For example, in the middle of the 18th century, many children in New France, some as young as five or six, were forced to work as live-in household servants. Were these children no longer part of a nuclear family?

The point is that uniform, fixed definitions of family are incomplete, and even misleading. Increasingly, the family is regarded as "a social construct."[6] Consequently, the mutability of the structure of family must be an important component of its definition. Later in this chapter, we will discuss the ways in which the structure of the Canadian family has adapted to its changing socioeconomic environment.

The Dimensions of Family Life: An Inclusive Approach

If the traditional definitions of family are incomplete, how do we at least determine some key characteristics of "family"? Margrit Eichler attempted to answer this question by developing a series of concepts she referred to as "internal dimensions."[7] We will focus here on five of these: the procreative dimension, the socialization dimension, the residential dimension, the economic dimension, and the emotional dimension. Box 10.1 describes some of the main elements of each.

This list is by no means complete, but even the five dimensions included here do provide a framework for analyzing the diversity of Canadian families. When we compare Eichler's dimensional approach with the traditional definitions offered by Murdock, Coser, and even Canada's census, we see an important distinction: Eichler viewed families as a set of relationships, whereas Murdock and Coser defined them as a fixed set of characters. Eichler's perspective allows us to look at "the family" as a dynamic institution that can account for a wide range of configurations. We will therefore use a dimensional analysis to understand Canadian families and the changes they have undergone, particularly since the 1950s.

Next, we will examine some theoretical perspectives that sociologists have used to study families. These theories enable us to determine how families affect, and are affected by, other social institutions and social relationships.

> **BOX 10.1**
>
> *Five Internal Dimensions of the Family*
>
> **Procreative Dimension**
> - Does the couple have children?
> - If so, are they from the current relationship or a previous one?
>
> **Socialization Dimension**
> - Are both parents, one parent, or neither parent involved in childrearing?
>
> **Residential Dimension**
> - Do all family members live under the same roof?
> - Does one family member live in a separate dwelling only?
> - Does one family member have an additional separate dwelling?
>
> **Economic Dimension**
> - Is one family member solely responsible financially for the other members?
> - Do two or more family members share the financial responsibility for other members?
> - Are family members financially responsible for themselves?
>
> **Emotional Dimension**
> - Is the positive involvement of family members mutual or one-sided?
> - Is there a lack of involvement from family members?
> - Is there mutual negative involvement?
>
> *Source:* Margrit Eichler, *Families in Canada Today* (Toronto: Gage, 1983). Reproduced with permission of the author.

THEORETICAL PERSPECTIVES ON UNDERSTANDING DIVERSITY IN FAMILIES

Why do we consider theory? Why not just "look at the facts" and forget about theoretical questions and debates? The answer is that theories are useful models that help us make sense of the complexities of the real world. As Emily Nett explained: "To omit theory would be to reduce knowledge about families to a series of simple statistics or journalistic accounts at one extreme, and a compendium of value judgements, like a sermon, at the other. . . . Theory relates concepts and provides a basis for asking questions, finding answers, or doing both."[8]

Having acknowledged the importance of theory, we are faced with a dilemma. Given the limitations of space, how do we cover the many theorists who have written about families? In fact, we cannot. As a result, we have selected some of the theories that have appeared consistently in the literature on the family.

In this chapter, we divide theories of family into two groups, based on the analysis by David Cheal, a family sociologist:[9] (1) unification theories emphasize the universality of family structures and the benefits of family over its costs, (2) liberation theories emphasize both the diversity of families and the limitations that families place on individual members, as well as the strategies used to overcome these constraints. The debate between unification theorists and liberation theorists is central to our discussion about the nature of family.

Unification Theories

Following World War II, sociological analysis in the United States and Canada, including analysis of the family, was dominated by several schools of thought that were all variations of the unification approach. Although some of these theories had appeared earlier in the 20th century, it was during the prosperous decades of the postwar boom that they enjoyed immense popularity.

Unification theories of the family hold the view that the family is an adaptive unit mediating between the individual and society; the family meets the needs of individuals for personal growth, development, and physical and emotional integrity. To the extent that the family meets an individual's social needs, it is seen as "functional."

Since the 1960s, however, competing schools of thought have put forward alternative perspectives on family life that have generated a lively debate on the family and its role in modern society. We will describe in detail two of the predominant sociological perspectives within the unification school—structural–functionalism and general systems theory—and offer brief summaries of two other perspectives that have been used by unification theorists.

Structural–Functionalism

Talcott Parsons, G.P. Murdock, W.F. Ogburn, W. Goode, and B. Schlesinger are well-known and influential structural–functionalists who wrote about families. **Structural–functionalism** argues that the family is an important institution that maintains social stability. Family members have many and diverse normatively prescribed activities that become integrated into a dynamic system called family. That is to say, family members behave according to the prescribed family norms of the society in which they live. Talcott Parsons, the sociologist most closely associated with this theory, saw the family as performing three particular functions that contribute to social harmony and integration: reproduction, socialization of new members, and emotional support.

As society changes, institutions, including the family, adapt to new social realities. For example, before industrialization and urbanization, families played a central role in educating their own children. They performed as a "unit of production" as well as a "unit of consumption." These families grew much of their own food, produced some of the implements used in their daily lives, and in some cases built their own houses. With all these responsibilities, extended families were important for a family's well-being.

As industries grew and people moved off the land to the cities, families became essentially units of consumption only. Institutions such as schools began to take over some of the social functions of families. A highly mobile family unit meets industrial capitalism's need for a flexible workforce. Therefore, according to Parsons, the smaller nuclear family was ideal. The extended family declined in importance in favour of the more independent nuclear family. Fewer familial commitments, argued Parsons, meant that workers were less preoccupied with concerns of family members and more with productivity. Thus, the breakdown of the extended family, in structural–functional theory, served the needs and interests of the modern capitalist economy. The family adapted. The isolated **conjugal family**, which superseded the extended family, was based on marriage, and it included a "breadwinner," usually the father, a homemaker, usually the mother, and the children conceived from that marriage. In the view of the structural-functionalists, this family type became the norm. It was, to use their terminology, "functional."

If such a familial arrangement is seen as functional, however, then those arrangements that do not conform to this model, or attempt to alter it, must, by definition, be dysfunctional. Structural–functionalists have not dealt with conflict and change, nor have they considered the dynamic nature of interpersonal relations. Change is seen as disruptive, and individual opposition to social pressure as "deviance." But, regardless of the criticisms that have been levelled at it, structural–functionalism has played a major role in determining how we conceive of families in our society.

General Systems Theory

General systems theory, also referred to simply as systems theory, analyzes the family as a total system that has an impact on all its members. This theory became popular in the 1960s, and although some ideas associated with it have been severely criticized since the 1970s, it is still being used in psychiatry, psychology, and family therapy. Of the many sociologists who have made important contributions to family studies using systems theory, the most influential is Reuben Hill.[10] According to Hill four qualities make it possible to study the family as a system:

1. Family members occupy various interdependent positions, that is, a change in the behaviour of one member leads to a change in the behaviour of other members.

2. The family is a relatively closed, boundary-maintaining unit.

3. The family is an equilibrium-seeking and adaptive organization.
4. The family is a task-performing unit that meets both the requirements of external agencies in the society and the internal needs and demands of its members.

Qualities 3 and 4 illustrate some of the similarities between structural–functionalism and systems theory. With reference to the first quality, Maureen Baker explained the emphasis of systems theory on recurring behaviour that is triggered by similar and interdependent conditions or responses. Baker gave the example of the cyclical nature of family violence, pointing out that children who grow up in a violent home environment are more likely to be abusive when they are parents.[11]

With respect to the second quality, systems theorists do not universally accept the idea that a family is relatively closed, although there is agreement that it is a "boundary-maintaining unit." In a study by Montgomery and Fewer, it was argued that families differ along a continuum of "relative openness" and "relative closure" that acknowledges some diversity in family behaviour. A family's relative openness "refers to the degree to which a mindful system is receptive to information."[12] For systems theorists, then, the boundaries that families maintain are in most cases permeable.

Perhaps the most controversial belief of systems theorists is their rejection of the idea of "causes of behaviour." As Montgomery and Fewer explained: "In systems theory, there is no cause, since behaviour is interactional and processual and has no discernible beginning."[13] Rather than cause, systems theory looks at behaviour in terms of "fit." Causal explanations of behaviour are viewed as inadequate because of three assumptions they make: (1) there is only one possible response to a given action, (2) the receiver of a particular action is incapable of generating alternative responses, and (3) the receiver has no impact on the person who commits the action. This line of reasoning ignores other factors that have an impact on behaviour. The following example will illustrate this deficiency.

If Driver A is cut off by an aggressive and inconsiderate Driver B, Driver A might respond by giving Driver B "the finger," to use the vernacular. However, according to systems theory, Driver A has other options. He or she could simply ignore Driver B. (If they are driving on a freeway in Los Angeles, Driver A might pull out a gun and shoot Driver B, as has actually happened on occasion.) Driver A's response, however, is not caused simply by Driver B's action. The choice made by Driver A might be based on his or her own background and personality. Furthermore, Driver B's action might be prompted by something that Driver A has done, driving too slowly, for example. Behaviour is too complex to be attributed to a single cause.

As already mentioned, systems theory is applied regularly in family therapy, as well as in psychiatry and psychology. But sociologists from other schools of thought have criticized this approach. For example, systems theory's approach to spousal abuse, as illustrated by Montgomery and Fewer, suggests that the abused are at least partly to blame for their own abuse, because there is no "cause" that can be attributed solely to the abuser. The "circular pattern" of behaviour identified with systems theory means that "a problem within a family is not attributed to one individual as its instigator, but rather the problem is seen as being sustained by a continuous, circular interaction process among all family members."[14] Consequently, the victim must accept some of the blame, which draws attention away from the abuser. As you can imagine, this analysis has drawn considerable fire from researchers and frontline workers dealing with the problem of abuse, usually that of women by men. We will refer to this problem when we discuss liberation theories.

Liberation Theories

Liberation theories of family emerged in the 1960s and 1970s as people became increasingly skeptical of perceived "traditional" family roles. This growing skepticism of the traditional family was part of the general demand for changes in many social institutions and values, changes that invariably meant more freedom from the restrictions of traditional norms and values.

Traditional family patterns are inherently restrictive for some family members, as liberationist theorists point out. If the family is functional, it is only to the extent that it helps maintain unequal relationships that are based on the power of some individuals, or classes of individuals, over others. Liberation theorists, therefore, question the assumption that the family merely adapts to social change in a way that meets the personal needs of individual members.

Another important distinction between the unification and liberation approaches is that sociologists of the latter school do not limit their analyses to the identification of inequalities in the family. Instead, they extend their theories to consider alternative strategies to overcome those inequalities, up to and including the establishment of new relationships. Diversity, then, becomes the norm in the study of family, which makes dimensional theories, such as the one used by Eichler, useful tools for the analysis of families. We will describe two approaches that dominate liberation theories: conflict theory and feminist theory.

Conflict Theory

Conflict theory had its beginnings in the writings of Karl Marx and Friedrich Engels. During the 1960s, Marxist and neo-Marxist analyses were increasingly used by sociologists to explain relations within families as well as the family's role in capitalist society. Many feminist theorists have been influenced by conflict theory, although as the discussion of feminism will show, Marxists and feminists differ in some respects on the reasons for the subordination of women in families and in society as a whole. More recent theorists associated with conflict theory include D. Smith, W. Seccombe, and E. Zaretsky.

According to Marx and Engels, the character of social institutions and relationships, including the family, are determined by the economic system, or **mode of production**. As a result, families have evolved throughout history to serve the economic needs of each historical period. These historical periods have been characterized by a class struggle between oppressors and oppressed, including that between slave owners and slaves, feudal lords and serfs, and capitalists and workers. Now let us illustrate the impact of the mode of production and class struggle on the family.

If we examine the evolution of the family from prehistoric times, two major transformations stand out. First is the gradual change from natural tribal societies where group marriages were the norm, and the entire community was considered a family, to an increasingly monogamous (single-partner) marital arrangement that led to a smaller, more independent family structure. Second is the rise of **patriarchal** cultures and the subsequent decline and virtual disappearance of **matriarchy** as the social and religious norm. On the basis of the literature and the archaeological evidence, this fundamental shift is thought to have begun at approximately 2400 B.C.E. as a result of invasion and conquest.

In Roman times family had an entirely different connotation. Friedrich Engels explains that the Latin word *famulus* means "household slave," and the plural *familia*, refers to "the totality of slaves belonging to one individual."[15] Under the Romans, the

---

BOX 10.2

Two Other Approaches to Unification Theory

Symbolic Interactionism
Interactionist theorists define the family as "a unity of interacting personalities" and believe that through the interactions of its members, a family develops a conception of itself. Proponents of this approach include E. Burgess, C. Cooley, G. Mead, and S. Stryker. Social psychologists have used this approach in small-group laboratories to study parent–child and husband–wife interactions.

Family Life Course Perspective
According to life-course theorists, most families pass through a series of four stages: (1) a childless couple, (2) a couple with children, (3) **empty nesters**, when the children leave home, (4) a widow or widower. T. Hareven, G. Elder, and P. Uhlenberg are key proponents of this theory.

family developed into an institution where the male had absolute power over the rest of the family. Such an arrangement allowed him to ensure that his inheritance was bequeathed according to his wishes, which were stated in his will.

Obviously, families have come a long way since the days of "swords and sandals." Nevertheless, to appreciate the etymological roots of the word "family," consider that from Roman times until this century, a father's inheritance invariably went to his sons. If not slavery, then modern family relations have at least clung to a clearly defined hierarchy. Only in the latter part of the 20th century was this custom successfully challenged in some societies.

On the basis of these two changes, from large natural families to small independent families, and from matriarchal to patriarchal families, Engels reached the conclusion that monogamy was the first family form based "not on natural but economic conditions, namely on the victory of private property over original, naturally developed common ownership."[16] He also concluded that monogamy led to antagonism between men and women.

The antagonism in capitalist society can be illustrated by examining the dual role that the family plays in the service of industry. On the one hand, the family consumes the outcome of production, which is consumer goods, and on the other, it produces the workers employed by the industries that produce the goods. The patriarchal character of families results from the division of labour between men and women. As already discussed, women play a larger role in meeting the domestic needs of their families, a role for which they are not paid, and one that often compromises their wage-earning capacity. Although there is evidence that women are playing a more significant role in the workplace, as we shall see later in this chapter, in domestic affairs women are still the main providers. Conflict theorists argue that this creates a relationship of dependence and is the basis of the antagonism between male and female within the family.

Ironically, there is a similarity between conflict theory and the structural–functionalism of Talcott Parsons. It was Parsons, after all, who decreed that the nuclear family was best suited to industrial capitalism. Marxists would not disagree; particular family structures are tailored to meet the needs of the mode of production. However, the crucial difference between Parsons and conflict theorists is that whereas the former sees the ties between family and the economy as beneficial and positive, the latter see them as oppressive and conflictive.

Engels argued that the victory of workers over the ruling, or capitalist, class will lead to the elimination of other forms of oppression. Hence, with the disappearance of private property, the oppressive character of monogamy will also disappear, and a new monogamy, one based on mutual respect and equality, will emerge. This will result in a fundamentally different kind of family.

Sociologists from the feminist school who also identify the subordination of women with the present social order have utilized conflict theory. However, as we will see in the next section, some feminists see Marxist theory as incomplete because, although it demonstrates the inequality between male and female partners, it ignores the female's role as mother; the focus here is on "production," rather than on "reproduction." Consequently, there is a series of issues related to reproduction, the intervention of governments, for example, that conflict theory does not take into account. These differences have led to a feminist–Marxist dialogue that has clarified some important issues for both sides.

Feminist Theory

Feminist theory of the family differs from all other approaches, including conflict theory, because it uses gender, rather than the individual, the family unit, or class as the most important factor in analyzing families. For feminists, gender is not a concept to be taken for granted. Despite the biological differences between men and women, gender is regarded as a "complex social construction with multiple dimensions that bear on the dynamics of families and other institutions."[17] Gender shapes our individual identities, which are played out in school and work, as well as in the family.

Gender is essentially about "power" as it determines what is expected of males and females. These expectations reinforce a patriarchal hierarchy of relations that are based on the domination and oppression of women by men and that play an important role in shaping "family norms." The family, in return, is the primary organizational institution for gender relations. It is "the place where the sex/gender division of labour, the regulation of sexuality, and the social construction and reproduction of gender are all rooted."[18]

To understand the concept of family, feminists argue that we must focus on aspects of familial relationships that in many ways challenge the idea of "harmony" associated

---

CRITICAL THINKING BOX 10.2

What kinds of stress do modern families endure as a result of economic hardship? To what extent do these stressors affect family stability? If these stressors were eliminated, how do you think familial structures and relationships might differ from their present forms? Think of the kinds of conflicts that might be reduced. Are there some tensions you would expect to persist even in a world where basic economic needs were met?

with traditional family life. Feminist theorists draw four conclusions from their analysis of these relationships:

1. Families are "arenas" in which individuals struggle to pursue different social and economic interests.
2. Families are founded on relationships in which men dominate; hence, they are patriarchal systems.
3. Families are systems where women generally accept their subordinate position, resulting in the **ideological legitimation** of inequality.
4. The definition of families as unified groups promotes **familism**, an ideology that presupposes "traditional" family norms and values.

The first two conclusions are consistent with the views of conflict theorists, who also view families in terms of competing interests and power relationships. As stated earlier, according to Engels, the antagonism between male and female led to the first class oppression—that of men over women. Class conflict, then, can take on a patriarchal form. The second two conclusions raise an important distinction that merits further discussion, namely the distinction between the **ideology of the family** and the family as it exists in modern society. Here is how the sociologist Meg Luxton explains the distinction:

> To understand "the family" we have to differentiate between ideology and the actual ways in which people interact, co-reside, have sexual relations, have babies, marry, divorce, raise children, and so on. In other words, "the family" exists in two quite different forms: as "familism," a widespread and deeply embedded ideology about how people ought to live; and as economic and social groups which in fact organize domestic and personal life.[19]

We have already discussed theoretical approaches that focus on how families "ought" to be structured. The patriarchal basis of the nuclear family dictates a "natural" structure where men dominate women and children and where men have an independent identity outside the family. However, by emphasizing the complementary functions performed by men and women in the family, we keep the power of men over women hidden or "obscured."

The ideology of family extends beyond the family itself to define women's "proper place" in the economy. Because motherhood is seen as women's primary vocation, their labour outside the home is assumed to be of secondary value to that of men, who traditionally have had the role of earning enough money to provide for their entire family. Furthermore, women are seen as most qualified for occupations that resemble their roles as wife and mother. "Suitable" careers for women include caring for and teaching the young, nursing, clerical and service work, as well as producing and selling food and clothing. As a result, familism has reinforced the economic exploitation of all women.

Another important element of familism is the belief that the family shelters us from an increasingly impersonal world. The harmony that characterizes nuclear families contributes to a more stable society. If we deviate from this model, we risk undermining social stability. Feminists, on the other hand, argue that such harmony is in many cases an illusion maintained by ignoring the gender and age basis of family violence. Traditional theorists sometimes view family violence "as a series of individual assaults or else a pathology of 'family systems.' These views ignore a crucial fact ... violence runs along the lines of power, with adult men and women abusing children, and men abusing women, much more than the reverse."[20]

In response to their oppressive environment, feminists such as Marlene Mackie urge women to take action. First, women must examine more carefully the social basis of their roles in reproduction, parenting, and the gender division of labour. Next, they must reject the ideology that underlies the traditional model of family and begin to examine how familial relationships can be a source of conflict and violence, rather than of harmony. In the end, feminists believe, women will recognize that "family" is not experienced the same way by all family members and that its hierarchical divisions can produce conflict.[21] Such an examination enables women to create a more satisfying life for themselves, both inside and outside their families.

Feminists have also demonstrated an awareness of some of their own shortcomings by carefully examining some of their own biases. For example, the multicultural character of our society has led many feminists to be more conscious of their own white, middle-class assumptions. In addition, the demands of gays and lesbians to have their family experiences recognized and accepted has forced feminists to broaden their analysis of the social norms that determine the reproductive and childrearing choices permitted in society.

Such analysis leads to the conclusion that family patterns and structures are subject to social rules and constraints, not just to biological ones. Consequently, for feminists, as for liberation theorists in general, it is the diversity of families that enables individuals to engage in meaningful and fulfilling relationships.

We have described some of the most commonly used theoretical approaches to the study of families. Each theory provides at least a partial picture of the structure of families, the roles played by their members, and how they are related to society as a whole. We have also attempted to account for diversity in families from the various theoretical approaches. Next, we will examine how families in Canada have changed and the extent to which these changes reflect a growing diversity in what we call "families."

THE CHANGING PATTERN OF CANADIAN FAMILIES

Depending on your point of view, families are doing one of two things: they are either evolving or deteriorating. For those whose image of family is based on two parents—one male and one female, married, with 2.2 children, the changes in the last several decades indicate the deterioration of what they regard as the traditional family. On the other hand, for those with a more inclusive perspective on family, the current changes are evidence that the family is a dynamic institution that has known change throughout history and that will continue to evolve regardless of what is considered a traditional family at any given point in time.

We will now identify some of the key changes that have taken place in contemporary family life, using a dimensional framework. Then we will present a brief summary of our conclusions.

Procreative Dimension
Childbirth: Not So Soon, Not So Many

Several important transformations have occurred in families that are related to procreation. One of the most dramatic is the decline in the birth rate since the 1950s. To say "They don't make them like they used to" is no overstatement when we examine the changes illustrated in Table 10.1. As you can see, the birth rate peaked in the middle of the 20th century and then declined by approximately 53 percent by the beginning of the 21st century. The most recent data shows that the rate dropped another 1.5 percent in 2002 from the previous year, which was the 11th decline in the past 12 years.[22] Thus, the downward trend in Canada's birth rate continues into this decade.

One contributing factor to the lower birth rate is that women are choosing to have their children later in life. Table 10.2 shows that for women in their 30s, the birth rate has steadily increased since the late 1960s; whereas for women in their 20s, the general trend has been downward (see table). This upward trend in the average age of women giving birth continued into 2003, when it reached 29.6 years, up from 26.9 years in the 1980s.[23]

Because of these changes, families in Canada are becoming smaller, on average, than they have ever been, and the age gap between parents and their children is growing as families put off having children until later in life. Future studies may provide insights into the impact of this evolving family structure on relationships within families.

Childless and Childfree Couples

So far in this section, our focus has been on families with children. However, some families either choose not to have children or are unable to have them. In addition, there are families whose children have grown up and left home, the so-called empty nesters.

Table 10.1 Canadian Births per 1000, 1901–2001 (population and growth components, 1901–2001 censuses, thousands)

Period	Census Population at End of Period	Total Population Growth[a]	Births	Deaths	Immigration	Emigration
1901–1911	7 207	1 836	1 925	900	1 550	740
1911–1921	8 788	1 581	2 340	1 070	1 400	1 089
1921–1931	10 377	1 589	2 415	1 055	1 200	970
1931–1941	11 507	1 130	2 294	1 072	149	241
1941–1951[b]	13 648	2 141	3 186	1 214	548	379
1951–1956	16 081	2 433	2 106	633	783	185
1956–1961	18 238	2 157	2 362	687	760	278
1961–1966	20 015	1 777	2 249	731	539	280
1966–1971	21 568	1 553	1 856	766	890	427
1971–1976	23 450	1 488	1 760	824	1 053	358
1976–1981	24 820	1 371	1 820	843	771	278
1981–1986	26 101	1 281	1 872	885	678	278
1986–1991	28 031	1 930	1 933	946	1 164	213
1991–1996	29 611	1 580	1 936	1 024	1 118	338
1996–2001	31 021	1 410	1 705	1 089	1 217	376

[a]Total population growth is the change in population numbers between two censuses.

[b]Beginning in 1951, Newfoundland is included.

Source: "Births per 1,000, 1901–2001," Adapted from the Statistics Canada website: http://www40.statcan.ca/101/cst01/demo03.htm

Among childless and childfree couples, the number of empty nesters tends to fluctuate the most. During good economic times, their number rises as the result of children becoming financially independent enough to leave home and in many instances start a family of their own. The departure of children is often seen as positive by the parents who gain a measure of freedom and independence for themselves once the responsibilities of childrearing are completed. However, during periods of economic uncertainty, some children either delay leaving home or return after encountering financial difficulty. In these

The Many Faces of Diversity

Table 10.2 Age-Specific Birth Rates (per 1000 women), 1961–1997*

	Age Group			
Year	20–24	25–29	30–34	35–39
1967	161.4	152.6	91.8	50.9
1977	102.9	125.5	65.4	20.2
1987	76.1	116.7	73.2	23.2
1997	64.1	103.9	84.4	32.5

*Newfoundland included only in 1997.

Source: Adapted from Statistics Canada, Table 2.9, *Births and Deaths*, Catalogue No. 84-210, 20 May 1997.

cases, the "empty nest" becomes the "cluttered nest." Consequently, as we learned from conflict theory, economic conditions play an important role in shaping the structure of families.

Increasing economic pressures on families, along with changing attitudes regarding the role of women in society, have led some couples to decide not to have children at all, as research by Rachel Schlesinger and Benjamin Schlesinger suggests (see Table 10.3). Most studies have indicated that there are four reasons for choosing to remain childless: greater opportunity for self-fulfilment is first, followed by the desire for a more satisfying marriage. Tied for third place are female career and monetary considerations. However, as the sociologist Emily Nett[24] pointed out, there are no studies indicating the extent to which these goals are achieved.

Table 10.3 Voluntarily Childfree Couples: Some Research Findings

- The women are well educated, career oriented, and have a less traditional view of the female role.
- The women earn more than the average woman's salary, and this gives them a greater financial role in the marriage.
- The voluntarily childfree males have less stereotyped gender role attitudes than other males.
- Childfree families have a high degree of gender equality in their marital relationship.
- Women are usually the first to consider not having children.

Source: From Rachel Schlesinger and Benjamin Schlesinger, *Canadian Families* (Toronto: Canadian Scholars' Press, 1992), p. 35. © 1992, Canadian Scholars' Press. Reproduced by permission of the publisher.

On balance, childless and childfree couples make up a growing percentage of families in Canada. According to the 2001 Census,

> As of May 15, 2001, married or common-law couples with children aged 24 and under living at home represented only 44% of all families in Canada. In 1991, they accounted for 49% of all families, and in 1981 they represented more than one-half (55%). At the same time, couples who had no children living at home accounted for 41% of all families in 2001, up from 38% in 1991 and 34% in 1981.[25]

Given these trends, as well as Canada's aging population, it is logical to assume that the proportion of empty nesters will increase, complemented by those families who choose not to have children in the first place. The growing number of childless and child-free households is further evidence that family structure in Canada continues to change.

Socialization Dimension

The socialization dimension focuses on who in families is responsible for parenting. With a majority of women working outside the home nowadays, stereotyped attitudes toward raising children have changed. Traditional role differentiation, which cast men in the role of "breadwinner" and women in the role of "nurturer," has gradually been replaced by a recognition that both parents have a role to play both in the financial stability of families and in the nurturing of their young. However, this rethinking of male–female roles has not been accepted universally, or to the same degree, so that the process of change has been uneven.

On average, although men are playing a much more significant role in childrearing than they ever have before, in many families women are still spending more time than men as the family caregivers. In addition, despite men's increasing involvement in household chores, women devote more hours per day to **domestic labour** than their spouses do, whether they—the women or the men—work outside the home. Next, we will address the issue of childrearing itself and then the division of other forms of domestic labour.

Balancing Family Responsibilities with the Rest of Our Lives

Statistics Canada reported that in families with two working parents, women spend much more time on childcare than men do. Table 10.4 compares the daily time allocation of married women and men with and without children. One thing is for sure: children take their toll on men and women alike. Having children results in spending more time on unpaid work for both parents, and less time for sleep, free time, and paid work.

However, as the table shows, that impact is not distributed evenly between mothers and fathers. Women spend more than an hour and a half more on unpaid work than their partners, and that extra time appears to be at the expense of other areas of their lives. It

Table 10.4 Time Allocation of People Aged 25–44 Employed Full-Time, 1998 (hours per day)

	Married Women Without Children	Married Men Without Children	Married Women with Children	Married Men with Children
Unpaid work	3.2	2.3	4.9	3.3
Paid work	6.2	7.1	5.5	6.9
Sleep	8.0	7.8	7.8	7.5
Free time	4.3	4.9	3.6	4.2

Source: Adapted from Statistics Canada, "Women in Canada: A Gender-Based Statistical Report," 2000, Catalogue No. 84-210, 14 September 2000.

is therefore clear that, despite the progress that has been made to date, equality between the sexes where family responsibilities are concerned is far from a reality. To further illustrate this point, only 20 percent of respondents to a national survey disagreed with the statement "When children are young, a mother's place is in the home."[26]

Women's time allocation of unpaid labour naturally has an impact on the time they have to earn a living. In fact, the gap between women and men in the time devoted to unpaid labour is almost identical to that devoted to paid labour. That is, women work roughly one and a half hours less per day outside the home than their partners, 5.5 hours versus 6.9 for men.

Another important conclusion is that women in childless families still do more unpaid labour per day than their spouses do. Thus, having no childcare responsibilities does not necessarily equalize the workload at home. Thus, while men are doing more domestic chores than they did in earlier times, families still rely on the unpaid contribution of women's time more than that of their male counterparts.

Gender roles have been transformed by changing economic conditions that have made two-income families the norm and by social values that have forced women and men to rethink their roles in day-to-day family life. Another factor that has an important impact on socialization within families is where family members actually live. In particular, the rise in the number of single-parent families and families formed from previous marriages has reshaped the way family members, particularly children, are socialized.

Residential Dimension

For unification theorists, the nuclear family, with parents and children living under one roof, is assumed to be the norm and is considered one of the foundations of a stable, functional society. However, families can be characterized by an increasing diversity of living arrangements. This diversity is the result of two factors. First, a higher divorce rate has led to a larger number of lone-parent families and families in which children spend time at the residences of both parents. Apart from divorce, the choice of single women to have children without getting married first has also contributed to the higher number of lone-parent families.

The Impact of Divorce

The liberalization of divorce laws through the 1968 *Divorce Act* and the revised *Divorce Act* of 1985 have had an enormous impact on the divorce rate in Canada. In 1968 there were 11 343 divorces, or 54.8 per 100 000 people. By 1987 it had peaked at 96 200, or 362.3 per 100 000. Since then it has fluctuated, peaking again in 1992 at 79 034 and dipping to 67 408 in 1997.[27] Regardless of the recent fluctuations, the increase between 1968 and 1997 was enormous—almost 600 percent! However, between 1999 and 2003 the rate stabilized, increasing at below 2 percent per year.[28]

Of course, as the divorce rate climbed, so did the number of lone-parent families. As shown in Table 10.5, lone parents made up 18.5 percent of the total number of families with children in 1996, in contrast with 9.0 percent in 1961. However, that dipped to 15.7 percent between 1996 and 2001, so it remains to be seen if this dip is a minor hiccup or a more permanent reversal of a four-decade trend.[29]

Table 10.5 Single-Parent Families, 1961–1996

	Families Headed by Women		Families Headed by Men		
	000s	As % of All Families with Children	000s	As % of All Families with Children	Women as % of Single Parents
1961	272.2	9.0	75.2	2.5	78.4
1966	300.4	9.0	71.5	2.2	80.8
1971	378.1	10.4	100.7	2.8	79.0
1976	464.3	11.6	95.0	2.4	83.0
1981	589.8	13.7	124.2	2.9	82.6
1986	701.9	15.5	151.7	3.3	82.2
1991	786.4	16.4	168.2	3.5	82.4
1996	945.2	18.5	192.3	3.8	83.1

Source: Adapted from Statistics Canada, Table 2.6, *Families: Number, Type and Structure*, Data Products, Nation Series, 1991 Census of Population, Catalogue 84-210, 6 July 1992.

Of the total number of lone-parent families, more than four-fifths are headed by women, and this rate has not changed much since 1961. Statistics from 2001 show a further increase of 12.7 percent in female lone-parent families, while the figure for males was even greater, at 27.8 percent.[30] The pattern of mostly female lone-parent families is reinforced by court custody decisions, whereby in 73.6 percent of the cases the children were awarded to the mother and in 11.8 percent to the father. In the remaining 14.3 percent of the cases, the decision was in favour of joint custody.

What effect do these figures have on the residential dimension of family life and the overall quality of that life? Carolyne Gorlick contrasted the traditional, or unification view of divorce with its reverse, what we are calling the liberation perspective. To the unification theorist, divorce is a deviant phenomenon, "characterized by stages of denial, mourning, anger, and readjustment."[31] Therapeutic and casework analyses, according to Gorlick, see divorce as a crisis, rather than as a necessary part of an adjustment period.

Those who disagree with the traditional family perspective argue that separation from a marriage that is not working can contribute to the individual's personal growth in a way that was not possible during the marriage. First, it can create opportunities for widening the circle of family ties through remarriage and through new friendships that become possible in a different residential setting. Second, for family members leaving a non-supportive family environment, divorce is a means of liberation from fear, anxiety, and physical and emotional abuse. Finally, the opportunities for family renewal are not necessarily dependent on remarriage. Lone custodial parents enjoy a measure of independence to make choices for themselves and their children that would not be possible otherwise. Particularly for parents fleeing domestic abuse, this is an important part of their liberation.

Gorlick pointed out that neither the unification nor the liberation perspective provides a complete picture of divorce and its impact on families. Separation initially creates stress for all family members, especially children, who often cannot understand the complexities of divorce. With time, however, divorce can be seen as the beginning of a new life and a new family, not merely as the end of an old one. Diversity in residential patterns then has a positive impact on family growth and renewal.

In addition to divorced women, there are those who choose to have children without getting married. The number of women in this category rose from 11.0 percent in 1981 to 24.2 percent in 1996.[32] These women made a conscious decision that the traditional family structure is not the most suitable for them or their children. As women's incomes increase, having children without being married becomes a more economically viable option. That, in part, may explain the findings in a recent study by Susan Crompton: in a survey of mature women, aged 29–54, 52 percent believed that having children was very important, even though they had no desire to marry.[33] Consequently, the lone-parent family is becoming a more common family structure for many reasons.

Economic Dimension
Women in the Workforce

The most dramatic economic transformation in family life is the sharp increase in the number of two-income families. Duxbury and Higgins reported that in 1961, of all two-parent families in Canada, only 20 percent were dual-income.[34] By comparison, the latest census shows that that figure is now 62 percent, a more than threefold increase in the number of women in the workforce.[35]

We have already discussed the impact of women working outside the home on the division of parental responsibilities in the areas of childrearing and domestic labour. The traditional "breadwinning" and "nurturing" roles are not adequate to characterize Canadian families in the early 21st century. This is a direct result of changes in the economy and the makeup of the workforce. For example, because the financial stability of families has become increasingly dependent on having two incomes, families have had to adjust to (1) both parents balancing both family and occupational responsibilities, (2) the growing need for external childcare, and (3) the emotional stress of job loss for both parents. In addition, lone parents are faced with the task of raising a family on only one income, as well as meeting childrearing and occupational responsibilities without the emotional and physical support of a spouse.

As families have become more dependent on two incomes, they have also become more dependent on external childcare, which has, in itself, added to the expense of family life. To demonstrate, in 1987, two-parent families with preschool children spent 4.4 percent of their income on childcare, a sizeable sum. However, lone-parent families with preschool children fared much worse, spending 11.8 percent of their income on childcare.[36] The cost of childcare is just one of the pressures facing today's families in Canada. Next, we will focus on the broader issue of the financial pressures that are brought to bear on families.

CRITICAL THINKING BOX 10.3

Some people believe that if families would make do with less, one parent could afford to stay home with the children. As a result the "traditional family" would be preserved. Do you believe the current cost of living makes this option feasible, or has the dual-income family become a necessity?

Unemployment and Poverty in Canadian Families

One way that parents have adapted their jobs to their family responsibilities is by working part time. However, in keeping with what we learned about the socialization dimension, women have been most often the ones to do the adjusting. According to Statistics Canada, 29.1 percent of women in two-parent families choose to work part time because of the needs of the family, compared with 4.2 percent of men. For lone female parents, this figure drops to 20 percent because these women do not have a spouse with whom to share childrearing responsibilities.[37]

An obvious problem that affects a family's financial stability is unemployment. This is serious for all families, but the numbers indicate that it is particularly devastating for single-parent families. Statistics also indicate that considerably more female lone parents face unemployment than male lone parents. In 1992, for example, the unemployment rate for women who were single parents was more than double that of women in two-parent families, 19.2 percent versus 9.8 percent. That same year the figures for men were 13.9 percent and 8.3 percent, respectively.

Although the unemployment crisis of the 1990s improved by the end of the decade, poverty has persisted and the gap between rich and poor has increased. To measure the rate of poverty, the Canadian government uses a "low-income cutoff" according to which any family that spends more than 58.5 percent of its income on food, shelter, and clothing is considered poor. By this standard, the percentage of Canadian families living in poverty has grown significantly in the past decade. The sociologist Alfred Hunter reported that "according to the current criterion, 14 percent (or 3 800 000) of Canadians were poor in 1991, up from 11.8 percent in 1986 and 12 percent in 1981."[38] By 2000 the figure had improved slightly to 12.6 percent.[39] However, given that government cuts to spending on welfare and other social services during the 1990s remain largely in place, many Canadian families continue to struggle with poverty.

Which family types are most vulnerable to financial stress? Table 10.6 gives a breakdown of the percentages of low-income families by category, including elderly families, childfree couples, two-parent families, and female single-parent families. As you will see, female single-parent families are far more at risk of falling into the low-income category than other family types. Also, Table 10.6 illustrates that the percentage of low-income families started to climb again in 2002.

Emotional Dimension

Any discussion of the emotional dimension of family relations inevitably leads to the issue of domestic violence. We will briefly describe the kinds of abuse that are most common and how the high incidence of domestic violence affects our perception of the family.

Table 10.6 Low-Income Rates by Family Type

	1994	1996	1998	2000	2001	2002
Economic families, two persons or more	9.4%	10.7%	8.8%	7.9%	6.6%	7.0%
Elderly families	2.5	3.0	3.6	2.9	2.2	2.7
Married couples	6.3	7.1	5.6	5.8	5.0	5.5
Two-parent families with children	8.3	9.7	7.4	7.4	5.9	5.4
Female single-parent families	44.7	49.0	39.1	33.2	30.1	34.8
Unattached individuals	30.7	33.7	30.5	28.5	26.1	24.8

Source: Adapted from Statistics Canada, "Low-Income Rates (1992 Base After-Tax Income LICO) by Main Family Types," *The Daily*, 20 May 2004, Catalogue 11-001, Thursday, 20 May 2004, available <www.statcan.ca/Daily/English/040520/d040520b.htm>, accessed 6 May 2006.

Violence Against Female Partners

In 2004, according to Statistics Canada, of those women who reported incidents of violent spousal abuse, 23 percent reported either being beaten, choked, or assaulted with a knife or gun by their spouses, compared to 15 percent of male victims. Forty percent of women reported they were pushed, shoved, or slapped as opposed to 35 percent for men. Men, on the other hand, were more likely to be slapped or have something thrown at them (15 percent, compared to 11 percent for women). On the whole, 44 percent of women reported being injured as a result of spousal violence, in contrast to 19 percent of male victims.[40]

If we look at the most serious crime, homicide, we find that between 1993 and 2002, roughly two-thirds of female victims of solved homicides were killed by either their spouses or their ex-spouses, in contrast to 24 percent of male victims.[41] Sadly, spousal abuse of all types is a well-documented fact of life in Canada today.

Although violence against women has been traced back as far as ancient Greece and Rome, it has not always been seen as a problem. Marion Lynn and Eimear O'Neill quote from *The Rules of Marriage*, from the 15th century, on what were considered at the time the merits of wife assault: "When you see your wife commit an offence, don't rush at her with insults and violent blows. Scold her sharply, bully and terrify her. And if this still doesn't work . . . take a stick and beat her soundly, for it is better to punish the body and correct the soul. . . . Readily beat her not in rage but out of charity . . . for [her] soul so that the beating will rebound to your merit and her good."[42]

This tolerance of wife assault was still common in 19th-century Britain, where a husband was permitted to strike his wife with any instrument no wider in diameter than his thumb.

Thus, although today wife assault is illegal, for most of recorded history it has been accepted in patriarchal societies as a consequence of a man's authority over his wife and his children. (Remember the Latin origins of the word *family*, as quoted at the beginning of this chapter.)

Women are not the only victims of domestic violence. Children, too, bear the emotional scars of the dark side of family life. As Statistics Canada reports, "Of all violent crimes against children under 12 years of age and reported to police between 1988 and 1990, 41% were perpetrated by a member of their family: 24% involved a parent and 17% involved another family member."[43] What the statistics show is that some children have at least as much to fear from their families as they do from strangers.

Finally, assault against male partners, although comparatively rare, is also acknowledged in studies of domestic violence. In 1991, 43 percent of female victims of violence and 3 percent of male victims had been assaulted or murdered by their partners.

Effect of Violence on Family Relations

The high incidence of domestic violence runs counter to the traditional view of the family as a refuge from a cold, heartless world. It cuts across class lines, although it tends to be more visible among lower-income groups because people in these groups have more frequent interaction with relevant government departments, such as social services.

Feminist theorists have been at the forefront of research into the causes and outcomes of domestic violence. The subjects of their studies, be they women or children, are seen not only as victims, but also as "survivors." The pain suffered by victims is acknowledged, but so too is their capacity to overcome their circumstances through their adaptive capacities and strengths. Thus, the end of one negative family structure can lead to the emergence of a new positive one, although, of course, this does not happen without its own sacrifice and struggle. The emotional dimension, as Eichler concluded, "runs the gamut from the most tender, emotionally satisfying, positive involvements to the most frightening, abusive physically and mentally harmful relationships."[44] The diversity in these relationships must be the basis for further analysis of families.

SAME-SEX MARRIAGE: THE CONTINUING EVOLUTION OF CANADIAN FAMILIES

On 20 July 2005, Bill C-38, Canada's **same-sex marriage** law received royal assent. This bill, while legalizing marriage between gays and lesbians, also guarantees the right of religious groups to refuse to perform marriages that go against their beliefs. Gays and lesbians, as well as their supporters celebrated what amounts to the latest redefinition of family in Canada. Three other countries, Spain, the Netherlands, and Belgium, have also recognized same-sex marriage.

> **CRITICAL THINKING BOX 10.4**
>
> In this chapter we have analyzed the family using five "dimensions." Can you think of any other "dimensions" that might be added to those we have discussed? How would your additional dimensions enhance our understanding of families? What forms of diversity would they demonstrate?

This legal redefinition of marriage and, consequently, the family came after much heated and emotional debate. "Traditional family" proponents lobbied against the passing of C-38. One such group, United Families Canada, argued in a petition to the government that "[L]egalizing same sex marriage in Canada would undermine traditional marriage in our society, thereby undermining support for families as well . . ." and that ". . . same-sex marriage will hurt Canada, disenfranchise our children of their right to a parent of both genders, and drastically weaken the family."[45]

On the other side of the debate, organizations such as Equality for Gays and Lesbians Everywhere (EGALE) defend same-sex marriage as a human rights issue, as you can see from the following press release that came out after the passage of C-38: "In a generation, Canadians will look back on a time when lesbian and gay people were denied full citizenship, just as we look back on the days when women or Aboriginal people could not vote or times when Canadian citizens were interned because of ethnic origin."[46]

From the perspective of diversity, the broadening of the definition of marriage to include same-sex couples merely reaffirms that the structure of families in Canada continues to change as it always has. The intensity of the debate demonstrates that controversy, as always, is no stranger to the issue of defining "family" in our society.

CHAPTER SUMMARY

We began this chapter by referring to what some describe as a crisis in the family as a social institution. What we have attempted to show is that changes to the modern family do not necessarily constitute a crisis. Although the traditional ideology of the family has come under attack, families themselves are doing what they have always done. They are responding to political and economic changes in society by changing themselves, becoming more diverse as they adapt to their new circumstances.

We began by attempting to reach a meaningful definition of the term *family*—one applicable to the wide variation in the structure of Canadian families. As the process by which the definition is developed is as important as the definition itself, the introductory section of the chapter also examined how we arrive at our definition. In attempting to reach a suitable definition, we discovered that the notion of the "traditional family" is incomplete at best, as it is unable to transcend the constraints of popular middle-class culture.

Next, we looked at how families have been studied from several theoretical perspectives, which are broadly divided into two categories: unification and liberation theories. Each approach has its own rationale for the nature of relations among family members and of family structure. No one approach gives a complete picture of family life or reasons for changes that have taken place in Canadian families, but some approaches are better equipped than others to explain the diversity of modern families.

Change has always been an essential characteristic of families. Consequently, we should not be surprised that what is referred to as the "traditional family" has always been more of an abstract idea than a reality. The statistics provided in this chapter have clearly demonstrated the kinds of changes that have taken place in Canadian families, particularly since the 1950s. They reveal a social institution that is evolving, as it always has, in response to other changes in our environment, be they social, political, or economic.

Finally, we addressed the redefinition of families, including the Canadian government's passage of Bill C-38, which recognized the legality and sanctity of same-sex marriages. This latest development makes all the more important our contention that understanding the **diversity** of family life is essential if we are to address the stresses and challenges facing Canadian families.

KEY TERMS

conflict theory, p. 280
conjugal family, p. 278
domestic labour, p. 289
empty nesters, p. 281
extended family, p. 274
familia, p. 281
familism, p. 284
feminist theory, p. 283
general systems theory, p. 278
ideological legitimation, p. 284

ideology of the family, p. 284
liberation theories, p. 280
matriarchy, p. 281
mode of production, p. 281
nuclear families, p. 274
patriarchal, p. 281
same-sex marriage, p. 296
structural–functionalism, p. 277
unification theories, p. 277

DISCUSSION QUESTIONS

1. Conduct two interviews, one with someone you know who is from what the text defines as a traditional family, and someone from a nontraditional family. What similarities and differences can you identify in their familial relationships? To what extent do the concepts from this chapter apply to either family?
2. Review the content of Bill C-38. Then, conduct research on some of the groups who have argued for and against the recognition of same-sex marriage. Evaluate the basis of each side of the debate. Why do you think the government finally chose to legalize same-sex marriage?
3. Canadian families are based on patriarchal (male-dominant) authority. What would life be like in a family based on matriarchal (female-dominant) authority? How would it differ from a patriarchal family?
4. Which of Eichler's dimensions do you think has the greatest impact on family relationships? Explain your answer.
5. Do you agree that "traditional" family patterns of authority can create antagonistic relationships of dependence in families? Why or why not?
6. You are married and have a child. Both you and your partner work full time. Plan a schedule for a week that ensures that the domestic chores are divided equally.

NOTES

1. "New Mega-Trends Reflect Family Decline," *Today's Family News*, 7 January 2005, available FamilyFacts.ca <www.fotf.ca/familyfacts/tfn/2005/010705.html>, accessed 8 April 2006.
2. Dr. Darrel Reid, "Crisis or Opportunity? Eleven Practical Steps for Strengthening the Family," *Commentaries*, November 1999, available FamilyFacts.ca <www.fotf.ca/familyfacts/commentaries/110199.html>, accessed 8 April 2006.
3. George P. Murdock, *Social Structure* (New York: Macmillan, 1949), p. 1.
4. Rose Laub Coser, *The Family: Its Structure and Functions*, 2nd ed. (New York: St. Martin's Press, 1974), p. xvi.
5. Statistics Canada, *A Portrait of Families in Canada* (Ottawa: Vanier Institute, 1994), p. 7.
6. Brigitte Kitchen, "Family Policy," in Maureen Baker, ed., *Families: Changing Trends in Canada*, 2nd ed. (Toronto: McGraw-Hill Ryerson, 1990), p. 313.
7. Margrit Eichler, *Families in Canada Today* (Toronto: Gage, 1983), p. 8.
8. Emily Nett, *Canadian Families* (Vancouver: Butterworths, 1993), p. 24.
9. David Cheal, *Family and the State of Theory* (Toronto: University of Toronto Press, 1993), p. 4.
10. Reuben Hill, "Modern Systems Theory and the Family: A Confrontation," *Social Science Information* 10 (1971) (5): 12.
11. Baker, "Theories, Methods, and Concerns," p. 13.

12. Jason Montgomery and Willard Fewer, *Family Systems and Beyond* (New York: Human Sciences Press, 1988), p. 118.
13. Montgomery and Fewer, p. 21.
14. David Cheal, "Theoretical Perspectives," in G. Ramu, ed., *Marriage and the Family Today*, 2nd ed. (Scarborough: Prentice-Hall, 1991), p. 22.
15. Friedrich Engels, "The Origin of the Family, Private Property and the State," in Robert Tucker, ed., *The Marx-Engels Reader*, 2nd ed. (New York: W.W. Norton, 1978), p. 737.
16. Engels, p. 739.
17. Barrie Thorne, "Feminism and the Family," in Barrie Thorne and Marilyn Yalom, eds., *Rethinking the Family* (Boston: Northeastern University Press, 1992), p. 12.
18. Marlene Mackie, "Gender in the Family," in Nancy Mandell and Ann Duffy, eds., *Canadian Families* (Toronto: Harcourt Brace, 1990), p. 50.
19. Meg Luxton, "Thinking About the Future," in Karen Anderson et al., eds., *Family Matters: Sociology and Contemporary Canadian Families* (Toronto: Methuen, 1987), p. 238.
20. Thorne, p. 147.
21. Mackie, p. 50.
22. Statistics Canada, Health Statistics Division, "Birth Rate at All-Time Low," *Infomat: The Week in Review*, 27 April 2004, available <www.statcan.ca/english/freepub/11-002-XIE/2004/04/11804/11804_02p.htm>, accessed 8 April 2006.
23. Statistics Canada, "Births," *The Daily*, July 12, 2005 <www.statcan.ca/Daily/English/050712/d050712a.htm>, accessed 8 April 2006.
24. Nett, *Canadian Families*.
25. Statistics Canada, "Profile of Canadian Families and Households: Diversification Continues," *The Daily*, October 22, 2002, available <www12.statcan.ca/english/census01/products/analytic/companion/fam/contents.cfm>, accessed 8 April 2006.
26. Rick J. Ponting, *Canadian Gender-Role Attitudes* (Unpublished Manuscript, University of Calgary, 1986).
27. Statistics Canada, *Women in Canada 2000*, Catalogue No. 89-503-XPE, p. 43.
28. Statistics Canada, "Divorces," *The Daily*, March 9, 2005 <www.statcan.ca/Daily/English/050309/d050309b.htm>, accessed 8 April 2006.
29. Statistics Canada, "Lone-Parent Families as a Proportion of All Census Families Living in Private Households, Canada, Provinces, Territories, Health Regions and Peer Groups, 2001," November 26, 2003, <www.statcan.gc.ca/english/freepub/82-221-XIE/01103/tables/html/49_01.htm>, accessed 8 April 2006.
30. Ibid.
31. Carolyn Gorlick, "Divorce: Options Available, Constraints Forced, Pathways Taken," in Nancy Mandell and Ann Duffy, eds., *Canadian Families* (Toronto: Harcourt Brace, 1990), p. 212.
32. Statistics Canada, *Women in Canada 2000*, p. 43.
33. Susan Crompton, "Always the Bridesmaid: People Who Don't Expect to Marry," *Canadian Social Trends*, Summer 2005, available <www.statcan.ca/english/studies/11-008/feature/11-008-XIE20050017961.pdf>, accessed 8 April 2006.
34. Linda Duxbury and Christopher Higgins, "Families in the Economy," in Maureen Baker, ed., *Canada's Changing Families: Challenges to Public Policy* (Ottawa: Vanier Institute, 1994), p. 29.
35. Roger Sauvé, *Profiling Canada's Families III* (Ottawa: The Vanier Institute of the Family, 2004), available <www.vifamily.ca/library/profiling3/sample2.html>, accessed 8 April 2006.

36. Statistics Canada, *A Portrait of Families in Canada*, Catalogue No. 89-523E, 1993, p. 27.
37. Ibid., p. 29.
38. Alfred Hunter, "Social Inequality," in Robert Hagedorn, ed., *Sociology* (Toronto: Harcourt Brace, 1994), p. 275.
39. Statistics Canada, "Income of Canadian Families," *Analysis Series*, Catalogue No. 96F0030XIE2001014, 13 May 2003 <www12.statcan.ca/english/census01/products/analytic/companion/inc/contents.cfm>, accessed 8 April 2006.
40. Statistics Canada, "Family Violence in Canada: A Statistical Profile 2005," *The Daily*, 14 July 2005, available <www.statcan.ca/Daily/English/050714/d050714a.htm>, accessed 8 April 2006.
41. Maire Gannon, "Family Violence in Canada: A Statistical Profile," Statistics Canada, Catalogue No. 85-224-XIE, 2004, p. 35.
42. Marion Lynn and Eimear O'Neill, "Families, Power, and Violence," in Nancy Mandell and Ann Duffy, eds., *Canadian Families* (Toronto: Harcourt Brace, 1990), p. 285.
43. Statistics Canada, *A Portrait of Families in Canada*, p. 54.
44. Eichler, p. 13.
45. Canadian Citizens to Defend Marriage, Defend Marriage [website], 2003 <www.defendmarriage.ca>.
46. Canadians for Equal Marriage, "House of Commons Adopts Equal Marriage Bill by Decisive Margin" [press release], 28 June 2005, available <www.egale.ca/index.asp?lang=E&menu=20&item=1160>, accessed 8 April 2006.

PART III
THE TREATMENT AND PERCEPTION OF DIVERSITY

PART III

Part III examines how diversity is treated in the media and is perceived in literature. Continuing with the house analogy, if Part I is the structure and Part II the interior, then Part III is how people view the house—that is, how they assess it, value it, and treat it.

Chapter 11 looks at how the media portray diversity. It begins with a definition and overview of the history of mass media. The primary purpose of this chapter is to help the reader to understand the impact of mass media, especially the U.S. mass media, on the world in general and on diversity in Canada in particular.

Chapter 12 provides an exciting journey through the field of Canadian literature. At a general level, it looks at how contemporary Canadian writing reflects this country's continually evolving culture and diversity. At a specific level, it examines how Canadian literature illustrates the diversities covered in this text.

CHAPTER 11

The Medium Diversifies the Message: How Media Portray Diversity

Grant Havers and Paul U. Angelini

> *The medium is the message.*
> — Marshall McLuhan
>
> *Technology is our fate.*
> — George Grant

Objectives

After reading this chapter, you should be able to

- explain the meaning and uses of "mass media"
- explain the four major stages in the history of mass media
- understand the rise of the global village
- understand the influence of television on attitudes toward diversity
- understand the impact of American media on the world

The Medium Diversifies the Message: How Media Portray Diversity

INTRODUCTION: MEDIA AND DIVERSITY

Media have had enormous influence on human culture and are as old as civilization. They disclose a great deal about the cultures that employ them and the rich diversity within and among human cultures. How do mass media portray diversity?

Since the beginning of the electronic age, many observers have wondered whether mass media can encourage awareness of cultural diversity or, in fact, hinder such awareness. Mass media present the possibility of bringing the world together in peace and tolerance, because of their power to beam the news of the entire world in the living rooms of each family in every nation. However, this achievement does not guarantee greater awareness of diversity but instead presents diversity through distorted lenses. Whatever the implications, there is no question that representation of diversity by the media has an enormous impact on how we all see diversity.

DEFINING MEDIA AND "MASS" MEDIA

Perhaps it is easier to understand the meaning of "medium" than of "mass medium." A medium (*media* is the plural form) is a technique of communication. Language is the oldest medium of all. However, a **mass medium** is more than the means by which people speak and write to each other. If we understand the world as a "mass" of people, the term "mass media" suggests that media have enormous impact on the entire world. Two elements are important here.

First, mass media usually have "mass" audiences. All cultures in history have used various media (see Table 11.1), but these have not always been "mass" media. A "mass" medium is available to most people and cultures. It connects different parts of the entire world together.

Second, mass media vary in purpose. They may be sources of entertainment, such as television; information, such as newspapers; communication, such as telephones; or a combination of all three, such as computers. But they must always be available to a wide audience.

Throughout this chapter considerable discussion will be devoted to how successful mass media have been in educating the world about its diversity. One useful rule of thumb to remember is that every medium has an impact on its environment. Mass media, as Marshall McLuhan pointed out, are not simply tools that perform simple functions of informing or assisting human beings. A medium is more than a useful artifact; it affects human behaviour itself. To quote McLuhan's famous saying, "The medium is the massage"[1] (see Box 11.1). This claim can be interpreted in numerous ways, but only one will be offered here: every medium

Table 11.1 Which Civilizations Use(d) Which Media

Civilization	Main Medium Used
Sumerian	Pictographs, stone (3500 B.C.E.)
Egyptian	Papyrus, hieroglyphics (3000 B.C.E.)
Indian	Script (2400 B.C.E.)
Greek	Parchment, stone (700 B.C.E.)
Roman	Papyrus, parchment, stone (50 C.E.)
Chinese	Paper, script (105 C.E.)
Mayan	Script (50 C.E.)
Aztec	Script (1400 C.E.)
English	Paper (1200 C.E.)

Note: The dates here suggest approximately when the civilizations developed these media or made use of them. C.E. means "Common Era" and indicates the same time frame as A.D. refers to. B.C.E. means "Before the Common Era" and indicates the same time frame as B.C.

shapes the content of what it is conveying. This means that books, newspapers, radio, and television all can influence (or massage) the message that is delivered to its audience. As any consumer knows, how advertising portrays the product is usually more important than the product itself. Given this fact, it is centrally important to understand how mass media have shaped and influenced the meaning of diversity, for good or ill.

BOX 11.1

The "Prophet," Marshall McLuhan (1911–1980)

More than anyone else in this century, Marshall McLuhan understood the mass media. Born in Edmonton in 1911, he attended university in England in the 1930s, when he first studied the media. McLuhan became famous in the 1960s with his publications on the printing press, television, advertising, and technological change. *The Gutenberg Galaxy* and *Understanding Media* made him into a celebrity. It is ironic that by the time of his death in 1980, McLuhan himself had become a media phenomenon, appearing on the cover of major magazines, being asked for advice by politicians, and being hailed by many as a "prophet" for the television age.

A BRIEF HISTORY OF MEDIA AND MASS MEDIA

As already mentioned, throughout history civilizations of diverse origins have used media. People have always communicated with each other. But "mass media" have not always existed; that is, they have not always been available to large numbers of people. For example, writing was not always a mass medium; until the 15th century only elite groups in societies could write. By studying the history of media, we can begin to understand how media eventually became mass media.

The history of media can be divided into four stages (see Table 11.2), each of which marks a revolutionary turning point: (1) the oral stage, (2) the rise of the **alphabet** and writing, (3) the printing press, and (4) the electronic age, in which we now live.

The Oral Stage

The first stage was based on the condition of **orality**. Before the development of writing, the earliest human beings communicated almost entirely orally. Although human civilization has existed for perhaps six thousand years, the earliest writing appears to be not much more than five thousand years old. Until 3500 B.C.E. most people relied on the oral medium. It is estimated that the vast majority of languages spoken in human history have never been written down; little more than a hundred of them have a literature.[2] Human beings spoke words, sang poems, and gestured, but they did not write. Because they lacked any means of written communication, these oral cultures relied heavily on the constant memorization of their most cherished epics and folklore. It is important not to dismiss an oral culture as "primitive" on the basis of its lack of writing. Indeed, oral cultures were extremely sophisticated in their use of memory, which was their sole way of recording stories. Ancient Greek, African, Middle Eastern, and Aboriginal storytellers resorted to the oral transmission of their cultures' tales. This process required poets of each generation to pass down stories to the next through the spoken word.

The Rise of the Alphabet and Writing

The second stage began to emerge when oral cultures initiated the long transition toward written language. Around 3500 B.C.E. the Sumerians in Mesopotamia (modern-day Iraq) developed the **pictograph**, which consisted of pictures representing words or utterances. Other peoples, such as the Babylonians and Assyrians of the Middle East, also developed elaborate systems of pictographs. Around 3000 B.C.E. the ancient Egyptians developed their own system of pictographs, known as **hieroglyphics**.

By 1500 B.C.E. the second period of media had taken hold in the Middle East, which greatly accelerated the transition to written language: the alphabet had been invented. Human beings could now communicate using a system of signs that had letters, rather than symbols as in the case of pictographs.[3] This invention also enabled people around

The Treatment and Perception of Diversity

Table 11.2 Important Dates in the History of Mass Media

3500 B.C.E.	Sumerians develop pictographs.
3000 B.C.E.	Egyptians develop hieroglyphics, papyrus.
2000 B.C.E.	Hebrew begins to develop.
1500 B.C.E.	Phoenicians invent an alphabet.
105 C.E.	Chinese invent paper.
1455	Gutenberg invents the first printing press using moveable type.
1478	First printed ad appears.
1702	First English daily newspaper appears.
1741	First magazines appear in America.
1844	First telegraph line operates.
1873	Newspapers appear in the Middle East.
1876	Telephone is invented.
1877	Phonograph is invented.
1887	Gramophone is developed.
1895	Marconi develops the radio.
1896	First public motion picture shown in America.
1947	First TV news programs appear.
1979	Walkman is introduced.
1982	CD player emerges.
1985	Personal computer is established.
1990	Virtual reality becomes popular.
1998	Universities, industry, and government begin planning for Internet2.
2000	Napster file sharing upsets music industry.
2000	The Love Bug virus infects 45 million computers worldwide.
2000	British virtual "newscaster" Ananova joins other virtual performers on TV and the Net.
2000	Stephen King's novel *Riding the Bullet* becomes a bestseller via Net downloads only.
2001	Azerbaijan switches from the Cyrillic to the Latin alphabet.

the world to communicate with each other. Once it was discovered that materials such as papyrus could be employed to transmit the written word across great distances, various cultures became known to each other. The alphabet cannot be called a mass medium at this time, for until the invention of the printing press writing was still inaccessible to the vast majority of people.

Gradually, other cultures adopted the alphabet. The Phoenicians, a seafaring people of the Middle East, invented their alphabet about 1500 B.C.E. This reflected their desire to be able to trade goods using a flexible alphabet and enabled them to become independent of the dominant Egyptian culture of the time.[4] Eventually, the Greeks, in the eighth century B.C.E., took over the Semitic alphabet, which had 22 letters, and adapted it to their needs. One of the adaptations made by the Greeks was the introduction of vowels to the alphabet.[5]

As long as the alphabet remained the dominant medium, various ancient empires used it and expanded on it. Other cultures simply stuck to pictographs or combinations of symbols known as **script**. Whatever the type of medium used by a particular civilization, it reveals much about the priorities of that civilization. The Canadian historian Harold Innis shed light on these priorities by distinguishing **"time-oriented" media** from **"space-oriented" media**. Empires, such as Greece, that used durable media such as **parchment** (dried sheep- or goatskin) put great importance on the preservation of time, the recording of their history on materials that would last. Other empires that used papyrus, such as Egypt and Rome, and paper, such as China, viewed the conquest of space (or linkages to the outside world) as central. As a result, they cultivated media that could be transported across vast distances.[6] Because media such as stone and clay are impractical for transportation, it is clear that the empires that used papyrus, which is a very transportable medium, were most successful in acquiring knowledge about the rest of the world.

McLuhan introduced the idea that all technologies are "extensions" of human beings. Media "extend" a part of the human body by allowing that part to do something that it could not do by itself. For instance, print media—such as books and newspapers—are extensions of the eyes: they permit the eyes to see people, events, and cultures around the world that ordinary vision could not do by itself. Similarly, a computer extends the brain: the memory of the computer retains vast amounts of information that the brain cannot hold. The kinds of extensions a culture uses reveal which parts of the body that culture considers most important (see Table 11.3). That certain cultures adopted an alphabet suggests that the eye was very important to them, for the alphabet extends vision.

Although not all nations developed an alphabet (e.g., China did not), the alphabet became the dominant medium of western Europe, Russian Asia, India, and sections of the Far East (see Table 11.4). It has been said that every alphabet in the world is based on the Semitic alphabet.[7] The effects of the alphabet on the world increased with the third stage in the history of media.

The Printing Press

The third stage of history of media began in 1455, when Johannes Gutenberg (1390–1468), in Germany, invented the printing press with moveable metal type. This ushered

Table 11.3 Technology: The Extensions of Humanity

Technology	Part of Body It Extends
Club, hammer	Fist
Clothing	Skin
Knife	Teeth
Glasses, telescope, camera	Eyes
Writing, books, newspapers	Eyes
Refrigerator	Stomach
Wheel	Legs
Automobile	Legs
Radio	Ears
Television	Eyes, ears
Computers	Brain
Virtual reality	Whole body

Table 11.4 Some Major Alphabets of the World

Alphabet	Origin
Hebrew	Middle East
Ugaritic	Middle East
Greek	Mediterranean
Roman	Mediterranean
Cyrillic	Eastern Europe
Arabic	Middle East
Tamil	Southern India
Malayalam	Southwest India
Korean	Korea, East Asia

in a wave of social change that continues to shake the world today. (Before this time, the Chinese had developed a printing technique for wooden type.) Before the printing press with movable metal type, reading in the Western world was restricted mostly to the learned scribes and elites of the ancient and medieval worlds. Gutenberg's invention made it possible for millions to become literate, because it enabled the mass production of inexpensive books. Thus, the alphabet became a mass medium. Indeed, the age of mass media had arrived. The printing press had two major effects on the meaning of diversity.

First, it made people greatly aware of their diversity. Now it was possible to produce literature on a mass scale for a mass audience. The various peoples of western Europe became more aware of how different their cultures and languages were. The common, everyday languages or **vernaculars** of particular geographical areas could now be published using the printing press. Everyday French, German, and English people could now read the languages that they spoke. Nationalism, or love of one's nation, could not have developed without the printing press.[8] It made it possible for the various European peoples to read in their own languages. It also enabled them to develop languages that were truly "national," or common to all regions of a particular nation. Martin Luther translated the Bible into ordinary German by 1534. The different **dialects** (different versions of one standard language) of the various regions of Germany became one German language, because the printing press could put out books teaching different Germans how to conform to one national, printed language. Soon after 1455 cultural influences that had largely been restricted to Europe spread across the world.

Second, the press encouraged homogeneity or mass sameness. Once the printing press was used as the way of standardizing the language of a culture, everyone had to learn the same language. There was no longer room for individuals to create their own vernaculars—all had to conform to one language now. There was no such thing as "bad grammar," until the printing press made it possible to force one "good" grammar on everyone.[9] Indeed, Gutenberg's creation made compulsory education possible. Thanks to the printing press, everyone could now become literate, because everyone could be educated in the same language.

Thus, the print media of the alphabet and the printing press opened up the distinctive possibilities of diversity and homogeneity. This two-headed and contradictory process accelerated even further with the emergence of electronic media in the 19th century.

The Electronic Stage

The fourth stage, the electronic stage, began in 1844 when Samuel Finley Breese Morse's **telegraph**, a device that transmitted messages via electric wire, was installed between Baltimore and Washington, DC (see Table 11.5). Morse's first message on the telegraph was "What hath God wrought?"[10] It is not clear what Morse meant by this statement, but he may have been expressing surprise at the fact that his new invention

Table 11.5 Who Invented It?

Invention	Approximate Year	Who Got the Credit	
Printing press[a]	1455	J. Gutenberg	Germany
Telegraph	1844	S.F.B. Morse	United States
Telephone	1876	A.G. Bell	United States
Phonograph	1877	T. Edison	United States
Gramophone	1887	Emile Berliner	France
Radio[b]	1895	G. Marconi	Italy
Movies	1895	Lumière brothers	France

[a]This was the first printing press with moveable metal type; the Chinese had invented presses with *wooden* type almost 700 years before Gutenberg.

[b]There is no single inventor of the radio: credit must also be given to Heinrich Hertz (1857–94), a German scientist; Reginald Aubrey Fessenden (1866–1932), a Canadian, who experimented with radio transmission of voices in the early 1900s; and Lee de Forest (1873–1961), an American, who patented the vacuum tube in 1907.

had created a world whose diverse peoples could now communicate with each other almost immediately. The age of instantaneous communication through mass media had arrived.

Before the telegraph, messages and newspapers were transmitted as quickly as human beings could carry them—or move them—by horse. Now the telegraph could communicate messages across great distances in a relatively short time. The telegraph was the first of many inventions that were to be able to accomplish this feat: the telephone, radio, television, and computer followed in the 20th century. All of these encouraged the instantaneous transmission of messages from one part of the globe to another, allowing everyone everywhere to be involved in everyone else's business. Thus emerged the **global village**. This achievement was entirely the result of the discovery of electricity. This village produced **tribalism** among the human species, a feeling that, because of the linkages established by electronics to the outer world,[11] all human beings are part of one big group making up Earth.

Tribalism seems to foster awareness of diversity in an unprecedented manner. Thanks to electronic media, people around the world can now communicate with people in all the other parts, and they can do so instantaneously and effortlessly. For "space-oriented" cultures bent on developing linkages with the outer world, electronic media are a blessing, not a curse. But the question remains: Does this leap forward in communication translate into greater tolerance of diversity, or is the global village intolerant as well?

IS THE GLOBAL VILLAGE GOOD FOR DIVERSITY?

Television and Diversity

Media affect our behaviour. They are not only tools that perform the functions of communication and information; media "massage" their audiences: they can affect people's behaviour without people even being aware of this influence. Observers are increasingly putting media under the spotlight in case these media ignore, insult, stereotype, and misrepresent minorities of all kinds. As Michael Parenti observed, "The major distortions are repeatable and systemic—the product not only of deliberate manipulation but of the ideological and economic conditions under which the media operate."[12]

To answer the question "Is the global village good for diversity?" let us consider the effects of television, the dominant mass medium of our time and the technology credited as the most successful in expanding the global village. It is easy to observe why television is so successful, for it is one of the most widely available media of all time. Unlike print, television does not require the ability to read for it to be understood: no one requires special education to comprehend the messages of television. Long before they pick up a book to read, three-year-olds, for example, want the toys advertised on television.[13] In the global village, which consists of so many different cultures, how is life affected by the media? Four factors address this question.

First, television has encouraged many diverse communities and cultures to become aware of each other and themselves. Because television provides millions of people with access to a wider world, one of the effects has been to force many cultures to recognize their own isolation in the world. The feeling is that, if they do not show up on the nightly news, then they are not considered important. Although television has encouraged a feeling of isolation, it has also encouraged minorities and marginalized groups to demand an end to this isolation. Now that television can overcome the distances within the world, previously isolated groups want to be involved in the world's affairs. Having access to the world's information has encouraged this new demand. Indeed, it is significant that many political movements of minorities did not emerge until the television age.[14] Politics has been redefined in the media age (see Box 11.2).

Second, television has made it more difficult for the world to ignore the concerns and needs of these movements, especially when they have constantly appeared on the nightly news since, say, the 1960s. The plight of the poor around the world, for example, is increasingly difficult to ignore in the age of television exposure. The Live Aid concert organized by various musicians to fight starvation in Ethiopia in the mid-1980s is an example of how mass media exposure can make a huge difference to the success of a cause or movement—it succeeded in raising millions of dollars.

> ### BOX 11.2
>
> #### The Politics of Mass Media
>
> Mass media have influenced political movements for as long as politicians have existed. The printing press made possible the mass publication of pamphlets and newspapers, which are widely used by politicians.
>
> Television debates can also make and break a politician's career. Stephen Harper's performance on the television debates during the 2006 Canadian election helped his Conservative Party win a minority government. His performance in similar debates during the 2004 federal election campaign contributed to the defeat of his party. The medium does matter. In 1960, when John F. Kennedy debated with Richard Nixon for the U.S. presidency, those who watched the debate saw Kennedy, with his cool charm, as the winner. Those who listened to the debate on radio thought Nixon won. Evidently, television won over radio that year, because Kennedy defeated Nixon in the election by a narrow margin.

Third, television often distorts the meaning of diversity to suit its own aims and purposes.[15] Whereas it is true that television responded to the political movements of the 1960s with unprecedented exposure, this coverage had less to do with informing its audience than with providing entertainment. For example, at first glance it would appear that the media coverage of the various African-American and Vietnam protest movements in the 1960s represented an informative response to newsworthy events. It was hoped that this coverage would dramatize the plight of the African-American poor in the ghettoes of the United States, as well as the concerns of the antiwar protesters. What happened instead, according to some observers, was that the media decided to focus on those members of the movements who could provide the most entertainment or "news value." Typically, the media focused on leaders who packed the biggest dramatic punch, entertained the audiences, and looked physically attractive. Those leaders who lacked a dynamic visual image were ignored in the media coverage, even if they were important to their movements.[16] This was a sobering lesson about television coverage, and it raised a disturbing question: Does television inform about the diversity of the global village, or does it simply entertain?

Fourth, the nature of television is that it favours entertaining, fast-moving imagery that grabs the attention of audiences. For this reason, many political movements around the world choose television as the medium best able to get across their message.

CRITICAL THINKING BOX 11.1

Does TV Represent Diversity?

How often are characters from visible minority communities represented on your favourite television shows?

Entertainment is far more important as an objective to the producers of television than is providing copious detail on a subject. For this reason, television cannot inform as well as books or newspapers can. Yet television does not simply entertain. It can also misrepresent the reality of diversity. Although in recent years television networks have become more attentive to the need to reflect the minority population of North American society, their attempts have often been misleading.

The same holds true for Canadian media. In the history of television in Canada, there have been very few shows documenting or describing the lives of minority and Aboriginal communities. Canadian television, especially the CBC, has produced some fine documentaries on this subject, but very few visible minority and Aboriginal actors appear on mainstream dramas, comedies, and the like. *The Rez, North of 60,* and the *Aboriginal Television Network (APTV)* are attempts to remedy this. To make matters worse, visible minorities and Aboriginal peoples are often unfairly portrayed on the news as militant, violent, or unreasonable (see Box 11.3). This misrepresentation on television only serves to stereotype the Aboriginal and ethnic communities. It also results in coverage that is sorely lacking in information about the lives of these communities. For that matter, the media have only recently begun to employ women, as well as members of Canada's ethnic and Aboriginal communities, as journalists and broadcasters (see Box 11.4). In fact, in recent years there have been many incidents of racism in the media. This is especially true of issues in the fields of law enforcement and the treatment of minority groups.

CRITICAL THINKING BOX 11.2

Diversity and the News

Why do the media treat minority groups according to stereotypes? Is this due to the existence of racism, capitalism, or simple ignorance? Explain your answer in detail.

BOX 11.3

Indigenous Peoples and the Media

Until recently, Indigenous peoples in Canada were represented either unfairly or incompletely by the mass media. They suffered stereotypes that depicted them as criminally oriented, alcoholic, or lazy. Or their stories and histories simply were not seen or heard in the media. In the past 15 years, because of an increased awareness of the Aboriginal contribution to Canada's history, TV shows have included some Aboriginal actors, and the Aboriginal People's Television Network was founded in the early 1990s. Yet news programs still often portray Indigenous peoples as violent and prone to conflict when airing stories about land claims and treaty negotiations. The 2006 Native "occupation" of a Caledonia, Ontario partially finished subdivision is the latest example.

BOX 11.4

Journalists and the Canadian Supreme Court

The accuracy of court judgments is sometimes lost in the world of journalism, dominated as it is by time constraints (finding and tracking sources), deadlines, and collegial, and editorial expectations. Yet, accurately or not, journalists always have the "last word."

More generally, the courts and media have the power to set the parameters of debate, acceptance, and taste. In their own way each has the ability to inflict penalties on those who violate these norms.

Source: Florian Sauvageau, David Schneiderman, and Davis Taras, *The Last Word: Media Coverage of the Supreme Court of Canada* (Toronto: UBC Press, 2006). This text meticulously traces the media coverage of four cases that captured the attention of all Canadians: Delwin Vriend (gay rights); the Quebec Secession Reference (the right to leave Canada); Donald Marshall, Jr. (Aboriginal rights); and John Robin Sharpe (child pornography).

One way the media is racist in coverage is through their narrative on criminals and victims.[17] Minority crimes are generally analyzed as systemic ones meaning that entire groups are implicated and lumped together with the perpetrators. White lawbreakers, on the other hand, are usually individualized. This means their behaviour is reduced to some sort of individual shortcoming thereby separating them from their group. In short, when white people commit crimes their acts are random and individualized and are not symptomatic of all white people.[18] The best example of this type of narrative and reporting is the two high-profile crimes of the 1990s: the murder trial of Paul Bernardo and Karla Homolka and the Just Desserts shooting of 1994.

The Just Desserts reporting was extensive and high-profile. The media negatively implicated an entire group, Jamaican-Canadians,[19] and subsequently, Canadian immigration and deportations policies.[20] Their crimes are always the product of black culture and lifestyle; individual acts lead to the indictment of entire groups/cultures.[21] This is especially acute when the victims are white. Five months after this shooting, a man shot and killed two people at the Whip Burger Menu restaurant in Toronto's west end. The shooting disappeared from the media agenda within a couple of days.[22] Both the victims and shooter were black. Again, in March 1998, Christine Ricketts was canvassing door-to-door for the *Toronto Star*. The mother of two was strangled and found in the stairwell of a high-rise apartment. The *Toronto Star* paid serious attention, since she was in their employ, while it was a "yawner" for the rest of the media.[23] Ms. Ricketts was black and the strangler was white and just released from prison three months earlier where he served time for violently assaulting a prostitute.[24]

The reporting on the Bernardo and Homolka trial was arguably more intensive and emotional than the Just Desserts reporting. Yet these crimes were reduced to the actions of two individuals. The actions of both were explained by individual sicknesses and deficiencies[25] or individual mental illness.[26] No one asked "What's wrong with blue-eyed, blond-haired men of Italian descent?"[27] Regardless of how horrific and sickening the crime, the white "race" card was never played.[28] Similarly, no one equated the behaviour of Homolka with that of all blonde-haired white women!

Since the fall of the Soviet Union in 1989, the terrorist attacks of 11 September 2001, and the second American invasion of Iraq in 2003, the group that have most often been demonized by the media are people of Middle East descent, especially Muslims. This is especially the case with racial profiling (see Chapter 4). Muslims are continually portrayed as terrorists, violence-prone, technologically backward, incapable of understanding democracy, having a disdain for the idea of freedom, and hardcore fundamentalists. A six-month study of five major Canadian newspapers found that Muslims were routinely portrayed as barbaric fanatics.[29]

Media distortions do not stop here. As Michael Parenti has written,

> The basic distortions of the media are not ... random: rather they move in the same overall direction again and again, favouring management over labour, corporatism over anticorporatism, the affluent over the poor, private enterprise over socialism, Whites over Blacks, males over females, officialdom over protesters, conventional politics over dissidence ... national chauvinism over internationalism, US dominance of the Third World over revolutionary or populist national change."[30]

This consistency is made all the more likely by the behaviour of individual journalists. Individually, the phenomenon of **pack journalism** also plays a role. Reporters feed off one another, move in unison, and rush to be the first to receive accolades for "breaking" the story.[31]

A larger picture is provided by University of Windsor professor, James Winter, who uses the term **media think** to describe the way media owners, managers, and workers think, see, and represent the world around us.[32] In fact, he argues the "bias framework" provided by the media is "discernable and consistent" and reflects "their class, gender, race and corporate ties."[33] The media do not just represent corporate interests, they are corporate interests. Major newspapers, magazines, radio stations, and television stations are owned by corporate conglomerates. This is especially true of American television stations. For example, ABC is a wholly owned subsidiary of Time Warner, which owns, among others, Warner Brothers and MGM movie studios, CNN Radio, 30 magazines (including *Time*, *People*, and *Sports Illustrated*), book publishing companies, Internet companies (AOL), and cable television. Another example is NBC. During the early 1990s NBC never mentioned the fact that corporate giant General Electric was the target of a worldwide boycott that culminated with an Academy Award for the 28-minute documentary *General Electric: Deadly Deception*, which chronicled this very fact.[34] NBC is wholly owned by General Electric.

The most recent and visible example of media think and pack journalism is the Jessica Lynch "rescue" story of July 2003. The soldier had not been shot, tortured, or stabbed defending democracy in Iraq. Her injuries were a result of a truck accident during transport and the "heroic" rescue was rather a benign visit by U.S. troops to pick her up at the hospital in the city of Nasiriyah without encountering any resistance.[35] Her "heroic" rescue was a wonderful example of Pentagon and media manipulation. The BBC referred to it as "one of the most stunning pieces of news management ever conceived."[36] Nonetheless, Jessica Lynch received a hero's welcome on her arrival home.

The Power of American Media

One related issue here is whether mass media can accurately reflect the ideas of all societal groups. Most of the mass media in the world are controlled by American corporations whose aims may conflict with teaching about ethnic diversity. The United States has been accused of **television imperialism**, referring to the idea that it attempts to control other nations by using television as a means to dominate cultures.[37] In other words, it has been suggested that American media persuade people around the world to adopt American values by selling television shows and print media that portray American life, while paying little attention to life in other societies. How is this possible? Edward Herman and Noam Chomsky's *Propaganda Model* helps shed some light on mass media behaviour.

Herman and Chomsky's influential work, *Manufacturing Consent: The Political Economy of the Mass Media* (2002) dismisses the idea of an objective, independent, and truth-seeking media operating in democratic society. Rather, the media are part of a complex process of "managing" public opinion in a way that favours dominant groups at the expense of those that are not. Specifically, they describe the behind the scenes, invisible (to the public) forces that make the mainstream media a propaganda tool to transmit information to the public that is overwhelmingly biased in favour of ruling interests, capitalism, and other private interests. They call it the **propaganda model**. They use this model to explain how the media can "manufacture" the "consent" of the public by choosing what stories to ignore, what stories to cover, and how to cover them. By doing so, they control what the public is allowed to see, hear, and read.

The Propaganda Model is based on the idea that different "filters" eliminate unfavourable elements from news stories so the final message that reaches the public is one that supports the dominant interest. There are five interrelated filters.[38]

The first is the size, ownership, and profit orientation of the media. While media conglomerates are large, they are owned by a relatively few (see Box 11.5). Media companies are expensive to run. The cost of resources, technology, machinery, land, salaries, and buildings is enormous. This severely limits the number of people who can "set up shop."

BOX 11.5

CanWest Global

In 2000 the late Izzy Asper, owner of CanWest Global, purchased Southam newspapers and the *National Post* from former media mogul Conrad Black's Hollinger Corporation. Shortly thereafter, Asper sought to reverse Southam's longstanding claim of editorial independence. No local editorials were allowed to differ from those emanating from head office.

Source: Jeffery Klaehn, ed., *Filtering the News: Essays on Herman and Chomsky's Propaganda Model* (Montreal: Black Rose Books, 2005), p. 86.

Consequently, making money is essential to survival. This is best achieved by supporting monied interests. Only the wealthy need apply.

The second filter is advertising. Advertisers choose which media will survive when they choose where to spend their advertising dollars. The media will link a buying market (general public) to sellers (advertisers). This advertising distorts news, since the media are more accountable to advertisers that to the pursuit of the "truth."

Consistent news sources is the third filter. It is extremely expensive to attempt and cover the world with reporters, journalists, radio transmitters, cameras, and satellites. A far cheaper alternative is to accept the constant stream of information and analysis from government, corporations, and other "experts." Therefore, the kind, favourable, and timid media treatment of these sources translates into "a steady, reliable flow of the raw material of news."[39] The pursuit of the truth is the first casualty of this symbiotic relationship.

Flak and enforcers is the fourth filter. "'Flak' refers to negative responses to a media statement or program."[40] It can take many forms, including letters, phone calls, petitions, boycotts, lawsuits, demonstrations, and other actions that register disagreement. Those in positions of power can produce flak that impact on media operation. Threatening to pull advertising dollars and spending them elsewhere is one popular method that has a chilling effect on media behaviour. Corporations can also complain to their employees and shareholders and fund right-wing think tanks and interest groups that provide more friendly news and analysis. In Canada, the two most noteworthy groups are the Fraser Institute and the Canadian Council of Chief Executives (CCCE) run by Thomas d'Aquino. The CCCE represents 150 companies with collective revenues of over $600 billion and collective assets in 2005 of over $2.5 trillion.[41]

The fifth and final filter is the idea of anticommunism. Since the fall of the Soviet Union, Herman and Chomsky have replaced the term "anticommunism" with "dominant ideology." The dominant ideology "helps mobilize the populace against an enemy."[42] In 2006, anticommunism has been replaced by "globalization" and the "war on terrorism." These general terms are used to coerce the general population to accept lower wages, lower standards of living, the shredding of the social safety net, and the attack on civil liberties. The writing of Oakville, Ontario's Linda McQuaig provides the best Canadian example of this filter. She meticulously documented the 25-year use of "deficit hysteria" by the government, capitalist elites, and the groups that represent them to attack and dismantle Canada's extremely popular social safety net. Once federal and provincial deficits were under control, they moved on to using "globalization" as the justification to continue their attacks.[43]

Certainly the United States has a huge role in media markets far beyond its borders: in almost thirty nations, including Canada, U.S. companies control 70 percent of the market for films, and they control 35 percent of the cable television market in Canada.[44] Moreover, U.S. movie corporations are the largest in the world. Using Innis's terms, the

United States is a "space-oriented" culture that is intent on broadening its cultural and economic access. This effort on the part of the United States has many effects.

Indeed, the perception among many nations is that U.S. movies are subtle forms of political propaganda, aimed at flattering and spreading the influence of the United States.[45] Mass media often do celebrate the values and benefits of their country of origin, and American media are no exception here. McLuhan suggested that copies of American magazines and catalogues smuggled into the communist bloc posed a greater threat to its governments than any official political propaganda;[46] once the citizens of this bloc saw what consumer goods the average Westerner enjoyed, the people challenged their governments severely. A U.S. movie that portrays ordinary people in possession of cars, stoves, and refrigerators may be a revolutionary message to the poor of the developing world. In this context, force is not needed to support the American way.

American media have at least two major effects on non-American nations. First, the flood of U.S. programming into other nations discourages the latter from creating their own programs that might better reflect their own national identities. Often the United States is accused of promoting **ethnocentrism**, a feeling of cultural superiority, through mass media. In short, it is believed that U.S. dominance over mass media has weakened attempts to strengthen ethnic diversity. The cultures of Latin America and the Middle East have constantly struggled to create programming alternatives that reflect their national identities.[47] Even nations such as France, which began this century dominating the film industry and the arts, have now fallen under the influence of U.S. media. It has reached the point where these nations are concerned about preserving their own culture and language. For this reason, both France and Canada have occasionally imposed tariffs to limit U.S. media imports.[48]

Second, some observers argue that the success of U.S. media throughout the world is difficult to challenge because these media can appear to convey and represent ethnic diversity so well. For example, the film directors, comedians, and dramatic actors featured in U.S. programming are increasingly coming from diverse ethnic backgrounds (e.g., Jewish, Italian, and African-American), reflecting the multicultural nature of U.S. society. These diverse elements may explain the broad appeal of American media to various parts of the world.[49] If this is the case, however, this use of diversity is a new phenomenon, for American media have not always represented ethnic communities with the importance they deserve. For example, one staple of American popular culture, the type of movie known as the "western," still does not represent African-Americans in numbers reflecting their participation in the settlement of the West. Although one out of four cowboys was an African-American during the opening up of the Western frontier, Hollywood westerns still predominantly use white actors. Similarly, movies on Vietnam have still not fully represented the actual proportion of African-American troops who fought and died there.[50] The absence of characters belonging to minority groups on children's television

The Treatment and Perception of Diversity

programs has received comment.[51] Thus, it is dubious whether all media can be expected to provide serious data about the complexity of diversity when profits and entertainment are considered more important.

The Language of the Global Village

Another problem related to the representation of diversity in media is the dominance of the English language, which hinders non-English cultures from developing their own networks: English is the "media language."[52] This condition is becoming apparent on the Internet, too. Even though the Net has been described as an electronic, computerized global village (see Box 11.6), the dominant language or **lingua franca** (the language spoken by all cultures in a given area of business) on the Net is English. Despite the fact that almost half of Internet users live outside North America, most computers cannot translate non-English into English. Fortunately, a North American consortium is now developing a universal translation code, known as Unicode, to allow computers on the Net to represent the letters and characters of virtually all languages.[53]

Unicode has been successful in international technology platforms, which should drive it into more environments. Unicode's continued success will rely on new computer operating systems, beginning with Microsoft NT, which are compatible with Unicode. The dream of developing a truly multilingual World Wide Web is not yet a reality, but it has moved much closer.

There is some additional reason to hope for change in this area. To be sure, a white Anglo-Saxon establishment has controlled the airwaves in the past: in addition, the vast

---

BOX 11.6

Advertising and Consumerism

The mass media have always targeted the consumer through advertising. With the advent of the Internet, this process has accelerated rapidly; every good and service imaginable can be purchased around the world.

Yet consumerism teaches only one value—the desire to obtain the newest commodities; the new is heralded over the old. Some media observers (such as Benjamin Barber) have called attention to the triumph of "McWorld," or the transformation of the globe into one homogeneous consumer market, which threatens to undermine distinctive cultures and sacred traditions. Will the hunger for new and improved goods displace the need to preserve distinctive cultures, traditions, and roots?

majority of the world's media corporations are in the United States. But this establishment may not always have a monopoly. Historically, ethnic communities in Canada and the United States who feel ignored by the mainstream Anglo-Saxon press have taken to developing their own. Since the early 20th century African-Americans have produced and directed films depicting the historical contributions of their community in American culture, as a way of counteracting the absence of African-American themes in Hollywood movies. African-Americans have found it necessary since the 19th century to create their own newspapers to respond to racist attacks and stereotyping.[54] In Canada, newspapers have been set up by Indian, Asian, Jamaican, and many other ethnic publishers to serve their own communities.

This trend toward greater ethnic participation in the industry continues. The rise of ethnic diversity in North America has led to the demand for hundreds of cable channel systems, divided by language and culture. For example, 107 languages are now spoken in Southern California, and the media industry has responded with videocassettes, disks, and computerized banking geared to the traditions and needs of these various ethnic groups. Canada's magazine *Maclean's* has recently introduced a special edition in Chinese. Women's television networks are gaining popularity in North America. There has even been discussion of setting up several television networks to serve the needs of distinctive ethnic communities, instead of placing network programming into the hands of only a few media companies.[55] Observers concerned about the "Americanization" of the world through mass media predict that nations as diverse as India, Egypt, Mexico, and Brazil will enjoy the benefits of the Anglo-American mass media while developing their own distinctive programming.[56] Since the 1960s, for example, India has ensured that an official radio network exists for each of its dozen most-spoken languages.

Surviving in the Global Village: Canada and Brazil

Canadians should take particular interest in those nations that have been able to maintain their ethnic identity against the onslaught of U.S. media. Canada has produced many influential thinkers on the subject of media technology and how it is a powerful support for American media.[57] Perhaps this is because Canada, more than any other nation, has

---

CRITICAL THINKING BOX 11.3

Diversity and Education

How many of your textbooks discuss in detail the history and culture of ethnic communities in Canada?

been bombarded with U.S. programming ever since the rise of the media industry in the United States. The importation of American media into Canada has been seen by many Canadians as a threat to their identity. In a sense, this threat is serious: by 1900 Canadians read mostly American books, and the Canadian press was seen as modelled on American newspapers. Since the 1930s, the Canadian government has sometimes responded to this problem by creating agencies that favour Canadian broadcasting (such as the CBC) and by placing tariffs on U.S. magazines.[58]

Still, many Canadians are skeptical about the success of these efforts. George Grant argues that national identity cannot be reconciled with technological progress in the global village. In 1995 these concerns again became pressing, when Disney successfully acquired exclusive rights to use the symbols of the Royal Canadian Mounted Police (those rights have now reverted to the Mounted Police Foundation). It is simply impossible to maintain one's distinctive ethnic identity in an age driven by technical advancement.[59] In Canada, the Québécois want to maintain the language and culture of their French ancestors, and they also want to enjoy the benefits of U.S. technology and media. Some argue, however, that this effort is contradictory and doomed to failure, for the embrace of American technology and media might mean the disappearance of Québécois culture. Can a time-oriented culture survive in an age of space-oriented media?

The problem of keeping traditional cultures alive is compounded by the fact that modern mass media, especially television, tend to be so "present oriented." Media tend to focus simply on the present, at the expense of the past.[60] "There is no memory on TV," as the saying goes. In an age of instantaneous information achievable through electronic media, it is understandable that consumers would not want to study the past or tradition. Why focus on what is antiquated when the new is changing all the time? But there remains a problem for those concerned with the survival of diversity. Because our technological age tends to value the "new" or the most technically sophisticated as the "best," and the "old" as the inferior and irrelevant, can a traditional culture survive?

This last question is particularly important to the study of diversity, for if cultures cannot survive in the media age, how can diversity? Some thinkers, such as McLuhan, have contended that the lack of a strong national identity is advantageous in a deeply technological age. Because the rate of change is so great, it is not wise to develop an inflexible nationalism. Indeed, some praise Canada for its lack of such a strong identity, adding that this limitation makes it better able to adjust to the pace of technological development.[61] Still, what if cultures and communities want to maintain their differences? Can this be done in the age of the global village, where Anglo-American media seem so dominant?

There is reason to hope for an optimistic response to these questions. Consider the example of Brazil. It has one of the largest television audiences in the world.[62] It also has wide access to U.S. programming. Yet there is no hard evidence to show that Brazilians

> **CRITICAL THINKING BOX 11.4**
>
> *Canadian Content*
> Why do Canadian news stories show up so rarely on U.S. news programs?

are becoming "Americanized" because of this access. Indeed, despite the high level of violence in Brazilian society, there is no great demand for violent genres like the "slasher films," so popular in the United States. Indeed, according to one survey, Brazilians continue to admire the technological innovation represented by U.S. television, but they do not adopt the values of the United States.[63] Of course, Brazil has been able to enjoy these attitudes because its government is determined to maintain a high level of its culture's programming, to compete with the constant onrush of U.S. media.

As nations come to grips with the effects of television on their cultures, a new media revolution is already on its way. The emergence of virtual reality (computer-generated imagery that mimics reality) is said to herald this new era.[64] Can this new age of mass media create greater awareness of diversity? It will certainly accelerate the global village, for this age promises technologies that will link the world more instantaneously than ever before. The "teleputer," a combination of a computer and television that will permit a user to see, hear, and communicate with another user from a distant part of the globe, may also soon be online. The world will become "wired." Whether this new interconnectedness through computers, the Internet, virtual reality, or teleputers will create greater tolerance for diversity is another question. The record of mass media on this subject is mixed: television accelerated the global village, but this has not always produced greater harmony or understanding among cultures.

CHAPTER SUMMARY

Human beings have always employed techniques of mass communication. The use of mass media ranges widely from communication to information to entertainment. Mass media also affect our behaviour. They shape the message and content of what is being communicated to millions of people, especially the message and content of diversity.

In each of the four stages in the history of media, great changes in human behaviour have taken place. The oral stage demanded that people use their memories to "record" events; the rise of the alphabet extended the power of the eye, as ordinary people began to see what they were saying for the first time; the printing press encouraged nationalism; and the advent of the electronic age created the global village.

Electronic media have had by far the greatest impact on defining diversity in the modern world. Perhaps the most important effect of electronic media was turning the world into a global village, in which, because of the speed of communication via electricity, people of all nations feel interconnected. Since the emergence of the telegraph, this effect has only been strengthened by other electronic media.

Television, still the dominant medium in the world today, continues to accelerate this feeling of global interconnectedness. Whether television can represent diverse cultures in an accurate manner is an open question, for the success of television lies in focusing on the new, the entertaining, and the profitable. The record is mixed. Television can make people more aware of other cultures, but its presentation of diversity can be misleading and superficial.

To complicate matters further, most of the media markets are controlled by U.S. corporations. This control has led to the dominance of the English language in these markets, as well as the sustained influence of U.S. programs around the world, including Canada. Many nations have taken steps to control the influx of U.S. programming to protect their national identities.

One thing is certain. Mass media have changed every aspect of our world forever (see Box 11.7). There is no possibility of returning to an age when media technology was restricted to a privileged few, and people—the "masses"—knew little about other cultures in distant lands. For those concerned with emphasizing diversity, the permanence of mass

BOX 11.7

Media Facts: Did You Know?

- Canadians watch on average 25 hours of television weekly.
- European TV networks start broadcasting in the afternoon.
- Africa has 2000 language dialects.
- The Chinese language has 40 000 characters.
- By age 20, the average North American has seen 800 000 commercials.
- There are 180 languages spoken in India.
- The average American has more TVs than bathrooms.
- Brazil has the fifth-largest TV audience in the world.
- The best market for American movies (outside the United States) is Japan.
- The average American 12-year-old has spent 13 000 hours in school and almost 20 000 hours watching TV.

media is ultimately a good thing. After all, even if the media in the global village do not always foster peace among its members, they certainly encourage the need to be aware of the various people who live in this village.

KEY TERMS

alphabet, p. 307
dialects, p. 311
ethnocentrism, p. 321
global village, p. 312
hieroglyphics, p. 307
lingua franca, p. 322
mass medium, p. 305
media think, p. 318
orality, p. 307
pack journalism, p. 318

parchment, p. 309
pictograph, p. 307
propaganda model, p. 319
script, p. 309
space-oriented, p. 309
telegraph, p. 311
television imperialism, p. 319
time-oriented, p. 309
tribalism, p. 312
vernaculars, p. 311

DISCUSSION QUESTIONS

1. Discuss five examples of commercials that illustrate how the "medium is the massage."
2. Outline the characteristics of the "propaganda model"?
3. Is Canada being threatened by television imperialism? Compare the number of Canadian and U.S. TV shows on a typical programming day.
4. You have been asked to produce and direct a TV situation comedy that accurately portrays the multicultural makeup of our society. How do you do this?
5. The Canadian government has asked your advice on how to improve programming so that it reflects Canada's diversity but is also entertaining. What do you suggest?

NOTES

1. The title of Marshall McLuhan's book from which this quotation is taken is called *The Medium Is the Massage*, not "the message." McLuhan's point was that the form of the message (the medium) shapes (or massages) how the message is perceived. McLuhan's skill with language is evident in his choice of wording here.
2. Walter Ong, *Orality and Literacy: The Technologizing of the Word* (New York: Routledge, 1988), p. 7.
3. See Harold Innis, *Empire and Communications* (Toronto: Press Procépic, 1986), p. 32.
4. Innis, pp. 38–42.
5. Ong, p. 28.

6. Innis, p. 5.
7. Ong, p. 89.
8. Innis, p. 128. See also Marshall McLuhan, *The Gutenberg Galaxy: The Making of Typographic Man* (Toronto: University of Toronto Press, 1962), pp. 218–19.
9. McLuhan, *Gutenberg Galaxy*, p. 231.
10. Morse, quoted in Neil Postman, *The Disappearance of Childhood* (New York: Delacorte Press, 1982), p. 68.
11. See Marshall McLuhan, *Understanding Media: The Extensions of Man* (New York: New American Library, 1964), p. 156.
12. Michael Parenti, *Inventing Reality: The Politics of News Media* (New York: St. Martin's Press, 1993), p. 1.
13. Neil Postman, *The Disappearance of Childhood* (New York: Delacorte Press, 1982).
14. Joshua Meyrowitz, *No Sense of Place: The Impact of Electronic Media on Social Behaviour* (New York: Oxford University Press, 1985), p. 132.
15. See Augie Fleras and Jean Leonard Elliott, *Multiculturalism in Canada: The Challenge of Diversity* (Toronto: Nelson, 1992), pp. 233–48.
16. Todd Gitlin, *The Whole World is Watching: Mass Media in the Making and Unmaking of the New Left* (New York: Oxford University Press, 1980), pp. 152–53.
17. James Winter, *Media Think* (Montreal: Black Rose Books, 2002), p. 147.
18. Winter, p. 147.
19. Jeffery Klaehn, ed., *Filtering the News: Essays on Herman and Chomsky's Propaganda Model* (Montreal: Black Rose Books, 2005), p. 31.
20. Winter, p. 146.
21. Winter, p. 147 and Klaehn, p. 31.
22. Winter, p. 146.
23. Winter, p. 146.
24. Winter, p. 146.
25. Augie Fleras and Jean Lock Kunz, *Media and Minorities: Representing Diversity in Multicultural Canada*, (Toronto: Thompson Educational Publishing, 2001), p. 31.
26. Winter, p. 146.
27. Fleras and Kunz, p. 31.
28. Fleras and Kunz, p. 31.
29. Fleras and Kunz, p. 29.
30. Parenti, p. 8.
31. Parenti, p. 42.
32. Winter, p. xxvii.
33. Winter, p. xxvii.
34. Parenti, p. 36.
35. Klaehn, p. 41.
36. Klaehn, p. 36.
37. Fleras and Elliott, *Multiculturalism in Canada*, p. 238.
38. Edward Herman and Noam Chomsky, *Manufacturing Consent: The Political Economy of the Mass Media* (New York: Pantheon Books, 2002), pp. 3–35.
39. Herman and Chomsky, p. 18.
40. Herman and Chomsky, p. 26.

41. Maude Barlow, *Too Close for Comfort: Canada's Future within Fortress North America* (Toronto: McClelland and Stewart, 2005), p. 4.
42. Herman and Chomsky, p. 29.
43. Winter, p. xxvii. Read any of McQuaig's books. Her easy and direct style make them quite readable. To fully understand the use of "deficit hysteria" and how easily it was replaced with "globalization" as an excuse to assault Canada's social safety net, it is recommended that you read them in chronological order.
McQuaig, Linda. *The Quick and the Dead: Brian Mulroney, Big Business and the Seduction of Canada.* Toronto: Penguin Books, 1991.
McQuaig, Linda. *The Wealthy Banker's Wife.* Toronto: Penguin Books, 1993.
McQuaig, Linda. *Shooting the Hippo: Death by Deficit and Other Canadian Myths.* Toronto: Penguin Books, 1995.
McQuaig, Linda. *The Cult of Impotence: Selling the Myth of Powerlessness in the Global Economy.* Toronto: Penguin Books, 1998.
McQuaig, Linda. *All You Can Eat: Greed, Lust and the New Capitalism.* Toronto: Penguin Books, 2001.
44. Jeremy Tunstall, *The Media Are American: Anglo-American Media in the World* (London: Constable, 1977), p. 182. See also Mary Vipond, *The Mass Media in Canada* (Toronto: Lorimer, 1989), p. 134.
45. McLuhan, *Understanding Media*, p. 271.
46. McLuhan, *Understanding Media*, p. 152.
47. See Tunstall.
48. Shirley Biagi, *Media/Impact: An Introduction to Mass Media* (Belmont, CA: Wadsworth, 1994), p. 472; see also Vipond, pp. 24–25.
49. Alvin Toffler, *Power Shift: Knowledge, Wealth, and Violence at the Edge of the Twenty-First Century* (New York: Bantam, 1990), p. 451.
50. Richard Slotkin, *Gunfighter Nation: The Myth of the Frontier in Twentieth-Century America* (New York: Atheneum, 1992), pp. 526–27.
51. Biagi, p. 392.
52. Tunstall, p. 126.
53. See Andrew Pollack, "Cyberspace's War of Words," *The Globe and Mail*, 10 August 1995, p. A15.
54. Biagi, pp. 45–49.
55. Marshall McLuhan and Bruce R. Powers, *The Global Village: Transformations in World Life and Media in the Twenty-First Century* (New York: Oxford University Press, 1989), p. 88.
56. Tunstall, p. 274.
57. Arthur Kroker, *Technology and the Canadian Mind: Innis/McLuhan/Grant* (Montreal: New World Perspectives, 1984).
58. Tunstall, p. 104; Vipond, pp. 24–29.
59. See George Grant, *Lament for a Nation: The Defeat of Canadian Nationalism* (Ottawa: Carleton University Press, 1965), pp. 76–87.
60. See Neil Postman and Steve Powers, *How to Watch TV News* (New York: Penguin, 1992).
61. See McLuhan and Powers, p. 165.
62. Conrad Philip Kottak, *Prime-Time Society* (Belmont, CA: Wadsworth, 1990), p. 12.
63. Kottak, p. 93.
64. See Derrick De Kerckhove, *The Skin of Culture: Investigating the New Electronic Reality* (Toronto: Somerville House, 1995).

CHAPTER 12

Perceptions of Diversity in Canadian Literature

Maureen Coleman

Literature . . . binds one human being to the next and shortens the distance we must travel to discover that our most private perceptions are universally felt.
— Carol Shields

A country's literature is a crystal ball into which its people may look to understand their past and their present, and to find some foretaste of their future.
— Robertson Davies, Edinburgh, 1988

Objectives

After reading this chapter, you should be able to

- understand how literature is the voice and mirror of a culture
- understand the evolution of diversity in Canadian society through its literature
- appreciate the relationship between power and voice in Canada as revealed in the stories of Native Canadians, racial and religious minorities, immigrants, and refugees

- recall stories and other writing that reflects the diversity of cultures between, and within, the geographical regions of Canada, along with our international connections
- appreciate how contemporary Canadian writing strongly gives voice to our continually evolving cultural diversity in society's perceptions of family, sexuality, and gender

INTRODUCTION

The Role of Literature in Recording and Defining a Culture

Every society has its shared and unique experiences, which are shaped by the storyteller into song, poem, drama, comedy, myth, and legend. Literature is the **voice** of a culture. This chapter will describe how the spoken and written record of Canadian life has evolved and was passed from generation to generation.

The first and continuing stories of diversity in Canada were those of the "two solitudes" of English and French, with the resulting development of two distinct, parallel literatures, each written in its own language. This chapter will trace the changes in the story of Canadian society that came with the influx of people from many and various backgrounds and with the passage of time. We can see how these stories circulated within the Canadian community and combined with those of others to form a new vision. Perceptions of gender diversity can be found in the body of **feminist literature** that has filled an absence in the story of Canadian society. Northrop Frye referred to this "cross-pollination" of cultures as a process that invigorates and strengthens.[1] We will also examine how Canadian writing found its way onto the world stage. Readers will discover how the literature of and by Canadians provides a mirror that reflects our image, in all its complexity and diversity, not only to the world but also to ourselves. In *Survival: A Thematic Guide to Canadian Literature* (reissued 2003) Margaret Atwood showed how Canadian stories present a new culture. It is one that is separate not only from the European motherlands of England and France, but also from our neighbours in North and South America. In the thirty years since *Survival*'s publication, perceptions of diversity have changed dramatically. Acceptance and appreciation of the many people whose stories have interwoven continue to expand and reshape Canada's multicultural odyssey.

Literature has the power to personalize and humanize sociological data and historical events. The writer creates a character, living in a time and place with that era's particular values and attitudes, through whose eyes we can view that world. In a sense each writer freeze-frames a particular Canadian world for the reader to enter. As well, we begin to think

BOX 12.1

Culture and Story: A Changing Relationship

A comprehensive study of this revolution in thinking about the relationship between culture and story can be found in *The Empire Writes Back*, co-written by Bill Ashcroft, Gareth Griffiths, and Helen Tiffin of Australia. It has led to **postcolonial writing** "in cultures as various as India, Australia, the West Indies, Africa, and Canada . . . [that] challenges the existing canon and dominant ideas of literature and culture. . . . [P]ost-colonial texts . . . constitute a radical critique of the assumptions underlying Eurocentric notions of language and literature."[2]

about where each of us fits into the Canadian saga. Although it is true that our stories are shaped by the society we live in, it is also true that the society is influenced by its stories. We can identify with the **universal themes** explored in all literature from a uniquely Canadian viewpoint.

CANADIAN LITERATURE: PART OF A POSTCOLONIAL GLOBAL TREND

Voices from the Margins Move to the Centre

Canadian stories, which are part of this international postcolonial body of writing, are those of Native Canadians, African-Canadians, immigrants, refugees, and women. Also included are those previously silenced because of physical and mental diversities. In Canada, as elsewhere in the world, the stories of marginalized individuals and groups within the society were often ignored or suppressed. However, with more positive attitudes toward diversity, this **absence of voice** is being overcome. The retelling of the story of Canada's first settlers, the Acadians—their expulsion and cultural rebirth—is found in two recent works: *The Acadians: A People's Story of Exile and Triumph* by Dean Jobb and *A Great and Noble Scheme: New Brunswick: The Expulsion of the French Acadian* by John Mark Faragher.

Perceptions of Native People: The Nature of Cultural Imperialism

Only in the last quarter of the 20th century did Aboriginal peoples throughout the world have the freedom to tell their own stories. This is certainly true in Canada for Native

Canadians. From all parts of the country, they struggle to overcome centuries of political and cultural imperialism. Separating myth from reality is a difficult task. In an article entitled "Seeing Red over Myths," Drew Hayden Taylor, writer, actor, playwright, and filmmaker of Ojibway heritage, discusses "the myth of pan-Indianism.... It reveals a persistent belief that we are all one people. Within the borders of what is now referred to as Canada, there are more than 50 distinct and separate languages and dialects. And each ... has emerged from a distinct and separate culture."[3]

Best-Known Stories: Romantic Stereotypes

Initially, the European newcomers viewed Canada's Native peoples with a mixture of scorn and fear. Concerted efforts were made to impose European culture on the Native populations. Little attention was paid or value attributed to Aboriginal cultures. However, from the middle of the 18th century, the literature reveals the romantic stereotype of the "noble savage." Charles Mair, in his verse drama *Tecumseh*, wrote:

> There lived a soul more wild than barbarous;
> A tameless soul—the sunburnt savage free—
> Free, and untainted by the greed of gain:
> Great Nature's man content with Nature's food.[4]

In *Survival: A Thematic Guide to Canadian Literature*, Margaret Atwood pointed out that our perception differs from the American good Indian/bad Indian viewpoint. The Canadian version was more of a victor/victim version, or, as Leonard Cohen expressed it in his novel of the same name, Native peoples were *Beautiful Losers*. The stories tend to end with the Native willingly sacrificing his or her own ways to those of the newcomer.

Untold Stories: Realism and Legend, Big Bear, and Louis Riel

Canada has another untold story of poverty and the loss of self-respect that comes with being a "kept people." Generations passed before this picture in the Canadian family album was seen. In *The Temptations of Big Bear*, Rudy Wiebe tells of the only Cree chief, Big Bear, who resisted offers of the white man to give up Native land for the railway and settlement.

Another dramatic story of the fight against the tide of western expansion is that of **Métis** leader Louis Riel. Riel's people were rejected by both whites and Natives and denigrated as "halfbreeds." Riel's story has become an integral part of Canada's myth and legend. Early tellings of the tale presented Riel as a reactionary, a fool resisting progress, and finally a lunatic who could not accept defeat. Later Riel's story would appear in many forms—poetry, drama, novel, documentary, and film, and in the rewriting or revision of history. In these Riel is recast as a hero, a bold, courageous leader, a visionary willing to sacrifice his life for his people. Chester Brown's *Louis Riel: A Comic Strip Biography* tells the story in a unique, accessible form.

For generations, Riel and Big Bear were ghosts haunting the Canadian imagination and conscience. Literature was also important in defining the place of the Métis in Canadian culture, first by obscuring it and later by revealing its true nature. Maria Campbell's autobiography *Halfbreed* is one of the first books to do this. Both Native and white societies had a strong traditional resistance to acceptance of the Métis, Canada's first people of mixed cultures.

Native Voices of Today Speak for Themselves

Native peoples maintained their strong oral tradition, and thus stories were passed down through the generations. As we are drawn into the original story circle, it becomes a healing circle as well, one that is much needed by both groups. In 1990 Thomas King edited a collection, *All My Relations: An Anthology of Contemporary Canadian Native Fiction*, containing the writing of 19 authors, including himself. That King is of Cherokee, Greek, and German descent is part of what is meant by "contemporary Canadian." Most of these stories are influenced by various aspects of traditional Native storytelling and humour. Others centre on Natives in a contemporary world, old injustices, residential schools, women's status, and problems of identity. Poet and Mi'kmaq from Nova Scotia Rita Joe wrote in the Introduction to her third book of poems, *Lnu and Indians We're Called*: "One way for native people to experience the positive parts of their culture is through spiritual productiveness. This is what I try to convey—there is a great need to tell our story in 1991, more than ever."[5]

Personal Voices Move Closer to the Social Centre

Some stories address the condition of being "colonized" on a more personal level. Perhaps when *Funny Boy* by Shyam Selvadura won two literary awards in 1995—one Canadian, one Commonwealth—it indicated both a cultural and a literary maturity. The title contains a double irony. The protagonist, Arjie Chelvara Tuam, is a young man who is "funny" in two ways. First, he is of Sri Lankan background, and, second, he is homosexual.

CRITICAL THINKING BOX 12.1

Discuss Taylor's challenge in understanding our assumptions about the diversity among Canada's first peoples. What other myths are widely held about Native peoples? What are the sources of these beliefs? Are they supported by fact?

> **BOX 12.2**
>
> *Native Writers Tell Their Stories*
>
> - *An Anthology of Canadian Native Literature in English* (1998), edited by Daniel David Moses and Terry Goldie, includes stories from 57 authors representing 28 different tribes, five traditional Inuit songs, and seven examples of traditional orature from southern First Nations.
> - Drew Hayden Taylor's CBC docudrama *Redskins, Tricksters, and Puppy Stew* entertains and enlightens on Native humour and its healing effects. *Me Funny* (2006), edited by Taylor, is a compilation of stories by Canadian Native writers.
> - *The Mi'kmaq Anthology*, edited by Rita Joe and Lesley Choyce, includes 16 writers.
> - Three books by current Lieutenant Governor of Ontario James Bartleman, *Out of Muskoka, On Six Continents: A Life in Canada's Foreign Service*, and *Rollercoaster*, describe his life in politics from his unlikely beginning as the son of a Native (Majikaning) mother and white father.

To find his way and his place in a mostly Western, white, and heterosexual Canadian society, Arjie veils his true self in the clothing of clown and joker.

Another personal story of living with isolation and rejection is *Uproar's Your Only Music* by Brian Brett. Born an androgyne, now called *middlesex* (complicated by a physical condition in which the person never goes through puberty), he describes the confusion, anger, difficulties, and eventual acceptance of his life.

Two memoirs speak of mental illness. In *I Might be Nothing: Journal Writing*, Carole Itter reveals the journals of her daughter, Lara, from age 15 to 22 when she was successful at ending her life. It chronicles her depression, memories of abuse, and suicidal urges. In *The Voice Inside Me: A Memoir*, Elizabeth Ikiru (a pen name) tells her story. She was a married woman with two children who for ten years struggled with manic depression. The book, published by her partner Morris Wolfe, includes her poems and sketches, plus his and her two children's insights.

Maggie de Vries' book *Missing Sarah: A Vancouver Woman Remembers Her Vanished Sister* uncovers the life and death of her sister, whose disappearance from the city's east end met with little interest on the part of the authorities.

CRITICAL THINKING BOX 12.2

Is the official recognition of *Funny Boy* a "coming out" from denial of our own ethnocentrism? Is it a hopeful sign that must always be balanced against ultraconservatism and censorship on other fronts? What reasons may explain why the Canadian level of ethnocentrism is more or less than in another culture? What social factors fuel the difficult and sometimes divisive debate over gay marriage?

Stories of Race and Ethnicity: "Other" Voices Finally Heard
Black Voices

African-Canadians in Nova Scotia and Ontario have been part of Canadian society since Loyalist times, and in some cases earlier. Many artists' works fill in the blanks in the history, near and distant, of their communities. They also celebrate their strength and survival against racism, unofficial and official (e.g., the destruction of Africville), along with their broadening roles in province, nation, and world.

Nova Scotia Native poet, playwright, critic, and novelist George Elliott Clarke is at the forefront of rewriting lost stories, as well as raising awareness of the many other black writers who call Canada home. Clarke has written *Whylah Falls* (1990), edited *Eyeing the North Star: Directions in African-Canadian Literature* (1997), and written *Execution Poems: The Black Acadian Tragedy of George and Rue* (2001) and *Odysseys Home: Mapping African-Canadian Literature* (2003). Clarke's 2005 novel *George and Rue*, based on a real event, creates a world of physical and emotional poverty and isolation around the lives of two brothers, which led to murder and their deaths by hanging. Other voices from Nova Scotia include

- poet Maxine Tynes (*Borrowed Beauty*, 1987)
- filmmaker Sylvia Hamilton (*Black Mother Black Daughter*, 1989)
- a cappella singing group Four the Moment

Many writers have arisen among the black Canadians who have come from the English-speaking islands of West Indies mostly to Toronto, and to Montreal from former French colonies. Ayanna Black, in her poetry collection *No Contingencies*, writes of her colour in "A Sense of Origin," of being a woman in "Reflection," and of urban poverty and street people in "Bag Lady" and "In Memory of Lorraine." Austin Clarke, in

"Canadian Experience" and in many other stories, writes that immigration to Canada means more than cultural and climatic change; if you are a black person, it also means adjusting to a different, more subtle version of racism than the one left behind in Barbados. Rosemary Brown, social worker, politician, and international aid worker, writes of similar experiences in her autobiography *Being Brown*.

Dany Laferriere is a writer from the Haitian community in Montreal whose books include *An Aroma of Coffee*, *A Drifting Year*, and *Dining with the Dictator*. The themes of *No Crystal Stair* by Maeruth Sarsfield are, passing as white, and surviving as black. It is a coming of age story set in the Little Burgundy district of Montreal.

Poet and short story writer Dionne Brand states that if you are absent from literature or the forms in which you are included are only negative, then you must "write yourself." To that end she fights stereotyping by creating complex characters. Brand is coauthor of *Rivers Have Sources, Trees Have Roots—Speaking of Racism*. Her 2005 fiction, *What We All Long For*, is placed squarely in the city of Toronto as she reveals the lives of four friends in their twenties. Although they are from different backgrounds, they share an understanding of how their lives are so different from their parents. Their experience comes from not being firmly rooted in either culture.

Neil Bissoondath, a Trinidadian of East Indian ancestry, like some other writers mentioned here, came to Canada as a student. In an essay entitled "I'm Not Racist But . . ." Bissoondath defined the difference between racism born of hatred and that born of ignorance. Like Brand, he believes that the only way to change this is to write against stereotyping. Both writers believe that Canada's multicultural policies might be contributing to this by focusing on the differences of various groups of Canadians, rather than on their similarities. In his book *Selling Illusions: The Cult of Multiculturalism in Canada*, Bissoondath discusses these often-controversial ideas.

Countering the cultural imperialism of English language and literature is also an objective of Marlene NourbeSe Philip, lawyer turned writer. In *Harriet's Daughter*, her novel about a teenager growing up in Toronto, and in her poetry collection *She Tries Her Tongue: Her Silence Softly Breaks*, NourbeSe Philip consciously set out to explore and value all forms of language other than the imposed traditional ones. She edited *Frontiers*, essays on these topics.

Laurence Hill's books *Any Known Blood* and *Black Berry Sweet Juice: On Being Black and White in Canada* are told from the perspective of a mixed race family who moved from the United States in the 1960s.

York University's website African Canadian Online (www.yorku.ca/aconline) is a resource for culture, dance, film, theatre, literature, music, and visual art. For a longer list of selected writers including Afua Cooper, Archie Crail, Esi Edugyan, and Donna Bailey Nurse, see the website.

> **CRITICAL THINKING BOX 12.3**
>
> Should a writer who does not belong to a particular cultural group be able to tell a story through the eyes of a character who does? Some writers who belong to particular cultural groups, women, blacks, Natives, Asians, or people with physical disabilities such as blindness or deafness believe that such writing is not valid or honest. What are the arguments for and against this phenomenon, known as "appropriation of voice"?

Contemporary Writing Focuses on Diversity: Different Voices Break the Silence

Immigrant and Refugee Voices

When we read **immigrant and refugee stories**, and those of their children or grandchildren, we begin to understand not only the nature of Canadian diversity, but also its complexity. The process of finding one's place in a new culture is difficult, and in each case different. For an immigrant or refugee this is especially true if it means learning a new language. Mary di Michele, who came from Italy in 1955, writes in her poem "Luminous Emergencies":

> That my tongue has been un-
> Mothered. That my tongue has thickened
> with English consonants and diphthongs,
> mustard and horseradish. That burning.
> That burdened.
> While on my lips Italian feels
> almost free, like wind in the trees
> when the window's sealed shut
> and you're inside playing Scrabble.
> No—English is not so cosy!
> It's hypothermic. It's haunted
> by ghost letters & gnomic.[6]

The state of being between worlds is written about often. Authors describe the relationship between giving up the language that names and contains the meanings of the place that has been left, as the loss of part of themselves. Along with the difficulties of learning a new language come the feelings of powerlessness and frustration of being unable to communicate thoughts and feelings.

Eva Hoffman came to Canada at the age of 14 from Poland. In her 1989 autobiography, *Lost in Translation*, Hoffman describes this complex process of loss and gain. Another story that explores the issue of the power of language is "A Class of New Canadians," by Clark Blaise. The main character, a teacher, is a new Canadian from the United States, who teaches English. Readers may become uncomfortable with his paternalistic attitude toward his adult students, who, like him, have come to Canada to make a new life.

Himani Bannerji, who teaches sociology at York University in Toronto, was born in Bangladesh and educated in India, and arrived in Canada as a young woman in her late twenties. In her story "The Sound Barrier," Bannerji analyzes the communication gulf between herself and her mother. She draws the reader from a Western culture into an understanding of what "mother" means to her mother, from an Eastern culture. Bannerji shows how the sound barrier between them goes far beyond the language of the word "mother" to include thousands of years of cultural meaning. She compares this disparity to the gulf of understanding that exists between Canadians of Western backgrounds and those from Asian cultures. The story "Why My Mother Can't Speak English" by Gary Engkent, who came to Canada as a child, also asks and attempts to answer this same question.

Voices of younger writers are found in *Red Silk: An Anthology of South Asian Canadian Women Poets* edited by Rishma Dunlop and Priscila Uppal. The writers, such as Sandeep Sanghera, Shauna Sing Baldwin, and Proma Tagore, are concerned with the stresses between their traditional culture and mainstream Canadian values, questioning the extent of the separation and the time needed for a blending to occur. Hiro Boga's lyrics ask if we lose our identity when we embrace those of the world. Dunlop examines the meaning of belonging. Kuldip Gill uses English in an Indo-Persian form, the ghazall. Sonnet L'Abbe's poems express political anger at pressures to "think white" and to see a narrow view of beauty. Soraya Peerbaye employs powerful imagery to portray social insight. Sharanpal Ruprai "contrasts Western gender mores with the demands of Sikhism." Priscila Uppal remembers the pain of searching for her mother, who has run away.

Final Decree, by Hungarian-born George Jonas, is a mystery story based on culture shock. Josef Skvorecky was already a successful writer when he came to Canada from Czechoslovakia in 1969 after the Soviet invasion. Some of his works in English are *The Cowards*, *The Bass Saxophone*, and *The Engineer of Human Souls*. The stories by both Jonas and Skvorecky reflect the experience of refugees who have come to Canada because of war and violence. Civil and political justice are important concerns to them. Evelyn Lau ran away from home when her parents, Chinese immigrants, refused to let her write. *Runaway: Diary of a Street Kid* is a factual account of her life on Vancouver's streets. It was followed by two novels, *Fresh Girls* (1993) and *Other Women* (1995). Her new work *Inside Out* continues the theme of reflection on her life, relationships, and identity as a writer. For immigrants who did not speak English, often it was the third generation before a storyteller's voice broke through the sound barrier into the mainstream culture.

Some immigrants, such as Michael Ondaatje, had the benefit of having come from a former British colony, where the second language was English. These writers understood well the relationship of voice to power and numbers in a society, and this theme becomes a vital element in their novels. In Ondaatje's novel *In the Skin of a Lion*, the construction of Toronto's Bloor Viaduct represents an important image and symbol of cultures mixing, clashing, and yet quite often and unpredictably meshing. He portrays the frustration and isolation of the language barrier to these newcomers.

The sometimes humorous side of culture clash is found in Antanas Sileika's novel *Buying on Time*. With the attainment of the new language, the link to Canadian society, comes the inevitable loss of the old language and the culture to which it gives voice.

THE INTERRELATIONSHIPS OF DEMOGRAPHICS AND STORY

The Great Canadian Experiment or the Micro–Global Village

Canada has been called the first international nation. Except for Native peoples, all of us are from somewhere else. We are, or are descended from, immigrants or refugees, who arrived here, at various times and from diverse points in the "global village." Anthropologists use the term "low context" to describe a culture where diversity is more common than similarity. Many who came to Canada planned the move to gain a better life for themselves and a more positive future for their children. For others it was a refuge from war, economic upheaval, or natural disaster. For still others flight from imprisonment, or other forms of political repression, was the impetus to move. For some, finding a life they can understand is not possible. Among us are people whose stories are as diverse as can be imagined (see Table 12.1).

One group is the Hong Kong Chinese. They came to Canada in the face of unknown but definite change, when Communist China took over the former British colony in 1997. Their story presents some cultural similarities with that of their predecessors, who came to find adventure and fortune in the "golden mountains" of British Columbia and stayed to build, and died building, the railways. However, their acceptance into Canadian society is much smoother than that of the Chinese who came in the late 19th and early 20th centuries, who were exploited as cheap labour and then denied the right to bring

Table 12.1 Some Canadian Writers and Their Ancestry or Origins

Writer	Ancestry or Origin
Clark Blaise	American
Joy Kogawa	Japanese
Rohinton Mistry	Indian
Mary di Michele	Italian
Paul Yee	Chinese
Maxine Tynes	Black (Loyalist)
Drew Hayden Taylor	Ojibway and English
Dionne Brand	West Indian (Trinidad)
Lee Maracle	Métis/Salish
Alistair MacLeod	Scottish
Stephen Leacock	English
Gabrielle Roy	French
Morley Callaghan	Irish
Alberto Manguel	Argentinean
Michael Ondaatje	Sri Lankan

their families to Canada. They were also charged a prohibitive head tax, and suffered greatly from the prevailing attitudes that viewed them as the "yellow peril." Many years passed before word of this legalized racism was acknowledged.

Judy Fong Bates' novel *Midnight at the Dragon Café* is set in the fictional small town of Irvine just north of Toronto. Su-Jen (renamed Annie) and her mother arrive in the Canada of 1957. Over five years, from age seven, Su-Jen observes the turmoil in her family. Her young mother has an affair with her father's stepson, and bears a child; the family sends for a mail-order bride for her stepbrother; she observes her father's wisdom in dealing with the racism experienced from the restaurant's owner and customers. Fong Bates traces the girl's progress from naivety to wisdom, her understanding of the differences between her family and others in Irvine, and where she fits in in both.

One place the acknowledgment did appear was in Sky Lee's powerful novel *Disappearing Moon Cafe*, which tells a story from the point of view of a fourth-generation family member living in Toronto in 1987. Silence, both official and personal, kept this part of the Canadian story suppressed. It is symbolic of other omissions, denials, and neglects that signalled to third- and fourth-generation writers the need to revisit, uncover,

> **CRITICAL THINKING BOX 12.4**
>
> Is your part in the Canadian story represented in our literature? Is it part of an early and continuing chapter, such as that of the Acadians? Is it one with more contemporary immigrants or women, or perhaps both, at the centre? Perhaps it is a revised part of the story. If you are a Native person or a descendant of blacks, Japanese, or Chinese who have been Canadians for generations, how has your story evolved? Is it the story of flight from war or famine? Or is your chapter in the Canadian story yet to be told?

and reveal the complete, true—and often difficult—tale of the evolution of Canadian culture. Other realistic pictures can be found in the works of: Wayson Choy (*The Jade Peony*, *Paper Shadows: A Chinatown Childhood*, and *All That Matters*), a continuation of the lives of the Chen family; of Evelyn Lau; and of newer writers Denise Chong, Madeleine Thien, and playwright Betty Quan. In her memoir *A Leaf in the Bitter Wind*, Ting-Xing Ye writes of her 35 years living in China during the Cultural Revolution.

Two related works, Romeo Dallaire's *Shake Hands with the Devil: The Failure of Humanity in Rwanda* and Gil Courtmanche's novel *A Sunday at the Pool in Kigali*, are critical interrogations of the role of Canada and Canadians in the tragic civil war in Rwanda. Readers learn of events that led to genocides of both Tutsis and Hutus and destruction of their country, from which refugees came to Canada. Despite different points of view, Dallaire, as the general in charge of the United Nations troops, and central character Bernard Valcourt, a Canadian journalist, both convey the immeasurable harm done to the Rwandan people by all who were in a position to avoid the crisis. The authors point unflinchingly to the arrogance, indifference, self-interest, inaction, and utter lack of moral responsibility of outsiders. They also point out the courage of those who tried to intervene, at great cost, in a hopeless situation. The mask that hides the face of 21st-century colonization is stripped away and an accounting is demanded.

The Stowaway by Robert Hough tells a tale of a type of present-day refugees and their colonizers, involving Canada and international courts and justice. The event occurred when the Filipino crew of the tanker *Maersk Dubai* deserted when it docked in Halifax in 1996. They claimed the Taiwanese officers had thrown stowaways overboard during the trip. The fictional version shows the interlinking of global societies; injustice can prevail, as in both versions no one was convicted.

Montreal-born Shauna Singh Baldwin says of the heroine of her novel *The Tiger Claw* that "she had no geographical self just as I have no geographical self."[7] It is the life story of World War II resistance fighter Noor Inayat Khan, the daughter of an Indian Sufi mystic and an American mother, born in Moscow and raised in France, who fled to England when the Nazis invaded. Khan is a Muslim woman in love with a French Jew, spying for the British, India's colonizer. Life and death at the hands of the Nazis raises many questions about fascism, religious intolerance, love, and betrayal. Singh believes that many people, like Khan and herself, have the perspective of citizens of the world; this gives her hope.

BOX 12.3

Canadian War Stories: Changing Attitudes

Important works reflecting Canadian attitudes to war in the 20th century include

- World War I: Hugh MacLennan's *Barometer Rising*, Timothy Findley's *The Wars*, Alden Nowlan's poem "Ypres: 1914"
- World War II: Gabrielle Roy's *The Tin Flute* (1945), Timothy Findley's *Famous Last Words*
- Roch Carrier's *La Guerre, Yes Sir!*
- Rudy Wiebe's *Peace Shall Destroy Many* (1962)
- Raymond Souster's poems "That Morning in Brussels" and "The Nest"
- Farley Mowat's autobiographical *And No Birds Sang* (1979)
- Jane Urquhart's *The Stone Carvers*
- Francis Itani's *Deafening*

The Ash Garden, by Dennis Bock, examines the aftermath of World War II from an international perspective, as does Michael Ondaatje's *The English Patient*. In *Anil's Ghost*, by Ondaatje, the Canadian protagonist is a civil rights investigator caught in a civil war in her former homeland of Sri Lanka.

Execution by Colin McDougall presents the moral dilemma of World War II soldiers in executing orders to kill in questionable cases, especially regarding desertion. The execution of Private Harold Pringle shaped the plot and theme of the soldier's dilemma.

LITERATURE LINKS REGIONALISM, SOCIAL INEQUALITY, AND STRATIFICATION

Noah Richler's CBC radio series *A Literary Atlas of Canada* cites the country's storytellers as the ones who do "the extraordinary job ... of mapping, naming, and explaining the astounding variety of peoples, accents and histories and regions and points of view [almost] impossible to capture ...." The books discussed in this section give an overview of how writing from each region defined its culture and social structure. We can also see that the divisive effects of regionalism have lessened to a great degree, with travel, technology, and intermingling of peoples being important unifying factors. Our foibles and idiosyncrasies, what we like and don't like about each other, and what we have in common are the subjects of Will Ferguson's books: *How to Be a Canadian* (co-author Ian Ferguson), *Canadian History for Dummies*, *Beauty Tips from Moose Jaw*, and others. He comes to these topics as someone who was born in the North and lived, worked, and travelled extensively all over Canada. Ferguson also sees Canada from the outside, having travelled the globe and lived overseas for years. Canada and Canadians are presented with honesty, humour, and keen insight.

The Atlantic Region: The Geographical, Economic, and Sociopolitical Margins

Atlantic Canadian writers focus clearly on the poor working class. In many different styles, they describe this as the major difference between the Atlantic provinces in general and the part of Canada that is middle class. Pride of heritage, the power of the sea, and isolation are consistent themes. Atlantic Canada's people of French, English, Irish, and Scottish descent have strong oral traditions, as do Atlantic blacks and Native Mi'kmaq ("people of the dawn"). However, historically the latter two groups have been silenced. A revision of historical events and retelling of tales of "the others" are current stories, from the east coast.

Newfoundland

Our cross-country journey begins at Canada's most easterly point of land. We find ourselves windblown, perched on a cliff gazing over the north Atlantic through the eyes of E.J. Pratt, who in his poem "Silences" wrote

> There is no silence upon the earth or under the earth like the silence under the sea;
> No cries announcing birth,
> No sounds declaring death.[8]

In our imaginations we can travel offshore about 400 kilometres, back to 1912, and watch with horror as the ocean claims *Titanic* and about 1500 people. Pratt's epic poem "*The Titanic*" expresses the terrible power and beauty of the north Atlantic. Michael

Crummey's historical novel *River Thieves* engages us in the lives of early Newfoundlanders and their relationships with the Native Beothuk and Mi'kmaq.

We turn away from the sea and go inland to learn the story of *The Rowdyman*, written by Gordon Pinsent, a Native Newfoundlander. Narrator Will Cole's life contains the elements of the stories of so many from Atlantic Canada. He works in a pulp mill in a one-industry town, with little chance of future improvements, unless he goes away to Toronto to work. To leave or to stay is the big question for people from this area. Bernice Morgan explores these questions and other realities of life on "The Rock" in her historical novels *Random Passage* and *Waiting for Time*. The reverse phenomenon, the effect on the community when an outsider moves in, is related in Michael Winter's novel *The Big Why*.

Literary works from Canada's east coast are populated with characters who make the difficult choice. We witness the emotional wrench from family and community that is felt on both sides. We also see the determination of those remaining not only to survive, but also to nourish the community and cultures of home. Wayne Johnston's family saga *Baltimore's Mansions* is about the conflicts caused in communities and families leading up to Newfoundland's decision to join Canada as the tenth province. Other policies of then-Premier Joey Smallwood, particularly those involving uprooting whole communities from the isolated outports, became significant events in the story of Newfoundland. Donna Morrisey's novels *Kit's Law*, *Downhill Chance*, and *Sylvanus Now* are set in 1940s and 1950s outport Newfoundland in times of major change. She tells powerful tales of love, despair, and family secrets, women straining against a patriarchal environment and male tradition, and men struck down when their world collapses with the fishery. *Down in the Dirt* by writer, actor (TV show *Hatched, Matched and Dispatched*) Joel Hynes is set in present-day small-town Newfoundland. Keith Kavanagh is angry and self-destructive. He and his buddies are in that limbo between adolescence and maturity, preoccupied with booze, dope, love, and trouble. Hynes sees this rootlessness of youth as the great difference between his generation and his father's, when at 15 a young man was expected to go to sea. Derek O'Brien's exposé of sexual abuse of boys by priests and brothers, *Suffer the Little Children*, tells a tragic story with its far-reaching emotional, legal, and social scars on the whole island community. *The Long Run* by Leo Furey is a novel that tells the funny, heartbreaking story of Aiden Carmichael, an orphan who finally outruns the cruelty and abuse he suffers from the brothers at Mount Kildare Orphanage. Lisa Moore's characters' voices in her story collection *Open* are 21st-century urban ones, a couple in a troubled marriage and a Canadian visiting Cuba awakening to exploitation of the people by tourists.

Nova Scotia

Hugh MacLennan's novels, *Each Man's Son* and *Barometer Rising*, give a strong sense of the historical, economic, and social forces that shaped people's lives. Sheldon Currie's story "Glace Bay Miners' Museum" (film *Margaret's Museum*) continue this story.

"The Boat" from Alistair MacLeod's short story collection *The Lost Salt Gift of Blood* focuses on one family, as its members struggle with the desire for personal freedom, set against a strong sense of family obligation. In the community context, when it comes to survival there is no question about risking civil disobedience and outsmarting the Mounties: "After all, they're trying to enforce a law that was made in Ottawa by some guy that was never out in a boat in his life." MacLeod's award-winning novel, *No Great Mischief*, set in the present nationally and internationally, explores the themes of obligation and loyalty to one's people. The conflicts, and many subtle and complex ways in which the human spirit responds, are explored in the story.

Three other writers who give us a view of that world from the vantage point of having "gone down the road" are novelists Ann-Marie MacDonald (*Fall on Your Knees*), Lynn Coady (*Strange Heaven, Play the Monster Blind*), and playwright Michael Melski (*Hockey Mom, Hockey Dad, Miles from Home*). Although the settings in each case begin with a Cape Breton influence, the themes are universal. These stories include diversities that were, and remain, taboo subjects: mental illness, domestic violence, homosexuality, incest, and racism. All portray the evolution of attitudes regarding sexuality and women's understanding of their roles in the parental home, in relationships with men, as mothers, in the community, and the world. The changing roles of men as husbands and fathers are also realistically presented.

New Brunswick

Three authors whose writings are important in defining the culture of New Brunswick are Alden Nowlan, David Adams Richards, and Antoine Maillet. Maillet has become the voice of the Acadian people of New Brunswick, which since the Acadian cultural revival that began in the 1970s is recognized as the only Canadian province that is truly bilingual. *La Sagouine*, translated in 1979, and *Pélagie-la-Charette* were great successes worldwide. Her many stories and characters give voice to the geography, history, and people of Acadia with a passion for life and a sense of humour about human nature. Maillet believes that to recognize her writing is to recognize the people from whom she is descended. France Daigle is an Acadian author whose novel *Just Fine* has been translated into English. It is a touching, funny story of an agoraphobe and her neighbours as they dream of strange and fantastical ways to escape from their small town.

Alden Nowlan is considered the poet of the underprivileged. His immediate community motivates his poetry, novels, and short stories, especially the effects of poverty on the human spirit. In his best-known poem, "Britain St.," Nowlan writes of loneliness, violence, and despair with a gentle compassion. Nowlan was one of the first to write about unacknowledged social realities such as family violence and the plight of the single mother. His honesty compels us to look at the hypocrisy of denial.

There is a much harder edge to the stories of David Adams Richards about the harshness of life in the Miramichi. The characters in *The Coming of Winter*, *Blood Ties*, and *Lives of Short Duration* inhabit a world where subsistence is the norm, where every form of deprivation exists, and where social structures have collapsed. Hope is as rare a commodity as diamonds. Richards describes the moral decay that comes from consistent powerlessness as directly caused by economic exploitation of the land and its people, combined with general social breakdown. In *The Bay of Love and Sorrows* (1998), tension and rivalry develop between friends—privileged Michael Skid and farmhand Tommy Donnerall—ending in bitter misunderstanding and murder. In *Mercy Among the Children* a family becomes isolated, exploited, and abused because of the father's decision to never hurt another human being.

Larry Lynch's works, the short stories in *Learning to Swim* and his novel *An Expectation of Home* (2002), give us another version of life in the Miramichi. The former are tales told in the voices of several less than successful men as they attempt to master modern existence. *An Expectation of Home* challenges our expectations of home in the internal workings of the Gordon and Jolicoeur families, generations, genders, and classes. The Gordons and the Jolicoeurs are neighbours but not very neighbourly. In the small world of their pulp and paper industry town, and beyond, inevitably their lives intersect, overlap, and sometimes clash.

Prince Edward Island

Canada's smallest province, the mostly rural Prince Edward Island, is the setting for Lucy Maud Montgomery's prolific writing, especially her "Anne" and "Emily" stories. Perhaps because of Montgomery's idyllic descriptions of the island, the province has become a refuge for many artists who come "from away" to work in peaceful surroundings.

Alzheimer's disease is the subject of poet Hugh MacDonald's collection *Looking for Mother*. The poem of the title begins "It wasn't till mother lost her mind that I saw what I'd lost." It is a moving and honest search for his mother. He begins with childhood memories, interviews those who knew her best, and ends with her time spent in the Sacred Heart Nursing Home, lost even to herself.

Milton Acorn of Prince Edward Island is known as "The People's Poet." The titles of his works give an idea of the topics he was concerned with in his life and writing: *In Love and Anger*, *Against a League of Liars*, *I've Tasted My Blood*, *Dig Up My Heart*, *I Shout Love and Other Poems*, *Hundred Proof Earth*, and *The Island*.

Quebec: Cultural and Linguistic Isolation—Two Bodies of Literature Define the Two Solitudes

Quebec has produced a voluminous literature throughout its history and continues to do so. Québécois literature presents a picture both of how this region is different from the others and of internal diversities. Cultural and linguistic differences are central themes.

Roch Carrier's short story collection "The Hockey Sweater" gives a child's point of view. The Quebec–Canada difference is symbolized by the former's important cultural institutions of hockey (the Montreal Canadiens) and the Catholic Church, and the latter's federal government (the Post Office) and business (Eaton's). The stories of Carrier and Gabrielle Roy include people and events reflecting both French and English cultures. For others, such as Yves Thériault and Michel Tremblay, a personal or internal dichotomy acts as an analogy for cultural and political conflict. Thériault's futuristic short story, "Akua Nuten: The South Wind," contrasts north-south values and French-Métis differences. For Tremblay the issue was his homosexuality, evident in the play *Hosannah*. He uses a more aggressively separatist tone in his stories, movies, and plays. Michel Basilières' 2003 novel *Black Bird* has been described as a *Two Solitudes* for the 21st century, a fairy tale gone wrong. Anarchy, denial, secrets, and bizarre behaviour reign in the Desouche home, three generations of family members shaped by a mix of French and English influences. Language, gender, political, and educational differences cause family stress. The setting is Montreal in the uneasy time of the FLQ crisis. Filled with black humour, magical fantasy, farce, the bizarre, and the absurd, the story is a satirical take on everything that divides us, as it coexists with everything that unites us.

Prior to World War II the role of women in Quebec society was only visible from a masculine perspective. The strong patriarchies of church and state allowed only two choices for a woman: she could be a wife and mother or a nun. This applied to girls and women of all classes. Anne Hébert, Marie-Claire Blais, and Gabrielle Roy are writers who expanded and deepened that picture in works such as, respectively, *Kamouraska*; *Mad Shadows* and *A Season in the Life of Emmanuel*; and *The Tin Flute*. Whereas Roy came to Quebec as an adult, Hebert and Blais needed to leave Quebec to write.

The poems and stories of Irving Layton, Mordecai Richler, and Leonard Cohen speak of the Jewish community in Quebec, a minority within the English minority. Layton's poem "On Seeing the Statuettes of Ezekiel and Jeremiah in the Church of Notre Dame" presents the feeling of living in a cultural and religious isolation. Cohen's poem "Genius" describes the continuing presence of anti-Semitism in the world. Richler's last work *Barney's Version* is a funny yet sad fictional memoir set in a multicultural Montreal, particularly in the Jewish community. Their writings also include various aspects of contemporary Canadian life. In 1992, Richler published his controversial nonfiction work *Oh Canada! Oh Quebec! Requiem for a Divided Country*, on the matter of Quebec separation.

The timely new edition of *Earth and High Heaven* by Gwethalyn Graham is the tragic love story between Erica Drake, daughter of a WASP Westmount family, and Marc Reiser, a Jew. It is an honest portrayal of attitudes in 1942 Montreal/Canada, a valuable social history of the irrationality and corrupting effects of anti-Semitism.

Quebec Now: Urban Settings, National and Global Connections
Québécois Voices in Translation

Michel Basilières' *Black Bird* and Gil Courtmanche's *A Sunday at the Pool in Kigali* (movie *Hotel Rwanda*) have been previously mentioned.

Jacques Poulin's *Volkswagen Blues* takes us on a trip from Montreal to San Francisco with a writer named Jack. He picks up a hitchhiker, an Aboriginal woman, and her cat Chop Suey. As they travel their conversation ranges over a wide variety of topics, relating to characters—real and fictional—who had followed that path. The story of the French in North America and the history of Native peoples are revealed. It is a journey in which Jack learns about himself, with a new understanding of the complexity of personal identity.

In Monique Proulx's *The Heart Is an Involuntary Muscle*, narrator Florence is 25, a website designer, and smart. She both longs for and fears a romantic relationship with her colleague Zeno. She sets up technological barriers; to control an emotional situation, she simply logs off. Proulx uses a mystery format to examine the fears and pleasures of falling in love.

The title of Rejean Ducharme's novel *Go Figure* seems very appropriate for a story in which the narrator, Remi Vavasseur, sends his wife, Mammy, on a world tour with his volatile mistress, Raia. Mammy is recovering from a breakdown brought on by losing twins at birth. Remi goes into seclusion (as has Ducharme) and keeps a diary written to his wife. He seems to have the vain hope that Mammy's love and goodness will somehow stop that "madwoman" from making him lose his mind.

English Voices

Jeffrey Moore's *The Memory Artists* is a novel about how the brain processes memory. It asks the question: Is memory a blessing or a curse? Noel Burun, a chemist, has hyperamnesia (he forgets nothing) and synesthesia (one sense, sound, is experienced as another, colour). He cares for his mother as her Alzheimer's disease progresses, working in his basement lab to find a cure. He is joined by three others in this intense pursuit. All four are patients of Dr. Vorta, a neuropsychologist. His friend, Norval Blacquiere, an overconfident womanizer, is being treated for drug-induced synesthesia. Samira, a young Arab woman, is a victim of the date rape drug GHB, and suffers from short-term amnesia. Both are trying to escape painful memories. JJ Yelle, an alternative medicine enthusiast, has a condition induced by shock treatments that has him existing in a carefree, childlike state with only good memories.

In Emma Richler's *Sister Crazy* and *Feed My Dear Dogs*, Jem, the middle child of five siblings, takes us into the warmth and action of the Weiss family circle. The family personality is funny, creative, full of vitality, each member quite individual. She remembers all the details, books, movies, music, games of life in the safety of the family. In the background her present adult life breaks through with a note of sadness; she mentions her analyst. In *Feed My Dear Dogs* Jem's much loved older brother reminds her that the Weiss family is not the world, that she can't stay in the family forever. The world's stories of adventure, science, religion, philosophy, and history are intertwined with her own and her family's ever-changing worlds.

William Winetraub wrote *Crazy About Lili*, a coming-of-age story set in 1948 Montreal, but it might be any time or place. Richard Lippman, first-year McGill student, is introduced to the famous striptease artist Lili (St. Cyr) by his Uncle Morty. Morty's Montreal is a very different world from the Westmount of Richard's parents and friends. Not bound by religious or cultural constraints, Richard leaps into this life and the story relates the consequences of his obsessive behaviour. Weintraub's voice is satirical but honest, including an affectionate portrayal of his women characters.

Ontario: Largest and Richest, but Complex Internal Diversities

The vastness alone has led to diversity in stories about life in Canada's largest province, Ontario. They range from Morley Torgov's humorous stories of growing up Jewish in the steel town of Sault Ste. Marie, to Alice Munro's and Robertson Davies' differing portraits of conservative, small-town Ontario. Ontario's proximity to the pervasive American influence provides a theme for countless stories by Ontario writers. In an interview in *The Notebooks: Interviews and New Fiction from Contemporary Writers*, Catherine Bush said "[I think] . . . the point of being human is to try to imagine the Other somehow." Her novels *Minus Time* and *Rules of Engagement* do just that by placing women in traditional male careers. In the former the mother is an astronaut, in the latter a military scholar. She considers the impact of technology on morality, and the different ways of waging war. *Rules* contrasts the impersonal nature of war as waged by high-tech countries with the neighbour-versus-neighbour fighting in poorer places such as the Balkans and Africa. Bush's latest novel, *The Pain Diaries: A Neurological Mystery*, reveals the effects of chronic pain on the lives of three sisters. Questions about the need for drugs and survival arise when one sister disappears and another goes in search of her.

Hal Niedzviecki's writing is also concerned with how, as a society and as individuals, we are shaped by pop culture while we absorb it via all the modes of technology. *The Program* examines how one family has become obsessed with some element of mass media and how it is controlling their lives. The word "program" has many meanings. The family matriarch, Bubby Stern, is a TV addict, her way of blocking out past evils; her older son Maury writes advertising, and has written a book *Get with the Program: The Power of Slogans*; brother Cal chooses to follow this road to success, resulting in a constant tension with his natural inclinations. Maury's son Danny rebels against his father by joining a professor whose goal is to overturn the power of advertisers. The question is how innocent is this project. Danny writes his own program. In his novel *Muriella Pen*, set in the Toronto arts scene, Russell Smith uses wit and humour to skewer practitioners of political and artistic correctness.

The universal concerns of national and global political and civil justice, poverty, and the nature and source of violence are those not only of Atwood but also of Ontario writers Timothy Findley, Michael Ondaatje, Dave Godfrey, Hugh Hood, Richard Wright, and Alice Munro. Findley's *Not Wanted on the Journey* is a satirical retelling of the story of Noah's Ark from the perspective of the women and animals, rather than the patriarch, Noah. Stories by Ontario writers often include characters and events concerned with global issues. This is especially true of authors who, as immigrants and refugees, write from a viewpoint influenced by their places and cultures of origin. Nelofar Pazira's *A Bed of Red Flowers: In Search of My Afganistan* is one such story (also a CBC TV special).

The Prairies: Manitoba, Saskatchewan, and Alberta

Certain images and themes are common to the literatures of Manitoba, Saskatchewan, and Alberta and different from those of the rest of Canada. These images and themes are closely linked to the prairie landscape. The sense of isolation and loneliness in endless spaces of land and sky is background for Prairie stories reaching into tales of town, city, forest, and north of the tree line. Recently concern for the environment has become an important theme. Some of the first writers to describe the physical and emotional effects of the landscape are Margaret Laurence, Sinclair Ross, Max Braithwaite, Adele Wiseman, and Rudy Wiebe. Their stories reveal the diversities that exist within this region.

Saskatchewan

Our best guide to small-town life is W.O. Mitchell in his novels, especially *Who Has Seen the Wind*, and his *Jake and the Kid* radio and TV series. *The Last Cowboy* by Lee Gowan traces the lives of three generations of the MacMahon family. Sam Senior is frustrated with how his son is running the ranch, annoyed that he has a Chinese neighbour, and feels his life is a waste. His story is told in flashbacks to the 1970s, then to the 1930s, from the year 2000. His grandson Sam is a successful businessman with an unhappy home life. The book presents a social history of these years within the beautifully drawn landscape of southern Saskatchewan. Two of Sharon Butala's writings, *The Perfection of Morning: An Apprenticeship in Nature* and *Coyote's Morning Cry*, celebrate nature and how daily contact nourishes the spirit in us. She makes connections to the value of Native wisdom and to feminism, as well as her own religious insights.

Manitoba

Neepawa is the setting of Margaret Laurence's five related stories beginning with *The Stone Angel* and ending with *The Diviners*. She raised the blinds not only on the lives of women, but also on the town's hidden secrets, and prejudices regarding the Métis, the Chinese family, and the poor Bens. Carol Shields' *The Stone Diaries* and other novels of urban life are imbued with the history and geography of the West.

Contemporary writers Miriam Toews, Di Brandt, and Uma Parameswaran provide diverse insights into present-day Manitoba. In *Questions I Asked My Mother*, Brandt explores patriarchy within the Mennonite tradition. Teenaged Nomi, in *A Complicated Kindness* by Miriam Toews, wishes her mother and older sister had not run away from home leaving her and her father at a loss. Nomi's take on the contradictions of life, the strictures of religious and social norms, in her small Mennonite town are both funny and sad. Uma Parameswaran coined the term SACLIT, South Asian Canadian Literature and has written *SACLIT: An Introduction to South Asian Literature (Critical Essays)*, plays, poems and other works. Her novel *Mangoes on the Maple Tree* spans 20 days in the life of an Indo-Canadian family during the 1997 flood in Winnipeg. *Sisters at the Well* is a volume of poems marking the 15th anniversary of the crash of Air India Flight 182.

Alberta

Robert Kroetch, George Ryga, and Aritha Van Herk have all provided us with tales of Alberta. Three of Kroetch's books form his "Out West Trilogy": *The Words of My Roaring*, *The Studhorse Man*, and *Gone Indian*. They investigate four decades of social change, including the Depression years, post–World War II dislocations, and the good years of cattle farming in the 1960s and 1970s. A later novel, *Badlands*, has the reader accompany William Dawe, as he rafts up the Red Deer River into the heart of the Alberta badlands in his obsessive search for dinosaur bones. George Ryga was born in Alberta of Ukrainian background. He is best known for his play *The Ecstasy of Rita Joe*, the story of a young Indian girl who, on the one hand, is incited to rebellion by the white society around her and, on the other, is a victim of bureaucratic indifference. The plight of the Native person is a continuing concern throughout Ryga's plays and novels.

In Aritha Van Herk's stories the women characters are determined, independent, and adventurous risk takers—modern women who challenge the stereotypes about femininity. In *Judith*, the main character is fed up with urban living; she returns to the country to single-handedly operate a pig farm. In *No Fixed Address*, Arachne is a travelling saleswoman who weaves her way in and out of the small towns of Alberta and Saskatchewan selling her line of women's lingerie. The lives of these unconventional women, their relationships with neighbours, friends—male and female—and lovers are presented with energy and honesty.

British Columbia: The Pacific Coast, Gateway to the Orient

The unique geography of British Columbia, including the Rocky Mountains and the Pacific Ocean, has been significant in the shaping of its culture. The isolation and challenge of the mountains as a metaphor for life can be found in the poems of Earle Birney, particularly "Bushed" and "David." E.J. Pratt's narrative poem "Towards the Last Spike" presents the mountains as a barrier that must be overcome if Canada is to be united physically and politically. Many books of fact (Pierre Berton's *The Last Spike*) and fiction incorporate this event. However, the untold stories of the human cost of this achievement, especially the exploitation of Native peoples and Chinese immigrants, were cloaked in silence for decades.

The mountains are the setting for stories of the excitement and adventure of the gold rush days. Later, during World War II, the abandoned mining towns became the detention camps for Japanese-Canadians. The federal government confiscated property and separated families. Japanese-Canadians were banished from their homes in Vancouver, along the coast, and in the Fraser Valley, on suspicion of spying for Japan. *Obasan* (1981), by Joy Kogawa, and its sequel, *Itsuka* (1992), reveal this tragic tale and its aftermath from the perspective of the Japanese-Canadian community. It was not until the publication of *Obasan*, forty years after enactment of this policy, that the complete story came out. A docudrama by Dorothy Livesay, entitled "Call My People Home," tells the story in poetry form.

British Columbia and Victoria are names indicating British beginnings, but location also makes it Canada's gateway to the Orient. Sky Lee's novel *Disappearing Moon Cafe* (1990) takes us into the world of four generations of the Wong family. The first Wong arrived in the gold rush days and took part in the building of the railway. He, his wife, children, and grandchildren struggle to survive, and eventually build a successful business in Vancouver. It is the struggle of a community to form a Canadian identity despite isolation, racism, and culture clash. The main character is Kay, a Bay Street (Toronto) financial analyst and new mother. She revisits her family history, concluding that her forebears won a place for her and her son in Canadian society.

In Marilyn Bowering's novel *To All Appearances a Lady*, we join Robert Lam on a ship journeying up the coast of British Columbia determined to forget his past, to break family ties. Mysteries about his mother's death, his difficult stepmother, and even the true identity of his father must be faced as he becomes involved in the underworld opium trade. In *Intertidal Life* and other works, often set in the Pacific-coast islands, Audrey Thomas explores the complications of today's human relationships.

The inner-city life of Vancouver is at the centre of *The Dreamlife of Bridges*, a novel by Robert Strandquist, and *Colin's Big Thing*, the gritty memoir of journalist Bruce Serafin. Timothy Taylor's novel *Stanley Park* is a fictional imagining of events around the unsolved murder of two children referred to as "babes in the woods." He examines the city's loss of innocence, sense of collective guilt, and the creeping of fear into the general imagination.

The Treatment and Perception of Diversity

The North: Yukon and the Northwest Territories—The Romance and Realism of the Last Frontier

Many Canadians were introduced to the distant north of Canada through the poems of Robert Service ("The Cremation of Sam McGee"), the animal stories of Jack London, or *Klondike*, Pierre Berton's narrative of the gold rush. Even more of us entered that space with Farley Mowat, as described in *Never Cry Wolf* and *People of the Deer*, through Yves Thériault's novel *Agaguk*, and *Windflower* by Gabrielle Roy. Besides awakening us to the north's natural beauties and the unique culture of the Inuit, we learned how these were endangered by intrusions from the south. Mowat's controversial works pointed out his environmental concerns for the wildlife, but especially for the Native people, whose whole pattern of existence was being destroyed regardless of the consequences. Al Purdy's poems give us wonderful snapshots of the area's unusual natural beauty in "Arctic Rhododendrons," and sombre reflections on its long history of isolation in "Lament for the Dorsets."

Inuit writer Alootook Ipellie's story "Frobisher Bay Childhood" (Iqaluit) describes a time when his isolated community was in a time of transition. It began with the arrival of material goods from southern Canada, movies, and the establishment of an American military base nearby. In retrospect he can see it was the beginning of the erosion of a strong sense of community. Gradually, his people experienced the loss of independence that came with the traditional way of life.

CRITICAL THINKING BOX 12.5

How have the following elements of Canadian society promoted the recognition and acceptance of diversity within our culture?

1. We are a relatively new democratic society, from its beginnings made up of diverse peoples.
2. By law, all children must receive an education.
3. We have the transportation and communication means—railroads, airlines, the Trans-Canada Highway, CBC Radio and TV, the National Film Board, the Internet—to learn each other's stories, despite a vast geography and difficult climate.
4. We have developed a good standard of living.
5. We live under the *Charter of Rights and Freedoms*.

THE ROLES OF FAMILY, GENDER, AND SEXUALITY IN CANADIAN LITERATURE

Diversity Through a Child's Eyes

Probably the best-known Canadian stories through a child's eyes are told by Brian in *Who Has Seen the Wind?* by W.O. Mitchell, Anne Shirley in *Anne of Green Gables* by L.M. Montgomery, and Morag in *The Diviners* by Margaret Laurence.

Missy and Ruby in From Bruised Fell

Jane Finlay-Young's story is of two young sisters living at the end of the 20th century in an uncertain world of constant change. Even though their parents are living, they feel orphaned. Family life is constantly disrupted by their mother's mental illness, the parents' divorce, moves back and forth to England, their father's remarriage, and a series of relocations across Canada necessitated by their father's engineering work. The story is an honest and moving portrait of a family that is not so atypical. How Missy and Ruby respond to all the change in their lives, and their love and protection of each other, are at the centre of the story. Ruby, unable to accept the loss of her mother, finally succumbs to physical and emotional illness. The author depicts the inner lives and imaginations of these children in a sensitive and believable manner.

Tarcadia by Jonathan Campbell has Michael Chisholm, son of a union boss, recalling a summer of childhood fun with his brother Sid and friend Alex. Their "playground" is the neighbourhood of the now, infamous Sydney tar ponds, an atmosphere of nostalgia shaded with foreboding. They were modern-day Huck Finns, building a cabin on an old raft moored in the north pond, sailing a found kayak they made seaworthy, and occasionally falling out of it. Sometimes they noticed the oily surface and the bad smell, especially after a summer rain. They were as innocent of the dangers of their adventures, as the adults were of the price that was to be paid for decades of dumping the waste from the steel plant and raw sewage from the city into the water. For Michael, that summer, and childhood innocence ended in tragedy, when family tensions, long below the surface, erupt.

Literary Families Reveal the Diversity in Their Communities

The books and authors examined next are all about finding one's place in the world. Each story examines a different version of the patriarchal family.

Alice Munro's *Lives of Girls and Women* and Robertson Davies' *Fifth Business* are novels that begin in small-town Ontario. Jubilee is the name of the imaginary town created by Munro. *Fifth Business* is the first story of a trilogy set in the imaginary town of Deptford. The people are almost entirely white of Anglo-Saxon origin. The main differentiating feature is the type of Protestantism practised. Both stories underscore how childhood experiences affect the rest of our lives.

The gender of each author influences the gender of each text's central character. Munro wrote from the point of view of Del as a girl and a young woman, with the story ending the summer of high-school graduation. We are aware of her private, inner self in a way those in her everyday life are not. In *Fifth Business*, Dunny (Dunstable) Ramsay narrates his own tale. He is retiring from a lifelong career of teaching. Believing that people think his life has been dull and boring, he writes a story/letter to reveal the exciting secret life he led during holidays. The titles of these two books indicate that each is concerned about the roles we play in life and the opinions of others. Munro's title questions women's place in society. It is a novel of feminine experience. The meaning of the term "fifth business" is more obtuse, and therefore Davies prefaced the novel with a definition: it is a term used in theatre to refer to a character who is not a main character but acts as a vital catalyst in the plot.

In Munro's story we see the world through the narrator's eyes as she tries to understand herself and other girls and women. Del thinks about her relationship with her mother, the differences between herself and her friends, other women's behaviour, especially regarding their interactions with men. These reflections are all interwoven in the uneven process of Del's physical, emotional, intellectual, and spiritual growth. The development of Del's not always subtle, but decidedly female, sense of humour is an especially delightful aspect of the story. The importance of this novel, especially for women readers, cannot be overestimated. It was one of the first Canadian novels to outline the sexual awakening of a young girl within the context of the world of women. Female readers could only go so far in identifying with the experiences of male characters in stories of initiation into the adult world. In short story collections that followed, especially *Something I've Been Meaning to Tell You* and *Who Do You Think You Are?*, Alice Munro expands on how attitudes regarding women began to change in the second half of the 20th century. This happened partly because of stories like Munro's, which presented girls and women in a fuller light.

The setting in time is the second important difference between the two novels. *Lives of Girls and Women* provides an insightful analysis of the world of typical small-town Canadian women, over a wide age and social range, in the 1940s and 1950s. By contrast, Dunny grew up in pre–World War I days, when boys and men were constantly reminded of responsibilities to family and community. Another theme of Davies novel concerned the effects of external events or forces—fate, accident, timing—on the human psyche. The course of Dunny's life is altered when a snowball intended for him hits a pregnant woman. Davies, influenced by Jungian psychology, created and developed characters who cover the spectrum of human emotional possibilities.

Adele Wiseman's *The Sacrifice* (1954) is about a Jewish family, displaced from Russia, that finds its way to Winnipeg, where its members try to adjust to life in Canada. In the process conflict between the generations develops. A serious rift unfolds in the family that is the undoing of its patriarch, Abraham. The idea for Wiseman's later work *Crackpot* germinated from her own experience as a teenager, born in Canada into an immigrant family. She felt shame and embarrassment, and later guilt, to be seen with her grandmother, who dressed in dark clothes, covered her head in a "babushka," and spoke broken English.

The Apprenticeship of Duddy Kravitz (1959), by Mordecai Richler, takes place in the community of St. Urbain Street, Montreal. In this case the family is absent. Duddy's mother has died, his father has opted out, and his brother is away at university. Duddy learns his survival skills from the street's unsavoury characters. His immigrant grandfather impresses on him the importance of owning land, which becomes his driving ambition. Duddy succeeds at this, but realizes too late the cost was the betrayal of his girlfriend and friends and disappointing his grandfather.

Margaret Atwood is probably the best known among Canadian writers. In all of her work, she describes and analyzes how perceptions of women as objects change to views of women as subjects, worthy of consideration for their own sakes. Like other feminist writers, Atwood pointed out that, for this to happen, women themselves must be the first to change in the way they view themselves and their daughters. The story of Joan in *Lady Oracle* is a complex tale that traces her journey from an unhappy childhood of resistance to her mother's rigid expectations, to a series of relationships with men for whom she tried to appear in their feminine image. For years Joan leads a double life. Finally, she realizes that the only way to live is to tell the truth about herself. During this period Atwood's tone and style tended to be stark and unemotional. She made it her business to overturn expectations, to subject the reader to verbal shock treatment. Her short poem that opens *Power Politics* has this effect:

> you fit into me
> like a hook into an eye
> a fish hook
> an open eye [9]

Atwood often wrote from the point of view of alienated individuals. Some other themes that Atwood has explored are (1) inauthenticity, which she especially relates to women, (2) the dangers of a colonial mentality, with the resulting uncertainty about Canadian identity, and (3) the need to define one's own space as separate while still being related to others.

In the later phase of her work, covering from the late 1970s through the 2000s, Atwood "does employ a greater range of style and topics. She is by turns more lyrical and personal" as in this excerpt from "Variation on the Word *Sleep*," as well as "more satirical and political":[10]

> I would like to be the air
> that inhabits you for a moment
> only. I would like to be that unnoticed
> & that necessary.[11]

In some cases the elders in families and communities are important voices in Atwood's stories, revealing different roles for older people, how they adapt to changes in themselves and the world around them. *Handmaid's Tale* and *Oryx and Crake* are cautionary tales presenting readers with regressive futuristic societies that have resulted from war and environmental destruction.

Breaking the Silence About Matters of Gender and Sexuality

In Canada, as elsewhere, diversity was not accepted in certain areas of human life. Injustice, oppression, and all forms of human suffering were the result. As the social climate changed, society's secrets are revealed. One example is the TV docudrama *The Boys of St. Vincent* about pedophilia and breach of trust.

All types and styles of works written out of the feminine experience are an important part of the last quarter of 20th-century Canadian society and literature. Novels, such as the two mentioned below, with the lives of women at the centre of the story were rare before then. Carol Shields, in *The Stone Diaries*, created a story around the unsettled life of Daisy Goodwell. It is peopled with characters whose lives reveal differences in experiences of personal freedom, told from a woman's viewpoint. She writes of orphans, dysfunctional families, anti-Semitism, unfulfilled women, and the differing values in communities, as well as the effects of moving back and forth across the Canadian–American border. In *Unless*, Reta's daughter, Nora, drops out of university and sits on a Toronto street corner with a sign bearing one word, "GOODNESS." She has withdrawn into a void of spiritual and moral isolation, a silent protest against human indifference and injustice. In all of her stories, Shields values women's inner lives, their friendships, and their place in the community and the world; Shields writes without sentimentality but always with an honest, compassionate, and humorous view of the human condition.

Elly Danica's *Don't: A Woman's Word* and Sylvia Fraser's *My Father's House* are biographical works about being subjected to incest. When this knowledge became conscious for Fraser, a journalist and writer, she not only understood her own life better but also characters and events in her novels. Fraser's allusions to sexual politics between children, and between children and adults, which had been noted earlier by readers and critics, took on new meaning.

Poems, plays, and stories on the subject of homosexuality are now generally accepted. Jane Rule, in an article entitled "Lesbian and Writer: Making the Real Visible," discusses the conflicts that arise because of her honesty about her sexuality. She responds to opposing pressures from the gay community and the art world. Rule herself says, "I owe to my own art all the honesty and insight I have, not simply about homosexuals and artists, both of which I happen to be, but about the whole range of my experience as a member of a family, a community, a country."[12] Rule's best-known novel is *Desert of the Heart* (movie *Desert Hearts*). Gay and lesbian characters are part of Ann-Marie MacDonald's novel *Fall on Your Knees*, as is the matter of incest.

Themes being explored in recent novels are critiques of our image-conscious culture, sexual satisfaction in women, and risk taking among women characters, along with the constantly evolving man–woman relationships in a changing world. In *Waking Beauty* by Elyse Friedman, readers meet Allison Penny, whose life changes dramatically when she

---

BOX 12.4

Nonfiction

- *The Sexual Spectrum: Exploring Human Diversity* by Olive Skene Johnson. This work covers scholarly opinion and political debate around human sexual diversity. The Vancouver neuropsychologist rejects variation as a moral issue, using fact and theory to show diversity to be the norm.
- *Ten Thousand Roses: The Making of a Feminist Revolution* by Judy Rebick. This book is a collection of interviews with Canadian women active in promoting feminist causes from the 1960s and 1970s to the present. The title refers to the roses carried by women in the 1995 antipoverty march from Montreal to Quebec City. Included are the voices of women, representing small and large communities and varied economic and cultural groups across Canada. (Rebick publishes an online magazine, *rabble* (http://rabble.ca).
- *The Meaning of Wife* by Anne Kingston. The role of wife is traced from earliest descriptions to the 21st century. In her history, Kingston refers to both the famous and the infamous.
- *Mary of Canada: The Virgin Mary in Canadian Culture, Spirituality, History and Geography* by Joan Skogan. Skogan's "Mary map" records the presence of the Virgin Mary in Canada from its beginnings. Her factual research is accompanied by insights into Mary's role as a source of comfort and consolation to people in lonely or dangerous work.

meets her natural mother. Her very ordinary life is left behind for a modelling career. Her comparisons and insights into the rewards and punishments of each are playful and serious and shaded with a biting wit.

Eyehill by Kelly Cooper is a collection of connected stories of the relationships of men and women in central character Rhea's life. Characters' ways of communicating with each other are pivotal to tales of awakening desire, adultery, hypocrisy, self-delusion, and childlessness.

Arden, whom we meet in Lorna Jackson's novel *A Game to Play on the Tracks*, decides to live dangerously when she runs away from her husband with her baby son to work as a bar singer. Using five voices Jackson relates the differences of place, New Westminster and Vancouver Island, between the ways people parent, respond to loss, and approach the business of living, as she reveals Arden's story.

In Susan Swan's story *What Casanova Told Me*, we travel the world with two women. They are an unconventional archaeologist mother and her quiet, reserved archivist daughter in two very different eras, on a mysterious quest into history, distant and recent. She portrays the joys and freedom of travel, its power to change how we see the world, ourselves, and the gifts life has to offer.

By contrast, *Adultery* by Richard Wright examines the inevitable, and in this case drastic, consequences of an act of infidelity. What is different is the man's point of

---

BOX 12.5

The Voices of Experience

Some recent works focus directly on the themes of aging, illness, and death. The diversity of responses to these universal experiences reveals much about human nature and attitudes and assumptions regarding the elderly.

- Hagar in *The Stone Angel* by Margaret Laurence
- Eve in *The Book of Eve* by Constance Beresford Howe
- Barney Panofsky from Mordecai Richler's *Barney's Version*
- William McKelvey in Matt Cohen's novel *Elizabeth and After*
- The two grandfathers in Alistair MacLeod's *No Great Mischief*
- Kate's great-grandmother Lillian in Beth Powning's *The Hatbox Letters*

view, how his whole image of himself as a man, a husband, and father is shattered, and the painful struggle to repair it.

CHAPTER SUMMARY

Canada has a unique and complex culture, with diversity a defining characteristic. Using literature as our vehicle, we can tour Canada's social landscape. Readers can be drawn into the stories of the country's first peoples, our parents' and grandparents' times, as well as the experiences of more recent newcomers. Each presents strong ideas and feelings about a particular society. We experience the internal diversity, gaining understanding about relationships within the whole of Canadian society. As well, the diverse roles and attitudes of Canada and Canadians to global events is found in many contemporary writings. This applies equally to people who have been Canadians for generations and to newer members of the Canadian family.

In the first half of the 20th century, literature reflected a changing Canadian society. It revealed regional diversities, differences and changes in rural, small-town, and urban life, and social attitudes to women and children. Both world wars caused us to recognize that Canada's story is separate and distinct from European culture and other cultures from which its people originated. Although some diversities have diminished in importance, those between French- and English-speaking Canadians have persisted, and full acceptance of some cultural groups, continues to be a struggle.

Canadian writers are now at the forefront of a reversal in the relationship between power and voice. In the second half of the 20th century, voices of the dispossessed, marginalized, and devalued were acknowledged as telling the most vital stories being published in all parts of the world. In some societies, voices were stilled because of colonization and cultural imperialism, as with Canada's Native peoples. Feminist writings are also part of this global change in giving the power of voice to women's stories.

CRITICAL THINKING BOX 12.6

How has literature helped to change the way people think about diversity in Canada? Has it changed perceptions so that being different is viewed more positively than it was in the past?

KEY TERMS

absence of voice, p. 332
feminist literature, p. 331
immigrant and refugee stories, p. 338
Métis, p. 333
postcolonial writing, p. 332
universal themes, p. 332
voice, p. 331

DISCUSSION QUESTIONS

1. How does literature personalize and humanize sociological data and historical events?
2. Explain how the focus on diversity by many contemporary writers is part of a global trend of postcolonial writing.
3. Trace how stories about Canada's Aboriginal peoples moved from their portrayal as "noble savages," to images of rebels and reactionaries, to realism with a decidedly negative slant. Contrast these versions with those written today as many Native writers speak in their own voices about past and present.
4. Explain how Canada's literary history (voice) is linked to its political and historical development (power). How did this phenomenon lead to the evolution of two distinct bodies of literature, one in English and one in French?
5. In stories set in the various regions of Canada, certain themes recur that reflect their social and economic realities. Identify the themes that are important for: (a) Atlantic Canada, (b) Quebec, (c) Ontario, (d) the Prairie provinces, (e) British Columbia, and (f) northern Canada (Yukon, Northwest Territories, and Nunavut).
6. Why is it that if you are a person of colour, if you speak a language other than English or French, or you have come to Canada from an Eastern culture, it is only within the past thirty years that you had a chance of finding yourself reflected in Canadian literature?
7. Omissions and silences have led to writings that explore the effects, on individuals and communities, of being treated differently. What questions are being explored in the stories by and of Canadians with (a) black, (b) Chinese, (c) Japanese, and (d) certain European origins, as well as (e) in works written by more recent immigrants and refugees?
8. Some of the most controversial issues in Canadian society are directly related to the diversity of views regarding gender, sexuality, and the family. Discuss the differing viewpoints on one issue in each of these three areas of human diversity as written about in fact and fiction.

NOTES

1. Northrop Frye, "Northrop Frye's Canada," *Globe and Mail*, 15 April 1991, p. A13.
2. Bill Ashcroft, Gareth Griffiths, and Helen Tiffin, eds., *The Empire Writes Back* (London: Routledge, 1989), cover.
3. Drew Hayden Taylor, "Seeing Red over Myths," *Globe and Mail*, 8 March 2001, p. A15.
4. Charles Mair, "Tecumseh" [n.d.], in Elizabeth Waterston, *Survey: A Short History of Canadian Literature* (Toronto: Methuen, 1977), p. 22.
5. Rita Joe, *Lnu and Indians We're Called* (Charlottetown: Ragweed, 1991).
6. Mary di Michele, "Luminous Emergencies" [1955], in Russell Brown, Donna Bennett, and Nathalie Cooke, eds., *An Anthology of Canadian Literature in English* (Toronto: Oxford University Press, 1990), pp. 706–10.
7. Bruce Erskine, *War Tale Gave Writer Hope* (Halifax: The Chronicle Herald The Nova Scotian/Books, 2004), p. 1.
8. E.J. Pratt, "Silences" [1937], in Sandra Djwa and R.G. Moyles, eds., *E.J. Pratt: Complete Poems, Parts I and II* (Toronto: 1989). Reprinted with the permission of University of Toronto Press.
9. Margaret Atwood, *Power Politics* (Toronto: House of Anansi, 1973), p. 1. Reprinted with the permission of Stoddart Publishing.
10. Margaret Atwood, "Variation on the Word *Sleep*" [1981], in *True Stories*. © Margaret Atwood, 1981. Reprinted by permission of Oxford University Press Canada.
11. Russell Brown, Donna Bennett, and Nathalie Cooke, eds., *An Anthology of Canadian Literature in English* (Toronto: Oxford University Press, 1990), p. 584.
12. Jane Rule, "Lesbian and Writer: Making the Real Visible," in Melita Schaum and Connie Flanagan, eds., *Gender Images* (Boston: Houghton Mifflin, 1992), p. 175.

GLOSSARY

absence of voice: The situation in which the stories of marginalized individuals or groups in a society are ignored, degraded, and suppressed. This creates a slanted, inaccurate, and incomplete picture of the society. See **voice**.

age at first marriage: The average age at which men and women marry for the first time.

age-specific marriage rate: The number of people marrying in different age groups.

age-specific mortality rate: The number of people dying in different age groups.

aggregating: Summarizing demographic observations or information.

anomie: Etymologically the word means an absence, breakdown, or conflict over "norms" within a particular society or culture. Durkheim uses the word to describe what is almost an individual psychological state consisting of disorder, chaos, and subsequent feelings of meaninglessness (French, from Greek).

antidiscriminatory: Said of legislation or actions that attempt to ensure individuals or groups are not disadvantaged because of their **gender**, age, **disabilities**, **race**, or ethnicity.

ascribed status: A characteristic that people are born with and over which they have little or no control. An **ascribed status**, such as **race**, **sex**, social class, and age, significantly influences people's lives, affecting their chances of achieving educational, occupational, and financial success.

assimilation: The process whereby immigrants adopt the language, values, norms, and worldview of the host culture at the expense of their heritage culture.

band: A native political entity, ultimately defined by the federal government.

Beringia: A continent-sized landmass that linked Siberia and Alaska (also called the Bering Land Bridge).

bestiality: Human sexual interaction with non-human animals, especially farm animals and domestic pets. It is fiercely taboo in most human cultures, but it is a thriving subgenre of pornography.

"big R" religion: The type of religion one often reads about in introductory texts, as defined by either the religious specialist or the academic for a general audience. It ignores the specifics of the wide variety of religious traditions present at the local level. See also **"little r" religion**.

Bill C-31: An amendment to the *Indian Act*, passed in 1985, that enabled people who had lost their Indian status through marriage or through the marriage of their mothers to apply to be reinstated as **registered Indians**.

bisexual: The attraction to and sexual activity with members of both sexes, which may be simultaneous or serial, and may or may not be equal for both sexes.

bondage: A sexual practice that involves restraining one's partner through the use of rope, cuffs, handcuffs, latex/leather, hoods, etc. The restraint may be on the arms and/or legs only, but it may also involve complete immobilization.

butch: A category of identity that refers to those overtly masculine in appearance or demeanour. Commonly used in **gay** and **lesbian** communities to refer to masculine lesbians.

capitalist class: Karl Marx's term for those who own the **means of production**—the land, machinery, factories, and so on—required for the production of goods and services.

celibacy: The state of being unmarried.

census: A list compiled by the government of every person who is in a **population** at a given point in time.

civil partnerships: Official term for the legal bonding of same-sex partners with legal rights and responsibilities similar to **same-sex marriage**.

civil unions: See **civil partnerships**.

civil registry: A list, compiled by a government, of births, deaths, or marriages.

come out (of the closet): To disclose a truth about oneself, especially but not exclusively when this truth concerns **sexual orientation**.

community: A group of people connected by common cause or interest. Geographical **gay** and **lesbian** communities tend to appear in metropolitan areas, but communication technologies permit the geographically isolated to feel part of a sexual community.

comparative method: A method used to evaluate the quality of demographic sources; it involves calculating trends of vital events based on information from adjacent districts.

conflict theory: The perspective that sees society as consisting of many groups whose interests often conflict. The theory proposes that inequality stems from the **exploitation** and oppression of one section of society by another. Therefore, inequality should be reduced or abolished.

conjugal family: A family whose members are linked by blood ties.

conservative ideology: The belief that things are best left as they are. An example is the belief that biological causes determine male and female behaviour and, therefore, that attempts to change traditional **gender spheres** are futile.

crosschecking: A process used to evaluate the quality of demographic information; it involves using a variety of sources linked to people in order to verify the information obtained about them.

cross-sectional analysis: A method used to examine demographic patterns over the life cycle, involving dividing people into nonoverlapping age categories. Demographic patterns are then generalized to describe the demographic experience over a lifetime.

Glossary

Crow's Index of Selection: A method used to measure the potential for **natural selection** in human **populations**; it takes into account both deaths and births (**mortality** and **fertility**).

crude birth rate: The number of **live births** for every 1000 people in a **population**.

crude death rate: The number of deaths for every 1000 people in a **population**.

crude marriage rate: The number of marriages for every 1000 people in a **population**.

cults: Non-traditional religious movements, a "subspecies" of religion centred on a single person or principle. They can either be positive or negative organizations.

de-assimilation: The resegregation of a previously assimilated minority group, as when **gays** and **lesbians** withdraw from the **heterosexual** mainstream **community** to form distinct and visible communities. See also **assimilation.**

decennial census: The **census** recorded once every 10 years.

deinstitutionalization: A movement to discharge people from institutional settings and to place them in the **community**.

demography: The scientific study of human **populations**.

Department of Indian Affairs (DIA): The federal ministry or branch of the federal ministry responsible for Native people.

dependency ratio: The ratio of the number of people aged less than 15 and over 64 years to the number of people aged 15 to 64 years.

developmental disabilities: Intellectual development that has been delayed; formerly referred to as "mental retardation." See also **disability**.

dialect: A version of a language that is distinct from a standard version in terms of grammar, pronunciation, and vocabulary; it is usually restricted to a specific geographical area of a nation or territory whose regions share the same language.

dialectical process: The ongoing process of social discourse and negotiation between members of that society.

disability: According to the World Health Organization, "Any limitation (resulting from an **impairment**) in the ability to perform any activity considered normal for a human being or required for some recognized social role or occupation" (*International Classification of Impairments, Disabilities and Handicaps* [WHO, 1980]). In Canada, it has become common to use the term to represent the notion of disability, **handicap**, and impairment, and adjectives such as *mental, intellectual,* and *physical* are used with the word to denote which body part and/or function is affected.

discrimination: The unequal or unfavourable treatment of people because of their perceived or actual membership in a particular **ethnic group** which restricts their full participation in the social, economic, and political life.

disposable income: Income above that required for basic necessities, such as food, clothing, and accommodation.

documentary method: A method of evaluating the quality of demographic sources by using the opinions of informed colleagues to assess the accuracy and completeness of the records.

domestic labour: Work related to home and family maintenance. From feminist perspectives, usually associated with women's work in the home.

domestic partnerships: See **civil partnerships**.

ecclesiastical registries: A list of births, deaths, or marriages compiled by religious groups.

emigration: The movement of people out of a specific geographical area.

emigration rate: The number of people leaving an area for every 1000 people in the **population**.

empowerment: Obtaining the resources, such as physical or financial means, to take control of one's own life.

empty nesters: Parents whose children have left home to live on their own.

equality of opportunity: The condition that exists when all citizens, regardless of their **ascribed status**, have the chance to succeed educationally, occupationally, and financially.

essentialize: To attempt to reduce a phenomenon to some single common "essence" or universal definition. Most scholars now recognize that all descriptions and knowledge are conditional, and that definitions therefore cannot be compressed into one single explanation or classification.

ethnic group: A group of people who share norms, values, traditions, and ancestry, and thus are considered distinct.

ethnocentrism: Viewing or judging the world from the point of view of one's own culture. Two variations are the assumption that what is true of one's culture is true of other cultures and the belief that one's culture is superior to other cultures.

evolution: A change either in physical form or in the frequency of certain genes over time.

executive federalism: What exists when important and far-reaching political decisions are made by the prime minister, provincial premiers, federal and provincial Cabinet ministers, and senior bureaucrats. Some believe this method of decision making runs counter to democracy, or to the belief that citizens should be involved in government decision making. See also **federalism**.

expectation method: A method used to evaluate the quality of demographic sources, which involves calculating expected proportions of vital events on the basis of such factors as economic conditions, marriages, and **migration**.

exploitation: Karl Marx's term for the situation in which the **capitalist class** pays workers less than the real value of their work.

extended family: A family that comprises two or more nuclear families joined through blood ties. The classic example is a husband and wife, their unmarried children, their married children, and the spouses and children of their married children.

external migrants: Migrants who move from outside a specific area. The area can be defined at many different levels, such as the neighbourhood, province, or nation.

externalization: What occurs when human perceptions and understanding of the universe become externally manifest as representations in both the things that we make (objects, tools, art, music, institutions, culture, etc.) and the things we do with those "products" within the public sphere.

failed states, system of: According to fundamentalists, the nation-state of the modernist ideal, which is seen as morally bankrupt due to its secularization and the resulting corruption, social chaos, and meaninglessness.

familia: The total number of slaves in a **household** (Latin).

familism: An ideology that promotes the traditional view of the family as the norm: working father, stay-at-home mother, and children.

fecundity: The maximum number of children a woman can produce during her lifetime.

federalism: A system of governing a country that divides responsibilities between two levels of government. Each level is responsible for the same **population** and cannot abolish the other level.

feminist literature: Stories related from the point of view of women, where the feminine experience is central to the story and women's perceptions are differentiated from those of men. The stories often include issues and events omitted from previous literature.

feminist theory: The view that women are disadvantaged in society and therefore must seek equality with men.

feminization of poverty: A trend characterized by growth in the percentage of women living in poverty.

femme: A feminine **lesbian** who can pass as straight. Unlike **lipstick lesbians**, most femmes wish to be known as queer.

fertility: The number of **live births** in a **population**.

fertility rate: The number of **live births** per 1000 women aged 15 to 44.

fundamentalism: Generally, a description of a religious movement of those who return their focus to what they believe to be the fundamental truths and practices of a religion.

gay: An adjective applied most commonly to the male subset of the whole class of **homosexuals**. Less often, it refers to all homosexuals, including women. The term, as used in English, is borrowed from Old French *gai*, meaning happy, carefree, hedonistic—qualities historically associated with male homosexuals.

gay marriage: General term for the legal union between same-sex partners, no matter what official term is used in any given country.

Gay Pride Parade: The annual colourful parade that many **gay** communities organize to demonstrate and celebrate their status as valued and respected members of society.

gender: The cultural aspect of masculinity and femininity.

gender reassignment surgery: The procedures by which a person's physical appearance and the function of their existing sexual characteristics are changed to those of the other **sex**, from male to female or from female to male. Part of the treatment for problems of **gender identity**.

gender identity: The social role a person assumes, which is usually but not always masculine for men and feminine for women.

gender spheres: Areas of work, school, or recreation that are dominated by one or the other **gender** (e.g., engineering for men and secretarial work for women).

gender stereotypes: Generalizations about how men and women should behave, what their strengths are, and where they are best suited to work, learn, and play in society.

gender-role-bound: The constraint of behaving in accordance with what is expected of a particular social role related to the **gender** in question.

genealogical analysis: Recreation of family histories. Can be used to calculate the amount of relatedness (inbreeding) among people in the **population**.

genealogies: Family histories or trees.

general systems theory: A sociological approach that studies the family as a self-contained unit.

glass ceiling: The invisible **gender** barrier that keeps women at the bottom of the occupational hierarchy and prevents them from winning promotions to positions of power.

global village: A condition in which every part of the world is electronically connected to every other part, creating a feeling that every culture is involved in the affairs of all other cultures.

Golden Horseshoe: The narrow stretch of cities along Lake Ontario from Niagara Falls to Oshawa.

gross migration rate: The number of people who enter and leave an area for every 1000 people in the **population**.

group distinctiveness: The significant ways in which a group is different and can be differentiated from other groups.

"handicap": "Any resulting disadvantage for an individual that limits the fulfilment of a normal role or occupation" (*International Classification of Impairments, Disabilities and Handicaps* [WHO, 1980]). See also **disability**.

hate propaganda: A controversial and difficult-to-define term for speech intended to degrade, intimidate, or incite violence, or prejudicial action against an identifiable group. In Canada it relates to colour, **race**, religion, ethnic origin, and now also **sexual orientation**.

heteroerotic: Having sexual, but not necessarily affectionate, attraction to members of the other **gender**.

heterosexual: Preferring to develop romantic and sexual relationships with members of the opposite **gender**.

hieroglyphics: The pictographic system used by the ancient Egyptians.

homoerotic: Having sexual, but not necessarily affectionate, attraction to members of one's own **gender**.

homophobic: Denoting an aversion to, active hatred of, or even violence toward the fact of same-sex desire and those who experience it.

homosexual: Preferring to develop romantic and sexual relationships with members of one's own **gender** (from Greek *homos*, "same," not Latin *homo*, "man").

household: A group of people who live together.

ideological legitimation: An assumption or set of assumptions that attempts to justify a political, economic, or social relationship or system.

ideology of the family: See **familism**.

immigrant and refugee stories: Stories that examine themes concerning leaving a familiar world—the risk and complexities of becoming part of an unknown and different society while accepting the inevitable losses involved, such as coping with the distance from family and friends, learning a new language, and adjusting to new political and social institutions. One theme that appears often in stories by and about refugees and immigrants is attitudes toward the role of women.

immigration: The movement of people into a specific geographical area.

immigration rate: The number of people entering an area for every 1000 people in the **population**.

impairment: "Any abnormality of physiological or anatomical structure or function" (*International Classification of Impairments, Disabilities and Handicaps* [WHO, 1980]).

inclusive societies: Societies characterized by the involvement and participation of all of its members.

income: The flow of money received over a specified period.

***Indian Act*:** An act through which the federal government gave itself tremendous power over Native people in Canada.

individual discrimination: Acts of **discrimination** carried out by individuals.

infant mortality rate: The number of children dying under one year of age.

institutional completeness: A complete set of basic social services, so that a member of a given **community** might live virtually all his or her life and have all important needs filled within that community.

institutional discrimination: Discrimination that limits the full participation of minority groups in the social, political, economic, and educational institutions of Canada. May or may not be intentional.

internal migrants: Those migrants who move within a specific area. The area can be defined at many different levels, such as the neighbourhood, province, or nation.

internalization: Occurs when an individual or society creates objects of culture to represent an understanding of reality, then internalizes or reabsorbs into consciousness the objectified world in such a way that the structures of this world or culture come to determine the subjective structures of consciousness.

isonomy: A method used to estimate inbreeding by examining the frequency of marriages occurring between people who share a surname.

jiva: From the Sanskrit, "soul." Often used by the Jain tradition to describe the thing that differentiates living from non-living entities.

karma: The principle of universal causality resulting from action. It is the accumulated sum of all actions in which an individual has participated, and the subsequent results of those actions that "bear fruit" in some future existence.

kinship coefficients: A method used to estimate inbreeding by examining the proportion of surnames within a **population**.

language families: Groups of related languages.

language isolate: A language that has no known related language.

learning disability: A comprehensive term that describes limitations in one's capacity to learn. See also **disability.**

leather-fag: A term to describe a sexual subculture organized around leather clothing—caps, jackets, trousers or chaps, boots, wrist cuffs, belts—and the sexual practices they can be used for. The term *fag* has been reclaimed from the original pejorative to an affectionate term within the **gay community**.

lesbian: A female **homosexual** (after the Aegean island Lesbos, home of the homosexual woman poet Sappho of ancient Greece).

liberation theories: Approaches to the study of family that see conflict as an essential characteristic of families and change and diversity as a means of freeing some family members from an oppressive family environment.

life expectancy: The age to which most humans can expect to live (average age at death).

life span: The maximum age that a human has ever lived.

***lingua franca*:** The common language used among cultures with different languages to communicate with each other.

lipstick lesbian: A **lesbian** who is beautiful, stylish, or markedly feminine who can pass as straight, but, unlike a **femme**, is often "in the closet." See **come out (of the closet)**.

"little r" religion: The specific forms of religion passed from grandmother to mother to daughter, or from grandfather to father to son. See also **"big R" religion**.

live births: Babies born alive.

longitudinal analysis: A method used to examine demographic patterns over the life cycle that involves following a birth or marriage cohort through time.

mass medium: A medium available to and used by most people in the world.

matriarchy: A type of family in which authority is vested primarily in the female. See also **patriarchy**.

means of production: The land, machinery, factories, and other resources required for the production of goods and services. In a capitalist society, the means of production are owned by a small percentage of the **population**.

media think: In the terminology of James Winter, a professor at the University of Windsor, the way media owners, managers, and workers think, see, and represent the world around us.

Meech Lake Accord (1987): The constitutional accord agreed to by the Prime Minister and the 10 provincial premiers. The name of the Accord was taken from the Meech Lake cottage where the meeting took place. It died when it failed to receive the appropriate approval from the Manitoba legislature.

"mentally deficient": A term used in the past to describe a person whose intellectual development was limited.

meritocracy: A social system that rewards people in direct proportion to their merits (skills, talents, and abilities) rather than to their **gender**, **race**, or social connections. A meritocracy would remove all **systemic barriers** that block men and women from entering nontraditional spheres.

Métis: A person who is a descendant of a particular people of French-Cree heritage.

middle class: Those who own a small amount of **wealth** and are employed in relatively secure and high-paying occupations.

migration: The movement of people into and out of specific geographical areas.

mode of production: The means by which goods are produced in society; in Marxist theory, the defining characteristic of a society.

moksha: A primarily Hindu and Jain term from the Sanskrit language that means liberation or "release" from the cycles of birth and death in the traditions of South Asia.

morbidity: The number of people with a specific disease in a **population**.

mortality: The number of people in a **population** dying in a given period.

multiculturalism: The federal government's official commitment to furthering national unity by promoting the positive aspects of cultural differences and the English and French languages.

Multiculturalism Act: A federal act that officially sanctioned **multiculturalism**; became law on 21 July 1988.

National Policy (1879): A conscious attempt by the government of John A. Macdonald to build an economy based on manufacturing and to lessen Canada's dependence on resource exports.

natural selection: The preferential survival and reproduction of individuals in a **population** by virtue of possessing a genetic characteristic that gives them an advantage.

nature: Biological explanations for human behaviour and interaction. Sometimes referred to as *biological determinism*.

nature/nurture: See **nature** and **nurture**.

net migration rate: The increase or decrease in the size of a **population** for every 1000 people based on the number of people who enter an area minus those who leave.

nirvana: From the Sanskrit, "extinction." Primarily a Buddhist term to indicate a departure from the cycle of rebirths, and entry into a different mode of existence.

nominative records: In **demography**, sources of demographic information that list a person's name.

nomos: According to Peter Berger, the stable social environment that occurs when individuals freely identify and participate within the "social project" and feel they are not being forced to participate in a particular role.

nonrandom mate selection: A distinct preference in the choice of a mate or marriage partner.

normalization: The desire of people to live lives in as "typical" a way as possible or to function on a day-to-day basis in an average way.

nuclear family: A **household** that includes a married couple and their children.

nuptiality: The demographic **variable** that measures the incidence, rate, and other aspects of marriage in society.

nurture: Social explanations for human behaviour and interaction. Sometimes also referred to as *social determinism* or *social beaming* theory.

Glossary

objectivation: Occurs when the **products of the human cultural project** become the primary objects of our consideration and attention—that is, when we begin to interact with the representations we have created of reality as if they were the real universe itself.

omnibus bill: A government bill containing many separate items.

orality: The technique of communication that relies solely on the spoken or gestured word.

outed: Describes someone whose **sexual orientation** has been disclosed with or without consent, so that he or she is now "out of the closet." See **come out (of the closet)**.

pack journalism: A term used to describe the occupation of reporters who feed off one another, move in unison, and rush to be the first to receive accolades for "breaking" a story.

paleodemography: The study of prehistoric **populations** on the basis of their physical remains.

parchment: The dried skin of sheep and goats, used by the ancient Greeks for writing material.

patriarchal: Pertaining to a family type where authority is vested in the male. See also **matriarchal**.

pedophilia: Sexual love or desire directed toward children of either **sex** by adult men or women.

pictograph: A symbol or picture used in some writing systems to represent entire words or utterances, as opposed to a letter of an alphabet representing a vocal sound.

pluralism: The belief that ethnic conflict will always be a central part of modern, industrial societies and that ethnicity will always be a vital component of individual and group identity.

political power: The degree to which a person or a group can enforce its demands.

population: A group of people who live within a specific geographical or political boundary, who are genetically similar (i.e., they interbreed), and who share a cultural heritage during a certain time frame.

postcolonial writing: Writing that takes place in a period after colonization has ended. For example, many writers emerged in the West Indies after the islands gained independence from European colonizers.

Powley test: A set of 10 main considerations that must be taken into account in order for the courts to determine whether a **Métis** hunting without a provincial licence was doing so in accordance with his or her proper Aboriginal right. After Steve Powley, who with his son killed a bull moose in Sault Ste. Marie in 1993 (the court ruled in their favour).

prejudice: The attitude of judging people on the basis of statements, ideas, and beliefs that do not hold up under scrutiny.

products of the human cultural project: Things we make, such as objects, tools, art, music, institutions, culture, etc., to represent our understanding of the world we live in. Human beings project meaning into the universe by creating both a material and an institutional culture that reflects that meaning, and these products become the primary objects (**objectivation**) of our attention.

progressive ideology: The belief in using social change to improve society. For example, under the assumption that social causes determine male and female behaviour, changing how society raises children will liberate people from restrictive **gender spheres**.

propaganda model: A theoretical model used to explain how the media can "manufacture" the "consent" of the public by choosing what stories to ignore, what to cover, and how to cover them. By doing so, the media control what the public is allowed to see, hear, and read. Created by Edward Herman and Noam Chomksy and discussed in their work *Manufacturing Consent: The Political Economy of the Mass Media* (2002).

queer subcultures: Homosexual lifestyles of various types. These are viewed by some as a good foundation from which to rebuild a just, improved society. "Queer" originally represented the notion that homosexuality involved pathological forms of femininity in **gay** men and masculinity in **lesbian** women. From the 1960s on, the term was slowly expropriated by emergent lesbian and gay communities, both culturally as a "catchall" for non-**heterosexuals** and then politically as a means of turning heterosexist language upon itself and making a positive statement of identity: "Queer and proud."

race: An arbitrary system of classification that divides humans into different categories (races) based on differing physical characteristics, such as skin colour and eye shape. Biologically, humans are all of the same species or race; but sociologically, physical traits are important symbols.

racial profiling: Any action undertaken for reasons of safety, security or public protection, that relies on **stereotypes** about **race**, colour, ethnicity, ancestry, religion, or place of origin, or a combination of these, rather than on reasonable suspicion, to single out an individual for greater scrutiny or different treatment.

racism: Discrimination based on **race** and assumed behavioural and mental similarities or deficiencies. Racism usually takes the form of the belief in the superiority of one race to another.

regionalism: An attitude of the citizens of a certain region that they have not been given adequate recognition for their part in building Canada and have been penalized by the federal government in favour of another region.

registered Indian: Someone who is "legally" an Indian, according to the federal ***Indian Act***.

rehabilitation medicine: A field of medicine that aims to return or to restore people to a former state of health or well-being. The term now also refers to helping people to participate to their fullest potential in society by achieving the highest level of well-being possible.

rehabilitation team: A group approach to **rehabilitation medicine**, developed to combat the many physical, mental, or intellectual **disabilities** facing returning veterans after World War II, involving team conferences and team planning. The approach recognizes that treating the effects of disabilities includes social aspects such as education, vocational training, housing, and employment and that professional expertise and access to a variety of services are needed.

representation by population: The principle that allocates seats in the House of Commons to each province according to its share of the national population. For example, a province that has 10 percent of the population receives 10 percent of the seats.

reserve: An area of land that has been reserved for Native peoples' use.

residential schools: Church-run boarding schools for Native children that existed from 1910 to the 1960s.

royal commission: An information-gathering device used by the federal government to investigate issues deemed important to Canada. The commission travels across Canada, headed by people appointed by the federal government. A royal commission can only advise government; it cannot implement policy.

sacred canopy: Within the "social project" (the project of building a society) religion can be used to protect its cultural institutions from assault from within by its members. Religion can give authority and legitimacy to social institutions.

sadomasochism: Sadism (erotic love of cruelty or domination) and masochism (erotic love of pain and submission) together, with one person getting pleasure in inflicting pain or suffering and another person getting pleasure in being subjected to it. The word "sadism" derives from the writings of the Marquis de Sade; "masochism" from Leopold von Sacher-Masoch, known for a novel with masochistic themes.

same-sex marriage: The legal bonding of same-sex partners with the same legal rights and responsibilities as opposite-sex marriage.

same-sex parenting: Two adults of the same **sex** acting as parents to children in the context of a family.

***samsara*:** A word from the Jain religion used to describe the transient world and the cycles of birth and rebirth (reincarnation).

script: A system of combined symbols for words, as opposed to **pictographs** or an alphabet.

self-determination: Making free choices and acting without outside interference.

sentencing circle: An innovative Native justice forum based on traditional concepts of restorative justice that involves **community** members and not just legal professionals, and that provides ways of dealing with people charged with crimes that are alternatives to the choices usually available in the Canadian legal system.

sex: A man or a woman's biological sexual characteristics, as indicated by the reproductive organs and the hormonal system a person is born with.

sexology: The body of knowledge making up the science of **sex**, that is, the science of the differentiation and dimorphism of sex and of the erotic/sexual pair bonding of partners.

sexual inversion: Taking on the **gender** role of the opposite **sex**; a term used in the older model of homosexuality.

sexual orientation: One's erotic attraction toward and interest in developing loving relationships with members of either the other or one's own **gender**, independently of one's sexual behaviour.

sexual predator: A person who commits **sex** crimes, such as rape or child abuse, especially a repeat offender.

sexuality: All aspects of human constitution and behaviour related to **sex**, including disposition toward love and deep affection, sexual dysfunction, the quality of being sexual, and **gender**.

smudging: The burning of herbs, such as sweetgrass, to create a smoke bath. It is used for purifying people and ceremonial space, tools, and objects, much as incense is used by Catholics, Hindus, Jews and Buddhists. Smudging is a daily morning ritual for some.

social controls: The means society uses to ensure that men and women behave in **gender**-appropriate ways. May include laws, ridicule, and **discrimination**.

social inequality: The degree to which people have access to and control over valued resources, such as money, **wealth**, status, and power.

social movements: Major historical changes in the day-to-day lives of groups resulting from concerted social action. Examples include the women's movement and the civil rights movement.

socially constructed roles: Sets of connected behaviours, rights, and obligations as conceptualized by actors in a given social situation. A socially constructed role is mostly thought of as an expected behaviour in a given individual social status and social position rather than the individual's characteristics.

sociopsychological dimension to regionalism: Concerned with how people living in different regions feel and act toward each other, their **community**, and the federal government. See **regionalism**.

sodomy: A legally defined term variously applied to *zoophilia* (sex with animals) and to mouth-genital or anal-genital contact between human beings, especially males.

space-oriented: Said of media that facilitate delivery and communication of information across vast distances of territory or space. See also **time-oriented**.

statistically representative sample: A sample that is an accurate reflection of the population from which it was drawn.

stereotype: A collection of generalizations about a group of people, which are negative, exaggerated, and unable to be maintained when subjected to critical analysis.

straight trade: **Heterosexual** partners, usually men, who have **sex** with **homosexuals**, sometimes for money as prostitutes.

stratification: The state of a society when it is made up of groups of people who have differing degrees of access to and control over valued resources.

structural–functionalism: 1. A theory proposing that inequality serves a positive function by ensuring that the most functionally important occupations are carried out by the most talented people, thus preserving the stability and proper functioning of society. An example is the idea that strict divisions of labour between men and women reduce role confusion and ensure that necessary jobs are done by those best equipped to do them. 2. A sociological approach that views the family as a stabilizing force for its members and for society as a whole.

sweat lodge: A ritual "sauna" used by First Nations or Native peoples, in which a sauna-like sweat bath is prepared by pouring water on heated rocks. Also refers to a tradition-based practice in which people physically and spiritually cleanse themselves, often as part of their healing path and of (re)connecting themselves to their Native identity.

systemic barriers: Laws, discriminatory practices, and psychological roadblocks that prevent men and women from entering non-traditional spheres.

telegraph: A device invented in 1844 that transmits messages across distances using electric wire.

television imperialism: The power of television over other media; its tendency to be the dominant medium in a culture.

time-oriented: Said of media sufficiently durable to preserve the history and tradition of cultures across vast amounts of time. See also **space-oriented.**

transcendent God: A divine being existing above and independently of the material world.

transfer payments: The name given to the billions of dollars that the federal government gives to the provinces and territories to help them deliver services to their populations. Key types of transfer payments are the Canada Assistance Plan (CAP), Equalization, and Territorial Formula Financing.

transgender: A term describing a range of states in which an individual, through dress, behaviour, or body modification, displays attributes different from the social and cultural expectations of their **gender**. Transgender people usually do not opt for **gender reassignment surgery**.

transsexual: A person who experiences an incompatibility between his or her anatomical **sex** and his or her psychological **sex**. Most transsexual people seek to alter their bodies through hormone therapy and/or **gender reassignment surgery**.

transvestite: A man or woman who adopts the dress and the behaviour of the opposite **sex**.

treaty: In the context of Canada, a legal agreement signed between either the British or the Canadian government and one or more Native nations.

tribalism: The condition in which people are encouraged to act as members of a group rather than as individuals.

"Ultimate Reality": An all-pervasive force that is representative of the "Supreme Experience," the effulgent and universal Awareness, the ultimate and absolute that permeates the universe. This force is impersonal and transcends all ideas or concepts of personality; one cannot have a personal relationship with it.

unification theories: Approaches to the study of family that see families as adaptive units that mediate between individuals and society.

universal themes: Topics or subjects of a literary work that are common to all human beings, regardless of the work's setting in time or place or the nature of the society in which they live. Some examples include good and evil, family relationships, loyalty, a sense of belonging, and desire for a home.

upper class: Those who own a considerable amount of **income** and **wealth**. In Canada, the upper class constitutes about 5 percent of the population.

urban reserves: A new type of **reserve** being created in Saskatchewan, consisting of lands located in a municipality or Northern Administrative District whose main function is to provide central urban locations for Aboriginal businesses.

variable: A characteristic that differs or varies among groups.

vernacular: A form of speech or a **dialect** that is characteristic of a particular region or nation.

vital event: A demographic term referring to births, marriages, or deaths.

voice: The expression of one's opinion and experience in the spoken or written word, the opinion itself, or the right to express an opinion.

wealth: An accumulation of assets, such as a house, savings, or a car.

working class: Karl Marx's term for the vast majority of the population who must sell their labour in order to survive.

SELECTED BIBLIOGRAPHY

Chapter 1

Adamec, Robert. *Memorial Gazette* 37(9) (27 January 2005).

Archer, Keith, Roger Gibbins, Rainer Knopff, and Lesie A. Pal, eds. *Parameters of Power: Canada's Political Institutions*. Toronto: Nelson, 1995.

Black, Errol, and Jim Silver. *Equalization: Financing Canadians' Commitment to Sharing and Social Solidarity*. Canadian Centre for Policy Alternatives, March 2004.

Bowker, Marjorie Montgomery. *The Meech Lake Accord: What It Will Mean to You and to Canada*. Gloucester, ON: Voyageur Publishing, 1990.

Bradfield, Michael. *Regional Economics: Analysis and Policies in Canada*. Toronto: McGraw-Hill Ryerson, 1988.

Brodie, Janine. *The Political Economy of Canadian Regionalism*. Toronto: HBJ, 1990.

Brym, Robert. "Canada's Regions and Agrarian Radicalism." In James Curtis and Lorne Tepperman, eds., *Images of Canada: The Sociological Tradition*. Scarborough, ON: Prentice-Hall, 1990.

Conway, J.F. "Western Alienation: A Legacy of Confederation." In John A. Fry, ed., *Contradictions in Canadian Society*. Toronto: Wiley, 1984.

Davis, Jo, ed. *Not a Sentimental Journey: What's Behind the Via Rail Cuts, What You Can Do About It*. Toronto: Gunbyfield Publishing, 1990.

Dyck, Rand. *Canadian Politics: Critical Approaches*. Scarborough, ON: Nelson, 1993.

Gibbins, Roger. *Conflict and Unity: An Introduction to Canadian Political Life*, 3rd ed. Scarborough, ON: Nelson, 1994.

Harmer, Harry. *The Longman Companion to Slavery, Emancipation and Civil Rights*. Toronto: Pearson Education Ltd., 2001.

Hiller, Harry S. *Canadian Society: A Macro Analysis*. Toronto: Pearson, 2000.

Hurtig, Mel. *The Betrayal of Canada*. Toronto: Stoddart, 1990.

Johnston, Donald J., ed. *With a Bang, Not a Whimper: Pierre Trudeau Speaks Out*. Toronto: Stoddart, 1988.

Kilgour, David. *Inside Outer Canada*. Edmonton: Lone Pine Publishing, 1990.

Lithwick, N.H. "Is Federalism Good for Regionalism?" In Garth Stevenson, ed., *Federalism in Canada*. Toronto: McClelland and Stewart, 1989.

"Maritimes Kept Poor by Ontario." *The Hamilton Spectator*, 30 July 2001.

Mathews, Ralph. *The Creation of Regional Dependency*. Toronto: University of Toronto Press, 1983.

McQuaig, Linda. "Just One Sponsor, but Canadians Love CBC *People's History*." 21 December 2000. Straight Goods.com <http://goods.perfectvision.ca/ViewFeature.cfm?REF=23>. Accessed 8 February 2002.

Morton, Desmond. *A Short History of Canada*. Edmonton: Hurtig, 1983.

Phillips, Paul. *Regional Disparities*. Toronto: Lorimer, 1978.

Qualman, Darrin. *The Farm Crisis and Corporate Power*. Ottawa: Canadian Centre for Policy Alternatives, 2001.

Savoie, Donald J. *The Canadian Economy: A Regional Perspective*. Toronto: Methuen, 1986.

Statistics Canada. *The Daily*, 27 January 2005.

Statistics Canada. *The Daily*, 4 May 2005.

Swan, Neil, and John Serjak. "Analysing Regional Disparities." In James Curtis, Edward Grabb, and Neil Guppy, eds., *Social Inequality in Canada: Patterns, Problems, Policies*, 2nd ed. Scarborough, ON: Prentice-Hall, 1993.

"Top 300 Private Companies," Globeinvestor.com <www.globeinvestor.com/series/top1000/tables/private/2005>. Accessed 29 April 2006.

Trudeau, Pierre E. *A Mess That Deserves a Big NO*. Toronto: Roberston Davies Publishing, 1992.

Wien, Fred. "Regional Inequality: Explanations and Policy Issues." In James Curtis, Edward Grabb, and Neil Guppy, eds., *Social Inequality in Canada: Patterns, Problems, Policies*, 2nd ed. Scarborough, ON: Prentice Hall, 1993.

Young, Lisa and Archer Keith eds. *Regionalism and Party Politics in Canada*. Toronto: Oxford University Press, 2001.

Chapter 2

Abu-Laban, B. "Arab Immigration to Canada." In J.L. Elliott, ed., *Two Nations, Many Cultures: Ethnic Groups in Canada*. Scarborough, ON: Prentice-Hall, 1979.

Bilson, G. *A Darkened House: Cholera in Nineteenth-Century Canada*. Toronto: University of Toronto Press, 1980.

Bourbeau, R., and J. Légaré. *Évolution de la mortalité au Canada et au Québec, 1831–1931. Essai de mésure par génération*. Montreal: Les Presses de l'Université de Montréal, 1982.

Brookes, A.A. "The Golden Age and the Exodus: The Case of Canning, Kings County." *Acadiensis* 11 (1981): 57–82.

Brunger, A.G. "Geographical Propinquity Among Pre-famine Catholic Irish Settlers in Upper Canada." *Journal of Historical Geography*, 8 (1982): 265–82.

Cavalli-Sforza, L.L., and W.F. Bodmer. *The Genetics of Human Populations*. San Francisco: W.H. Freeman, 1971.

Charbonneau, H. "Jeunes femmes et vieux maris: la fécondité des mariages précoces." *Population*, 35 (1980): 1101–22.

Charbonneau, H., and A. LaRose, eds. *The Great Mortalities: Methodological Studies of Demographic Crises in the Past*. Liege: Ordina Editions, 1979.

Connell, K.H. *The Population of Ireland 1750–1845*. Oxford: Clarendon Press, 1950.

Crawley, R. "Off to Sydney: Newfoundlanders Emigrate to Industrial Cape Breton 1890–1914." *Acadiensis* 17 (1988): 27–51.

Cressy, D. "The Seasonality of Marriage in Old and New England." *Journal of Interdisciplinary History* 16 (1985): 1–21.

Crow, J.F. "Some Possibilities for Measuring Selection Intensities in Man." *Human Biology* 30 (1958): 1–13.

Crow, J.F., and A.P. Mange. "Measurement of Inbreeding from the Frequency of Marriages Between Persons of the Same Surname." *Eugenics Quarterly* 12 (1965): 199–203.

Darroch, A.G., and M.D. Ornstein. "Family and Household in Nineteenth-Century Canada: Regional Patterns and Regional Economies." *Journal of Family History* 9 (1984): 158–77.

Dixon, R.B. "Explaining Cross-Cultural Variation in Age at Marriage and Proportions Never Marrying." *Population Studies* 25 (1971): 215–33.

Dobzhansky, T. "Natural Selection in Mankind." In G.A. Harrison and A.J. Boyce, eds., *The Structure of Human Populations*. Oxford: Clarendon Press, 1972.

Donnelly, F.K. "Occupational and Household Structures of a New Brunswick Fishing Settlement: Campobello Island, 1851." In R. Chanteloup, ed., *Labour in Atlantic Canada*. Saint John: University of New Brunswick, 1981.

Elder, G.H., and R.C. Rockwell. "Marital Timing in Women's Life Patterns." *Journal of Family History* 1 (1976): 34–53.

Elliott, J.L. "Canadian Immigration: A Historical Assessment." In J.L. Elliott, ed., *Two Nations, Many Cultures: Ethnic Groups in Canada*. Scarborough, ON: Prentice-Hall, 1979.

Fogel, R.W., S.L. Engerman, J. Trussel, R. Floud, C.L. Pope, and L.T. Wimmer. "The Economics of Mortality in North America, 1659–1910: A Description of a Research Project." *Historical Methods* 11 (1978): 75–108.

Gaffield, C.M. "Boom and Bust: The Demography and Economy of the Lower Ottawa Valley in the Nineteenth Century." *Canadian Historical Association, Historical Papers*, 1982: 172–95.

Gee, E.M.T. "Early Canadian Fertility Transition: A Components Analysis of Census Data." *Canadian Studies in Population* 6 (1979): 23–32.

Gibson, J.R. "Smallpox on the Northwest Coast, 1835–1838." *BC Studies* 56 (1982): 61–81.

Gossage, P. "Absorbing Junior: The Use of Patent Medicines as Abortificants in Nineteenth-Century Montreal." *The Register* 3 (1982): 1–13.

Hajnal, J. "Age at Marriage and Proportions Marrying." *Population Studies* 7 (1953): 111–36.

Harney, R.F. "Men Without Women: Italian Migrants in Canada 1885–1930." *Canadian Ethnic Studies* 11 (1979): 29–47.

Harrison, G.A., and A.J. Boyce., eds. *The Structure of Human Populations*. Oxford: Clarendon Press, 1972.

Henry, L. *Population: Analysis and Models*. New York: Academic Press, 1976.

Kaprielian, I. "Immigration and Settlement of Armenians in Southern Ontario: The First Wave." *Polyphony* 4 (1982): 14–27.

Katz, M.B., M.J. Doucet, and M.J. Stern. "Population Persistence and Early Industrialization in a Canadian City: Hamilton, Ontario, 1851–1971." *Social Science History* 2 (1978): 208–29.

Kaye, V.J., and C.W. Hobart. "Origins and Characteristics of the Ukrainian Migration to Canada." In C.W. Hobart, W.E. Kalbach, J.T. Borhek, and A.P. Jacoby, eds., *Persistence and Change: A Study of Ukrainians in Alberta*. Toronto: Ukrainian Canadian Research Foundation, 1978.

Keyes, J. "Marriage Patterns Among Early Quakers." *Nova Scotia Historical Quarterly* 8 (1978): 299–307.

Kussmaul, A. "Time and Space, Hoofs and Grain: The Seasonality of Marriage in England." *Journal of Interdisciplinary History* 15 (1985): 755–79.

Landry, Y. "Mortalité, nuptialité et canadianisation des troupes française de la guerre de Sept Ans." *Social History* 12 (1979): 298–315.

Lavoie, Y. *L'Émigration des Québécois aux États-Unis de 1840 à 1930*. Quebec: Éditeur officiel du Québec, 1979.

Li, P.S. "Immigration Laws and Family Patterns: Some Demographic Changes Among Chinese Families in Canada, 1885–1971." *Canadian Ethnic Studies* 13 (1980): 58–73.

Li, P.S. "Chinese Immigrants in the Canadian Prairie, 1910–1947." *Canadian Review of Sociology and Anthropology* 19 (1982): 527–40.

Lloyd, S. "The Ottawa Typhoid Epidemics of 1911 and 1912: A Case Study of Disease as a Catalyst for Urban Reform." *Urban History Review* 8 (1979): 66–89.

Matwijiw, P. "Ethnicity and Urban Residence: Winnipeg, 1941–1971." *Canadian Geographer* 23 (1979): 45–61.

McGinnis, J.D.P. "The Impact of Epidemic Influenza: Canada, 1918–1919." *Canadian Historical Association, Historical Papers* (1977): 121–40.

McKeown, T. *The Modern Rise of Population*. London: Edward Arnold, 1976.

McLaren, A. "Birth Control and Abortion in Canada, 1870–1920." *The Canadian Historical Review* 59 (1978): 319–40.

McQuillan, K. "Economic Structure, Religion, and Age at Marriage: Some Evidence from Alsace." *Journal of Family History* 14 (1989): 331–46.

Medjuck, S. "The Social Consequences of Economic Cycles on Nineteenth-Century Households and Family Life." *Social Indicators Research* 18 (1986): 233–61.

Model, J. "The Timing of Marriage in the Transition to Adulthood: Continuity and Change, 1860–1975." In J. Demos and S. Boocock, eds., *Turning Points: Historical and Sociological Essays on the Family*. Chicago: University of Chicago Press, 1978.

Nam, C.B., and S.O. Gustavus. *Population: The Dynamics of Demographic Change*. Boston: Houghton Mifflin, 1976.

Norris, D.A. "Household and Transiency in a Loyalist Township: The People of Adolphustown, 1784–1822." *Social History* 13 (1980): 399–415.

Osborne, B. "The Cemeteries of the Midland District of Upper Canada: A Note on Mortality in a Frontier Society." *Pioneer America* 6 (1974): 46–55.

Parker, W. "The Canadas." In A. Lemon and N. Pollock, eds., *Studies in Overseas Settlement and Population*. New York: Longman, 1980.

Roth, E. "Historic Fertility Differentials in a Northern Athapaskan Community." *Culture* 2 (1982): 63–75.

Roychoudhury, A.K., and M. Nei. *Human Polymorphic Genes: World Distribution*. Oxford: Oxford University Press, 1988.

Sharna, R.D. "Premarital and Ex-nuptial Fertility (Illegitimacy) in Canada 1921–1972." *Canadian Studies in Population* 9 (1982): 1–15.

Shryock, H.S., and J.S. Siegel. *The Methods and Materials of Demography*. San Diego: Academic Press, 1976.

Statistics Canada. Internet site <www.statcan.ca>.

Statistics Canada. *2001 Census Handbook*. Ottawa: Minister of Industry, 2003.

Statistics Canada. *2001 Census of Canada*.

Statistics Canada. *Births, 2001*. Ottawa: Health Statistics Division, 2003.

Statistics Canada. *Deaths, 2000*. Ottawa: Health Statistics Division, 2003.

Statistics Canada. *Marriages, 2001*. Ottawa: Health Statistics Division, 2003.

Swedlund, A.C. "Historical Demography: Applications in Anthropological Genetics." In J.H. Mielke, and M.H. Crawford, eds., *Current Developments in Anthropological Genetics*, vol. 1. New York: Plenum Press, 1980.

Veevers, J.E. "Age Discrepant Marriages: Cross-national Comparisons of Canadian-American Trends." *Social Biology* 31 (1984): 118–26.

Weaver, J.C. "Hamilton and the Immigration Tide." *Families* 20 (1981): 197–208.

Willigan, J.D., and K.A. Lynch. *Sources and Methods of Historical Demography*. New York: Academic Press, 1982.

Wrigley, E.A. "Family Limitation in Pre-industrial England." *Economic History Review* 19 (1966): 82–109.

Wynn, G. "Ethnic Migrations and Atlantic Canada: Geographical Perspectives." *Canadian Ethnic Studies* 18 (1986): 1–15.

Chapter 3

Allahar, Anton L. and James E. Cote. *The Structure of Inequality in Canada*. Toronto: James Lorimer and Company Ltd., 1998.

Badets, Jane, and Tina W.L. Chui. "Focus on Canada's Changing Immigrant Population." Statistics Canada, Catalogue No. 96-311E. Ottawa and Scarborough, ON: Statistics Canada and Prentice Hall, 1994.

Clement, Wallace. *The Canadian Corporate Elite: An Analysis of Economic Power*. Ottawa: Carleton University Press, 1986.

Cooke-Reynolds, Melissa, and Nancy Zukewich. "The Feminization of Work." *Canadian Social Trends*, 72 (Spring 2004): 27. Statistics Canada, Catalogue No. 11-008.

Creese, Gillian, Neil Guppy, and Martin Meissner. *Ups and Downs on the Ladder of Success*. Ottawa: Statistics Canada, 1991.

Crompton, Susan. "Left Behind: Lone Mothers in the Labour Market." *Perspectives*, Summer 1994: 23.

Selected Bibliography

Davies, James B. "The Distribution of Wealth and Economic Inequality." In James Curtis, Edward Grabb, and Neil Guppy, eds., *Social Inequality in Canada: Patterns, Problems, Policies*, 3rd ed. Scarborough, ON: Prentice Hall Allyn and Bacon Canada, 1999.

Gee, E., and S. Prus, "Income Inequality in Canada: A Racial Divide." In M. Kalbach and W. Kalbach, eds., *Perspectives on Ethnicity in Canada*. Toronto: Harcourt Brace, 2000.

Grabb, Edward G. *Theories of Social Inequality: Classical and Contemporary Perspectives*. Toronto: Holt, Rinehart, and Winston, 1990.

Hou, Feng, and T.R. Balakrishnan. "The Economic Integration of Visible Minorities in Contemporary Canadian Society." In James Curtis, Edward Grabb, and Neil Guppy, eds., *Social Inequality in Canada: Patterns, Problems, Policies*, 3rd ed. Scarborough, ON: Prentice Hall Allyn and Bacon Canada, 1999.

Hunter, Alfred A. *Class Tells: On Social Inequality in Canada*. Toronto: Butterworths, 1981.

"Likelihood of Saving Increase with Income." *Infomat: A Weekly Review*, 20 July 2001, Cat. No. 11-002E.

Macionis, John J., Juanne Nancarrow Clarke, and Linda M. Gerber. *Sociology*. New Jersey: Prentice-Hall, 1993.

Marx, Karl. *Critique of the Gotha Programme*. Moscow: Progress Publishers, 1970.

Naiman, Joanne. *How Societies Work: Class, Power, and Change in a Canadian Context*, 2nd ed. Toronto: Irwin Publishing, 2000.

Picot, G., and A. Heisz. "The Labour Market in the 1990s." *Canadian Economic Observer*, February 2000. Statistics Canada, Catalogue No. 11-010-XPB.

Roy, F. "Social Assistance by Province, 1993–2003." *Canadian Economic Observer*, November 2004. Statistics Canada, Catalogue No. 11-010.

Statistics Canada. "Income Distributions by Size in Canada, 1997." Catalogue No. 13-207-XPB, 1999.

Turner, Bryan S. *Equality*. London: Tavistock Publications, 1986.

Chapter 4

Abella, Rosalie. *Report of the Commission on Equality in Employment*. Ottawa: Supply and Services Canada, 1984.

Barrett, Ralph V. "Pedagogy, Racism and the 'Postmodern Turn.'" *The College Quarterly* 1 (Fall 1994).

Berger, Peter, and Brigitte Berger. *Sociology: A Biographical Approach*. New York: Basic Books, 1972.

Berton, Pierre. *Why We Act Like Canadians*. Markham, ON: Penguin, 1987.

Bissoondath, Neil. *Selling Illusions: The Cult of Multiculturalism in Canada.* Markham, ON: Penguin Books, 2003.

Brodie, Janine, and Linda Trimble. *Reinventing Canada: Politics of the 21st Century.* Toronto: Pearson Education Canada, 2003.

Canadian Policy Research Network. "Populations Projections for 2017." Available <www.cprn.org/en/diversity-2017.cfm>. Accessed 29 April 2006.

Driedger, Leo. *Race and Ethnicity: Finding Identities and Equalities*, 2nd ed. Don Mills, ON: Oxford University Press, 2003.

Elliott, Jean Leonard, and Augie Fleras. *Engaging Diversity: Multiculturalism in Canada.* Scarborough, ON: Nelson, 2002.

Elliott, Jean Leonard, and Augie Fleras. *Unequal Relations: An Introduction to Race and Ethnic Dynamics in Canada.* Scarborough, ON: Prentice-Hall, 1992.

Fleras, Augie, and Jean L. Kunz. *Media and Minorities: Representing Diversity in a Multicultural Canada.* Toronto: TEP, 2001.

Fleras, Augie, and Jean Leonard Elliott. *Multiculturalism in Canada: The Challenge of Diversity.* Scarborough, ON: Nelson, 1992.

Gould, S.J. *The Mismeasure of Man.* New York: W.W. Norton, 1981.

Government of Canada. *The Canadian Multiculturalism Act: A Guide for Canadians.* Ottawa, 1990.

Haas, Jack, and William Shaffir. *Shaping Identity in Canadian Society.* Scarborough, ON: Prentice-Hall, 1978.

Hawkins, Freda. *Canada and Immigration: Public Policy and Public Concern.* Kingston, ON and Montreal: McGill-Queen's University Press, 1988.

Heisz, Andrew. Statistics Canada. "Ten Things to Know About Canadian Metropolitan Areas: A Synthesis of Statistics Canada's Trends and Conditions in Census Metropolitan Areas Series." Ministry of Industry, 2005.

Henry, Frances. *The Caribbean Diaspora in Toronto: Learning to Live with Race.* Toronto: UTP, 1994.

Henry, Frances, Carol Tator, Winston Mattis, and Tim Rees. *The Colour of Democracy: Racism in Canadian Society*, 2nd ed. Toronto: Harcourt Brace, 2000.

Hill, Daniel G. *Human Rights in Canada: A Focus on Racism.* Ottawa: Canadian Labour Congress, 1977.

Hiller, Harry H. *Canadian Society: A Macro Analysis.* Toronto: Prentice-Hall, 2000.

James, Carl E. *Seeing Ourselves: Exploring Race, Ethnicity and Culture*, 3rd ed. Toronto: TEP, 2003.

Selected Bibliography

Johnson, Walter. *The Challenge of Diversity*. Montreal: Black Rose Books, 2006.

Kalbach, Madeline A., and Warren E. Kalbach. *Perspectives on Ethnicity In Canada: A Reader*. Toronto: Harcourt Brace, 2000.

Kelley, Ninette, and Michael Trebilcock. *The Making of the Mosaic: A History of Canadian Immigration Policy*. Toronto: UTP, 1998.

Kennedy, K.A.R. *Human Variation in Space and Time*. Dubuque, IA: Brown, 1976.

Li, Peter S. *Destination Canada: Immigration Debates and Issues*. Don Mills: Oxford University Press, 2003.

Li, Peter S. *Ethnic Inequality in a Class Society*. Toronto: Wall and Thompson, 1988.

Montagu, A., ed. *The Concept of Race*. London: Collier-Macmillan, 1964.

Ontario Human Rights Commission. "Paying the Price: The Human Cost of Racial Profiling: Inquiry Report," 2004. Available <www.ohrc.on.ca/english/consultations/racial-profiling-report.shtml>. Accessed 29 April 2006.

Palmer, Howard. *Immigration and the Rise of Multiculturalism*. Toronto: Copp Clark, 1975.

Smith, Charles C. "Crisis, Conflict and Accountability: The Impact and Implications of Police Racial Profiling." *African Canadian Community Coalition on Racial Profiling*. March 2004.

Stoffman, Daniel. *Who Gets In? What's Wrong with Canada's Immigration Program and How to Fix It*. Toronto: Macfarlane Walter and Ross, 2002.

Wise, Tom. "Racial Profiling and Its Apologists." *Z Magazine*, March 2002.

Chapter 5

A good place to begin a search for information on Native culture is the local Native Friendship Centre. These are found in most Canadian cities. Other good sources of information are the Native Studies departments found in some community colleges and a few universities.

Barman, Jean. "Aboriginal Education at the Crossroads: The Legacy of Residential Schools and the Way Ahead." In D.A. Long and O.P. Dickason, eds., *Visions of the Heart: Canadian Aboriginal Issues*. Toronto: Harcourt Brace, 1996.

Bergman, Brian. "Dark Days for the Inuit." *Maclean's*, 4 March 1996, p. 67.

Dickason, Olive P. *Canada's First Nations: A History of Founding Peoples from Earliest Times*. Toronto: McClelland and Stewart, 1997.

Francis, Daniel. *The Imaginary Indian: The Image of the Indian in Canadian Culture*. Vancouver: Arsenal Pulp Press, 1992.

Selected Bibliography

Frideres, James S., and Rene Gadacz. *Aboriginal People in Canada: Contemporary Conflicts*, 6th ed. Toronto: Prentice-Hall, 2001.

Henslin, James, Dan Glenday, Ann Duffy, and Norene Pupo. *Sociology: Canadian Edition: A Down-to-Earth Approach*. Toronto: Allyn and Bacon, 2001.

Indian Treaties and Surrenders, vol. 1. Toronto: Coles Publishing, 1971. Reprint of federal government publication.

Knockwood, Isabelle. *Out of the Depths*. Lockeport, NS: Roseway Publishers, 1992.

LaRoque, Emma. "Three Conventional Approaches to Native People." In Brett Balon and Peter Resch, eds., *Survival of the Imagination: the Mary Donaldson Memorial Lectures*. Regina: Coteau Books, 1993. Pp. 209–18.

Lawrence, Bonita. *"Real Indians" and Others: Mixed-Blood Urban Native Peoples and Indigenous Nationhood*. Vancouver: UBC Press.

Purich, Donald. *The Métis*. Toronto: Lorimer, 1988.

Rice, Brian, and John Steckley. "Lifelong Learning and Cultural Identity: Canada's Native People." In Michael J. Hatton, ed., *Lifelong Learning: Policies, Practices, and Programs* (APEC pub. #97-HR01.5). Toronto: School of Media Studies, Humber College, 1997. Pp. 216–29.

Robinson, Angela. *Ta'n teli-ktlamsi Tasit (Ways of Believing): Mi'kmaw Religion in Eskasoni, Nova Scotia*. Canadian Ethnography Series, vol. 3. Toronto: Pearson Education Canada, 2005.

Smith, Donald. *Le Sauvage*. Ottawa: National Museum of Man, 1974.

Steckley, John. *Aboriginal Voices and the Politics of Representation in Canadian Introductory Sociology Textbooks*. Toronto: Canadian Scholars Press.

Steckley, John, and Bryan Cummins. *Full Circle: Canada's First Nations*. Toronto: Prentice-Hall, 2001.

Chapter 6

Berger, Peter L. *The Sacred Canopy: Elements of a Sociological Theory of Religion*. New York: Doubleday, 1967.

Buchignani, Norman. "South Asians in Canada: Accommodation and Adaptation." In R.N. Kanungo, ed., *South Asians in a Canadian Mosaic*. Montreal: Kala Bharati, 1984, pp. 157–180.

Buchignani, Norman. "Research on South Asians in Canada: Retrospect and Prospect." In Milton Israel, ed., *The South Asian Diaspora in Canada: Six Essays*. Toronto: The Multicultural History Society of Ontario (in cooperation with Centre for South Asian Studies, University of Toronto), 1987. Pp. 113–140.

Selected Bibliography

Buchignani, Norman, Dorren Indra, and Ram Srivastiva. *Continuous Journey: A Social History of South Asians in Canada*. Toronto: McClelland and Stewart, 1985.

Canadian Multiculturalism Act, R.S., 1985, c. 24 (4th Supp.). Available Canadian Heritage site <www.canadianheritage.gc.ca/progs/multi/policy/act_e.cfm>. Accessed 4 April 2006.

"A Tale of Perseverance: Chinese Immigration to Canada." *Life and Society*, CBC Archives, 2006. Available <http://archives.cbc.ca/IDD-1-69-1433/life_society/chinese_immigration>. Accessed 4 April 2006.

Chinese Immigration Act, S.C. 1900, c.32 S.6 and S.C. 1903, c.8 S.6.

Choquette, Robert. *Canada's Religions*. Ottawa: University of Ottawa Press, 2004. P. 145.

Citizenship and Immigration Canada. Archives. Available <www.cic.gc.ca/english>.

Clifford, James. *Routes: Travel and Translation in the late Twentieth Century*. Cambridge, MA: Harvard University Press, 1997.

Coward, Harold, and David Goa. "Religious Experience of the South Asian Diaspora in Canada." In Milton Israel, ed., *The South Asian Diaspora in Canada: Six Essays*. Toronto: The Multicultural History Society of Ontario (in cooperation with Centre for South Asian Studies, University of Toronto), 1987. Pp. 73–86.

Daniels, Roger. "The History of Indian Immigration to the United States: An Interpretive Essay." In Jagat Motwani, Mahin Gosine, and Jyoti Barot-Motwani, eds. *Global Indian Diaspora: Yesterday, Today and Tomorrow*. New York: Global Organization of People of Indian Origin, 1993. Pp. 439–445.

Dawson, Lorne L. *Comprehending Cults*. Toronto: Oxford University Press, 1998. Pp. 52–53.

Department of Justice of Canada website. <http://canada.justice.gc.ca/en/justice2000/libmin00.html>. Accessed 2005.

Durkheim, Emile. *The Elementary Forms of Religious Life*. New York: Free Press, 1995. P. 44.

Geertz, Clifford. *The Interpretation of Cultures*. New York: Basic Books, 1973. P. 90.

Johnson, Hugh. *The Voyage of the Komagata Maru*. University of British Columbia Press, 1989.

Kupferschmid-Moy, Denise. *Across the Generations: A History of the Chinese in Canada*. <http://collections.ic.gc.ca/generations/index2.html>. Accessed 29 April 2006.

Petros, C.I. "Indo-Canadians." In J. Motwani, M. Gosine, J. Barot-Motwani, eds. *Global Indian Diaspora: Yesterday, Today and Tomorrow*. New York: Global Organization of People of Indian Origin, 1993. Pp. 475–484.

Ramcharan, Subash. "South Asian Immigration: Current Status and Adaptation Modes." In R.N. Kanungo, ed., *South Asians in a Canadian Mosaic*. Montreal: Kala Bharati, 1984. Pp. 33–48.

Sampat-Mehta, R. "First Fifty Years of South Asian Immigration: A Historical Perspective." In R.N. Kanungo, ed., *South Asians in a Canadian Mosaic*. Montreal: Kala Bharati, 1984. Pp. 13–32.

Shohat, Ella. *Talking Visions: Multicultural Feminism in a Transnational Age*. Cambridge: MIT Press, 1998.

Sinclair, K. "Women and Religion." In M.I. Dudley and M.I. Edwards, eds., *The Cross-Cultural Study of Women: A Comprehensive Guide*. New York: City University of New York (The Feminist Press), 1986. Pp. 107–124.

Smart, Ninian. *The World's Religions*. Cambridge: Cambridge University Press, 1998. p. 10.

Smith, Huston. *Why Religion Matters*. New York: HarperSanFrancisco, 2001.

Statistics Canada. *The Daily*, February 17, 1998. Available <www.statcan.ca/Daily/English/980217/d980217.htm>. Accessed 26 April 2006.

Thomas, David. *Transcultural Space and Transcultural Beings*. Boulder, CO: Westview Press, 1996.

Wagle, Iabal. "South Asians in Canada, 1905–1920." In Milton Israel and N.K. Wagle, eds., *Ethnicity, Identity, Migration: The South Asian Context*. Toronto: University of Toronto Centre for South Asian Studies, 1993. Pp. 196–216.

Weber, Max. *The Protestant Ethic and the Spirit of Capitalism*. New York: Charles Scribner's Sons, 1958.

Weber, Max. *The Sociology of Religion*. Boston: Beacon Press, 1964. P. 1.

Chapter 7

Accessibility for Ontarians with Disabilities Act. June 14, 2005.

Alford, Glen, ed. *The Advocate*. Toronto: Ontario March of Dimes.

Bickenbach, Jerome. *Physical Disability and Social Policy*. Toronto: University of Toronto Press, 1993.

Bowland, A., C. Nakatsu, and J. O'Reilly, eds. *The 1995 Annotated Ontario Human Rights Code*. Toronto: Carswell, 1995.

Canadian Human Rights Act, R.S.C., 1985.

Driedger, Diane, and Susan Gray, eds. *Imprinting Our Image: An International Anthology of Women's Disabilities*. Charlottetown: Gynergy, 1992.

Eisenberg, Myron G., Cynthia Griggins, and Richard J. Duval, eds. *Spring Series on Rehabilitation: Vol. 2. Disabled People as Second-Class Citizens*. New York: Springer, 1982.

Findley, Timothy. *The Piano Man's Daughter*. Toronto: Harper Collins, 1995.

Higgens, Paul. *Masking Disability: Exploring the Social Transformation of Human Variation.* Springfield: Charles C. Thomas, 1992.

Hornberger, Chris and Peter Milley. *Final Report, Diversity Planning for Inclusive Employment.* Halifax, Nova Scotia: Halifax Global Incorporated, 2005.

Human Rights Legislation: An Office Consolidation. Toronto: Butterworths, 1991.

Ministry of National Health and Welfare: Disabled Persons in Canada. Ottawa: Ministry of National Health and Welfare, 1981.

Office for the Disabled Persons. *The Needs and Attitudes of Disabled Ontarians.* Toronto: Environics Research Group, 1989.

Rioux, Marcia, and Michael Bach, eds. *Disability Is Not Measles: New Research Paradigms in Disability.* North York, ON: Roeher Institute, 1994.

Rogers, Patricia. "Atlanta Olympics Take Aim at Barriers to the Disabled." *Toronto Star*, 16 July 1996.

Rubin, Josh. "Wheelchair Racers Preview Olympic Dash." *Toronto Star*, 14 July 1996.

Special Committee on the Disabled and Handicapped, First Report (Obstacles). Ottawa: 1980.

Speech by the President of the Treasury Board of Canada to the Disabled People's International Summit. September 10, 2004.

Chapter 8

Baker, Maureen. "Gender and Gender Relations." In R. Jack Richardson and Lorne Tepperman, eds., *An Introduction to the Social World.* Toronto: McGraw-Hill Ryerson, 1987.

Bly, Robert. *Iron John: A Book About Men.* Reading, MA: Addison-Wesley, 1990.

Carey, Elaine. "Women Still Two Steps Behind Men." *Toronto Star*, 9 August 1995, p. A15.

Colombo, Robert. *The 1994 Canadian Global Almanac.* Toronto: Macmillan Canada, 1994.

Fillion, Kate. *Lip Service: Challenging the Sexual Script of the Modern Woman.* Toronto: HarperCollins, 1995.

Friedan, Betty. *The Feminine Mystique.* New York: Dell, 1974.

Jones, Charles, Lorna Marsden, and Lorne Tepperman. *Lives of Their Own: The Individualization of Women's Lives.* Toronto: Oxford University Press, 1990.

Lorenz, Konrad. *On Aggression.* New York: Harcourt Brace and World, 1966.

Mackie, Marlene. *Exploring Gender Relations: A Canadian Perspective.* Toronto: Butterworths, 1982.

Miles, Rosalind. *The Women's History of the World.* Paladin: London, 1989.

Tannen, Deborah. *You Just Don't Understand: Men and Women in Conversation*. New York: Ballantine, 1990.

Wolf, Naomi. *The Beauty Myth*. Toronto: Random House, 1990.

"Women's Ranks Thin in Politics." *Toronto Star*, 28 August, 1995, p. A3.

Chapter 9

Bailey, J. Michael. *The Man Who Would Be Queen: The Science of Gender-Bending and Transsexualism*. Washington, DC: Joseph Henry Press, 2003.

Bailey, J. Michael, Michael P. Dunne, and Nicholas G. Martin. "Genetic and Environmental Influences on Sexual Orientation and Its Correlates in an Australian Twin Sample." *Journal of Personality and Social Psychology* 78(3) (March 2000): 524–36.

Foucault, Michel. *Foucault Live (Interviews, 1961–1984)*. Sylvere Lotringer, ed., Lysa Hochroth and John Johnston, trans. New York: Semiotext(e), 1996 (1989).

Foucault, Michel. *The Use of Pleasure*, vol. 2 of *The History of Sexuality*, Robert Hurley, trans. New York: Vintage Books/Random House, 1990 (1985).

Greenberg, A.S., and J.M. Bailey. "Do Biological Explanations of Homosexuality Have Moral, Legal, or Policy Implications?" *Journal of Sex Research* 30 (1993): 245–51.

Hamer, D., and P. Copeland. *The Science of Desire: The Search for the Gay Gene and the Biology of Behavior*. New York: Simon & Schuster, 1994.

Hatfield, Elaine, and Richard L. Rapson. *Love and Sex: Cross-Cultural Perspectives*. Toronto: Allyn and Bacon, 1996.

Klein, Fritz. *The Bisexual Option*, 2nd ed. New York: Harrington Park Press/Haworth Press, 1993.

Laumann, E.O., J.H. Gagnon, R.T. Michael, and S. Michaels. *The Social Organization of Sexuality: Sexual Practices in the United States*. Chicago: University of Chicago Press, 1994.

Laumann, Edward O., Stephen Ellingson, Jenna Mahay, Anthony Paik, and Yoosik Youm, eds. *The Sexual Organization of the City*. Chicago: University of Chicago Press, 2004.

LeVay, Simon. *Queer Science: The Use and Abuse of Research into Homosexuality*. Cambridge, MA: MIT Press, 1996.

Michael, Robert T., John H. Gagnon, and Gina Kolata. *Sex in America: A Definitive Survey*. Toronto: Little, Brown (Canada), 1994.

Murray, Stephen O. *American Gay*. Chicago: University of Chicago Press, 1996.

Nevid, Jeffrey S., with Fern Gotfried. *Choices: Sex in the Age of STDs*. Toronto: Allyn and Bacon, 1997.

Selected Bibliography

Ridley, Matt. *The Red Queen: Sex and the Evolution of Human Nature*. Penguin Books, 1995.

Suggs, David N., and Andrew W. Miracle. *Culture and Human Sexuality*. Pacific Grove, CA: Brooks/Cole, 1993.

Vatsyayana. *The Kama Sutra*, S.C. Upadhyaya, trans. London, UK: Watkins Publishing, 2004.

Zucker, K.J. and S.J. Bradley. *Gender Identity Disorder and Psychosexual Problems in Children and Adolescents*. New York: Guilford Press, 1995.

Chapter 10

Anderson, Karen, et al. *Family Matters: Sociology and Contemporary Canadian Families*. Toronto: Methuen, 1987.

Baker, Maureen, ed. *Canada's Changing Families: Challenges to Public Policy*. Ottawa: Vanier Institute, 1994.

Baker, Maureen, ed. *Families: Changing Trends in Canada*, 2nd ed. Toronto: McGraw-Hill Ryerson, 1990.

Canadian Citizens to Defend Marriage, Defend Marriage [website], 2003 <www.defendmarriage.ca>.

Canadians for Equal Marriage. "House of Commons Adopts Equal Marriage Bill by Decisive Margin." Press release, June 28, 2005. Available <www.egale.ca/index.asp?lang=E&menu=20&item=1160>. Accessed 29 April 2006.

Cheal, David. *Family and the State of Theory*. Toronto: University of Toronto Press, 1993.

Crompton, Susan. "Always the Bridesmaid: People Who Don't Expect to Marry." *Canadian Social Trends*, Summer 2005. Available <www.statcan.ca/english/studies/11-008/feature/11-008-XIE20050017961.pdf>. Accessed 8 April 2006.

Eichler, Margrit. *Families in Canada Today*. Toronto: Gage, 1983.

Gannon, Maire. *Family Violence in Canada: A Statistical Profile*. Statistics Canada, Catalogue No. 85-224, 2004.

Hagedorn, Robert, ed. *Sociology*. Toronto: Harcourt Brace, 1994.

Mandell, Nancy, and Ann Duffy, eds. *Canadian Families*. Toronto: Harcourt Brace, 1995.

Montgomery, Jason, and Willard Fewer. *Family Systems and Beyond*. New York: Human Sciences Press, 1988.

Nett, Emily. *Canadian Families*. Vancouver: Butterworths, 1993.

"New Mega-Trends Reflect Family Decline." *Today's Family News*, 7 January 2005. Available FamilyFacts.ca <www.fotf.ca/familyfacts/tfn/2005/010705.html>. Accessed 8 April 2006.

Selected Bibliography

Ramu, G., ed. *Marriage and the Family Today*, 2nd ed. Scarborough, ON: Prentice-Hall, 1991.

Reid, Dr. Darrel. "Crisis or Opportunity? Eleven Practical Steps for Strengthening the Family." *Commentaries*, November 1999. Available FamilyFacts.ca <www.fotf.ca/familyfacts/commentaries/110199.html>. Accessed 8 April 2006.

Sauvé, Roger. *Profiling Canada's Families III*. Ottawa: The Vanier Institute of the Family, 2004. Available <www.vifamily.ca/library/profiling3/sample2.html>. Accessed 8 April 2006.

Schlesinger, Rachel, and Benjamin Schlesinger. *Canadian Families in Transition*. Toronto: Canadian Scholar's Press, 1992.

Statistics Canada. *A Portrait of Families in Canada*. Ottawa: Minister of Industry, Science and Technology. Catalogue No. 89-523E, 1993.

Statistics Canada. "Family Violence in Canada: A Statistical Profile 2005." *The Daily*, 14 July 2005. Available <www.statcan.ca/Daily/English/050714/d050714a.htm>. Accessed 8 April 2006.

Statistics Canada. "Income of Canadian Families." Analysis Series, Catalogue No. 96F0030XIE2001014, 13 May 2003. Available <www12.statcan.ca/english/census01/products/analytic/companion/inc/contents.cfm>. Accessed 8 April 2006.

Statistics Canada. "Birth Rate at All-Time Low." *Informat: The Week in Review*. Ottawa: Health Statistics Division, April 27, 2004. Available <www.statcan.ca/english/freepub/11-002-XIE/2004/04/11804/11804_02p.htm>. Accessed 29 April 2006.

Statistics Canada. *Women in Canada 2000*. Ottawa: Minister of Industry, 1995.

Statistics Canada. *The Daily*. 20 May 2004.

Statistics Canada. "Births." *The Daily*, July 12, 2005. Available <www.statcan.ca/Daily/English/050712/d050712a.htm>. Accessed 8 April 2006.

Statistics Canada. "Divorces." *The Daily*, March 9, 2005. Available <www.statcan.ca/Daily/English/050309/d050309b.htm>. Accessed 8 April 2006.

Statistics Canada. "Lone-Parent Families as a Proportion of All Census Families Living in Private Households, Canada, Provinces, Territories, Health Regions and Peer Groups, 2001." November 26, 2003. Available <www.statcan.gc.ca/english/freepub/82-221-XIE/01103/tables/html/49_01.htm>. Accessed 8 April 2006.

Statistics Canada. "Profile of Canadian Families and Households: Diversification Continues." *The Daily*, October 22, 2002. Available <www12.statcan.ca/english/census01/products/analytic/companion/fam/contents.cfm>. Accessed 8 April 2006.

Thorne, Barrie, and Marilyn Yalom, eds. *Rethinking the Family*. Boston: Northeastern University Press, 1992.

Tucker, Robert, ed. *The Marx–Engels Reader*, 2nd ed. New York: W.W. Norton, 1978.

Selected Bibliography

Chapter 11

Barlow, Maude. *Too Close for Comfort: Canada's Future within Fortress North America.* Toronto: McClelland and Stewart, 2005.

Biagi, Shirley. *Media/Impact: An Introduction to Mass Media.* Belmont, CA: Wadsworth, 1994.

Barber, Benjamin R. *Jihad vs. McWorld: How Globalism and Tribalism Are Reshaping the World.* New York: Ballantine Books, 1996.

De Kerckhove, Derrick. *The Skin of Culture: Investigating the New Electronic Reality.* Toronto: Somerville House, 1995.

Dyson, Rose A. *Mind Abuse: Media Violence In An Information Age.* Montreal: Black Rose Books, 2000.

Fleras, Augie and Jean Lock Kunz. *Media and Minorities: Representing Diversity in Multicultural Canada.* Toronto: Thompson Educational Publishing, 2001.

Fleras, Augie, and Jean Leonard Elliott. *Multiculturalism in Canada: The Challenge of Diversity.* Toronto: Nelson, 1992.

Herman, Edward and Noam Chomsky. *Manufacturing Consent: The Political Economy of the Mass Media.* New York: Pantheon Books, 2002.

Innis, Harold. *Empire and Communications.* Toronto: Press Procepic, 1986.

Klaehn, Jeffery ed. *Filtering the News: Essays on Herman and Chomsky's Propaganda Model.* Montreal: Black Rose Books, 2005.

Kottak, Conrad Philip. *Prime-Time Society: An Anthropological Analysis of Television and Culture.* Belmont, CA: Wadsworth, 1990.

Kroker, Arthur. *Technology and the Canadian Mind: Innis/McLuhan/Grant.* Montreal: New World Perspectives, 1984.

McLuhan, Eric, and Frank Zingrone, eds. *Essential McLuhan.* Concord, ON: Anansi Press, 1995.

McLuhan, Marshall. *The Mechanical Bride: Folklore of Industrial Man.* New York: Vanguard, 1951.

McLuhan, Marshall. *The Gutenberg Galaxy: The Making of Typographic Man.* Toronto: University of Toronto Press, 1962.

McLuhan, Marshall. *Understanding Media: The Extensions of Man.* New York: New American Library, 1964.

McLuhan, Marshall, and Quentin Fiore. *War and Peace in the Global Village.* New York: Bantam, 1968.

McLuhan, Marshall, and Bruce R. Powers. *The Global Village: Transformations in World Life and Media in the Twenty-First Century.* New York: Oxford University Press, 1989.

McQuaig, Linda. *All You Can Eat: Greed, Lust and the New Capitalism*. Toronto: Penguin Books, 2001.

McQuaig, Linda. *The Cult of Impotence: Selling the Myth of Powerlessness in the Global Economy*. Toronto: Penguin Books, 1998.

McQuaig, Linda. *Shooting the Hippo: Death by Deficit and Other Canadian Myths*. Toronto: Penguin Books, 1995.

McQuaig, Linda. *The Wealthy Banker's Wife*. Toronto: Penguin Books, 1993.

McQuaig, Linda. *The Quick and the Dead: Brian Mulroney, Big Business and the Seduction of Canada*. Toronto: Penguin Books, 1991.

Meyrowitz, Joshua. *No Sense of Place: The Impact of Electronic Media on Myths*. Toronto: Penguin Books, 1995.

Ong, Walter. *Orality and Literacy: The Technologizing of the Word*. New York: Routledge, 1988.

Parenti, Michael. *Inventing Reality: The Politics of News Media*. New York: St. Martin's Press, 1993.

Postman, Neil. *The Disappearance of Childhood*. New York: Delacorte Press, 1982.

Postman, Neil, and Steve Powers. *How to Watch TV News*. New York: Penguin, 1992.

Rodman, George. *Mass Media in a Changing World*. New York: McGraw Hill, 2006.

Sauvageau, Florian, David Schneiderman, and Davis Taras. *The Last Word: Media Coverage of the Supreme Court of Canada*. Toronto: UBC Press, 2006.

Tunstall, Jeremy. *The Media Are American: Anglo-American Media in the World*. London: Constable, 1977.

Vipond, Mary. *The Mass Media in Canada*. Toronto: Lorimer, 2000.

Winter, James. *Media Think*. Montreal: Black Rose Books, 2002.

Winter, James. *Democracy's Oxygen: How Corporations Control the News*. Montreal: Black Rose Books, 1997.

Chapter 12

Aitken, Johan Lyall. *Masques of Morality*. Toronto: Women's Press, 1987.

Ashcroft, Bill, Gareth Griffiths, and Helen Tiffin, eds. *The Empire Writes Back*. London: Routledge, 1989.

Atwood, Margaret, ed. *The New Oxford Book of Canadian Verse in English*. Toronto: Oxford University Press, 1982.

Atwood, Margaret. *Survival*. Toronto: McClelland and Stewart, 1972/2003.

Selected Bibliography

Atwood, Margaret. *Power Politics*. Toronto: McClelland and Stewart, 1971.

Atwood, Margaret. *Lady Oracle*. Toronto: McClelland and Stewart, 1976.

Atwood, Margaret. *Bodily Harm*. Toronto: McClelland and Stewart, 1981.

Atwood, Margaret. *The Handmaid's Tale*. Toronto: McClelland and Stewart, 1985.

Atwood, Margaret. "What Do Canadians Want?" In Sarah Norton and Nell Waldman, eds., *Canadian Content*. Toronto: Holt Rinehart and Winston, 1988.

Atwood, Margaret. *Negotiating with the Dead: A Writer on Writing*. Cambridge, UK: Cambridge University Press, 2002.

Basilières, Michel. *Black Bird*. Toronto: Alfred A. Knopf Canada, Vintage Canada Edition, 2004.

Braithwaite, Max. *Why Shoot the Teacher?* Toronto: McClelland and Stewart, 1974.

Brand, Dionne. *What We All Long For*. Toronto: Alfred A. Knopf Canada, 2005.

Brett, Brian. *Uproar's Your Only Music*. Toronto: Exile Editions, 2005.

Broughton, Kathryn MacLean, ed. *Heartland*. Toronto: Nelson, 1983.

Brown, Russell, Donna Bennett, and Nathalie Cooke, eds. *An Anthology of Canadian Literature in English*. Toronto: Oxford University Press, 1990.

Butala, Sharon. *The Garden of Eden*. Toronto: HarperCollins, 1998.

Butala, Sharon. *The Perfection of the Morning*. Toronto: HarperCollins Ltd., 1994.

Callaghan, Barry. "Canadian Wrye." In Sarah Norton and Nell Waldman, eds., *Canadian Content*. Toronto: Holt Rinehart and Winston, 1988.

Campbell, Jonathan. *Tarcadia*. Kentville, NS: Gaspereau, 2004.

Campbell, Maria. *Halfbreed*. New York: Saturday Review Press, 1973.

"Canadian Studies: A Guide to the Sources." International Council for Canadian Studies (ICCS) site, 5 April 2006. Available <www.iccs-ciec.ca/blackwell.html>. Accessed 26 April 2006.

Choyce, Lesley, ed. *The Cape Breton Collection*. Porter's Lake, NS: Pottersfield Press.

Clarke, George Elliott. *Odysseys Home: Mapping African-Canadian Literature*. Toronto: University of Toronto Press, 2003.

Courtmanche, Gil. *A Sunday at the Pool in Kigali*. Tr. Patricia Claxton. Toronto: Vintage Canada Editions, 2004.

Dunlop, Rishma, and Priscila Uppal. *Red Silk: An Anthology of South Asian Canadian Women Poets*. Toronto: Mansfield Press, 2005.

Djwa, Sandra, and R.G. Moyles, eds. *E.J. Pratt: Complete Poems, Parts I and II*. Toronto: University of Toronto Press, 1989.

Ferguson, Will, and Ian Ferguson. *How to Be a Canadian*. Toronto: HarperCollins, 2003.

Ferguson, Will. *Beauty Tips from Moose Jaw*. Toronto: Knopf Canada, 2004.

Fowke, Edith. *Folklore of Canada*. Toronto: McClelland and Stewart, 1976.

Geddes, Gary, ed. *15 Canadian Poets × 2*. Toronto: Oxford University Press, 1988.

Geddes, Gary, and Phyllis Bruce, eds. *15 Canadian Poets*. Toronto: Oxford University Press, 1970.

Goh, Maggie, and Craig Stephenson, eds. *Between Worlds*. Oakville: Rubicon, 1989.

Henry, Frances, Carol Tator, Winston Mattis, and Tim Rees. *The Colour of Democracy*. Toronto: Harcourt Brace, 1985.

Hébert, Anne. *Kamouraska*. Toronto: Moussen, 1973.

Herberg, Dorothy Chave. *Frameworks for Cultural and Racial Diversity: Teaching and Learning for Practitioners*. Toronto: Canadian Scholar's Press, 1993.

Hutcheon, Linda. *The Canadian Postmodern*. Toronto: Oxford University Press, 1988.

Hutcheon, Linda, and Marion Richard, eds. *Other Solitudes*. Toronto: Oxford University Press, 1990.

Hynes, Joel. *Down to the Dirt*. Toronto: HarperCollins Canada, 2006.

Jobb, Dean. *The Acadians: A People's Story of Exile and Triumph*. New York: John Wiley & Sons Inc., 2005.

Karpinski, Eva C., and Ian Lea, eds. *Pens of Many Colours*. Toronto: Harcourt Brace Jovanovich, 1993.

Keefer, Janice Kulyk. *Under Eastern Eyes*. Toronto: University of Toronto Press, 1987.

King, Thomas, ed. *All My Relations*. Toronto: McClelland and Stewart, 1990.

Kroetch, Robert. *Badlands*. Toronto: New Press Canadian Classics, 1975.

Lai, David Chuenyan. "A 'Prison' for the Chinese Immigrants." *The Asianadian* 2(4): 1983.

Lecker, Robert, and Jack David, eds. *The New Canadian Anthology*. Toronto: Nelson, 1988.

MacLeod, Alistair. *No Great Mischief*. Toronto: McClelland and Stewart, 1999.

Moore, Lisa. *Open*. Toronto: House of Anansi Press, 2002.

Morrisey, Donna. *Kit's Law*. Penguin Books, 2000.

Moses, Daniel David, and Terry Goldie, eds. *An Anthology of Canadian Native Literature in English*, 2nd ed. Toronto, Oxford University Press, 1998.

Munro, Alice. *Lives of Girls and Women*. Toronto: McClelland and Stewart, 1971.

Munro, Alice. *Who Do You Think You Are?* Toronto: McClelland and Stewart, 1978.

Norton, Sarah, and Nell Waldman, eds. *Canadian Content*. Toronto: Holt Rinehart and Winston, 1988.

Nurse, Donna Bailey. *What's a Black Critic to Do? Interviews, Profiles and Reviews of Black Writers*. Toronto: Insomniac Press, 2003.

"Post-Survival Canada: 12 Writers in Search of a Paradigm." *The Globe and Mail*, 30 June 2001, p. D2.

Reaney, James. *The Donnellys: Sticks and Stones, The St. Nicholas Hotel, and Handcuffs*. Victoria and Toronto: Press Porcépic, 1983.

Richler, Noah. Introduction. *A Literary Atlas of Canada*. Available <www.cbc.ca/ideas/features/literary_atlas/index.html>. Accessed 26 April 2006.

Schaum, Melita, and Connie Flanagan, eds. *Gender Images*. Boston: Houghton Mifflin, 1992.

Shields, Carol. *Unless*. Toronto: Random House, 2002.

Shields, Carol. *The Stone Diaries*. Toronto: Viking Penguin, 1994.

Sileika, Antanas. *Buying on Time*. Toronto: University of Toronto Press, 1997.

Sullivan, Rosemary, ed. *Poetry by Canadian Women*. Toronto: University of Toronto Press, 1989.

Taylor, Drew Hayden. "Seeing Red over Myths." *The Globe and Mail*, 8 March 2001.

Taylor, Drew Hayden, ed. *Me Funny*. Vancouver: Douglas & McIntyre, 2006.

Thériault, Yves. *Aguguk*. Montreal: L'Actuelle, 1971.

Waterson, Elizabeth. *Survey*. Toronto: Methuen, 1973.

Weaver, Robert, ed. *Canadian Short Stories*. Toronto: Oxford University Press, 1960.

Wiebe, Rudy, ed. *The Story Makers*. Toronto: Gage, 1987.

Williamson, Janice. *Sounding Differences: Conversations with Seventeen Canadian Women Writers*. Toronto: University of Toronto Press, 1993.

Winks, Robin W. *The Blacks in Canada: A History*, 2nd ed. Montreal: McGill-Queen's University Press, 2003.

Winter, Michael. *The Big Why*. Toronto: Anansi, 2004.

Ye, Ting-Xing. *A Leaf in the Bitter Wind*. Toronto: Doubleday, 1997.

BIOGRAPHIES

Paul U. Angelini is the operations coordinator for the General Arts and Science Program at Sheridan College Institute of Technology and Advanced Learning. He has developed and delivered curriculum at Sheridan since 1988 in the fields of politics, sociology, human diversity, philosophy, and social movements. Paul completed his master's degree in political studies from Queen's University and his combined honours bachelor of arts in political science and sociology at McMaster University.

Michelle A. Broderick completed her Ph.D. in Biological Anthropology at the University of Toronto. Since graduating in 1994, she has taught a variety of courses at Sheridan College, McMaster University, and the University of Toronto. Michelle is currently working in Institutional Research at the University of Toronto.

Leslie Butler has been a community college professor for 11 years. She has taught English and general education courses, including social work, journalism, and graphic design. She received a master's degree in English from the University of Waterloo and a master's degree in journalism from the University of Western Ontario.

Maureen Coleman is a writer and reviewer, and also leader of Booktalk, a book club for reading and discussing works by international authors. She is retired from Sheridan College, where she taught courses focusing on cultural diversity through Canadian and international literature. She holds bachelor's degrees in History and Education from St. Francis Xavier University and a Master's in teaching English from the University of Toronto.

Eddie Grattan is an elementary teacher in Mississauga. He is a founding member of the Diversity Committee for the Mississauga South Family of Schools, and is especially interested in the relationship between educational practices and social class.

Brigitte Guetter has developed and taught courses in human sexuality, personality, abnormal, and social psychologies; sociology; cultural diversity; critical thinking; and communications since 1989 at George Brown College, Seneca College, Nova Scotia Community College, and Sheridan College. She is currently with the Multicultural Council of Windsor and Essex County. She received her bachelor of science and bachelor of education degrees from the University of Toronto.

Grant Havers received his Ph.D. in Social and Political Thought from York University in 1993. He is currently assistant professor of Philosophy at Trinity Western University in British Columbia. Grant's current research involves a study of political ideologies in the postmodern age.

Nancy Nicholls is a professor in the Social Service Worker program at Centennial College and was the 2000 recipient of the Board of Governor's Award for Excellence in Valuing Diversity. She holds a master's degree in social work from the University of British Columbia and an honours bachelor of arts from York University.

Geoff Ondercin-Bourne has taught communications, ESL, and General Education at Mohawk College for the past five years. Before that, he taught global issues, Canadian politics, sociology, literature, communications, and ESL at Sheridan College for 15 years. Geoff received a master's degree from McMaster University and an honours B.A. from the University of Guelph.

Mikal Austin Radford is a professor of Philosophy and Religion in the School of Community and Liberal Studies at Sheridan College (Davis Campus) and teaches courses in Asian and South Asian traditions at Wilfrid Laurier University (Waterloo, ON). His research specializes in religion, eastern philosophy, multiculturalism, and transnational identity formation. Mikal is in the final stages of his Ph.D. dissertation at McMaster University. He received both his master's degree and honours bachelor of arts in the Religion and Culture Department of Wilfrid Laurier University. His publications include "*Sallekhana, Ahimsa*, and the Western Paradox," "Role Models of Jaina Citzenship in the Western World," and "(Re)Creating Transnational Religious Identity with the Jaina Community of Toronto."

John Steckley has been teaching at Humber College since 1983, and has taught Anthropology at Memorial University of Newfoundland and at Trent University, and Native Studies at Laurentian University. His published books include *Beyond Their Years: Five Native Women's Stories* (1999), *Aboriginal Voices and the Politics of Representation in Canadian Introductory Sociology Textbooks* (2003), and *De Religione: Telling the Seventeenth-Century Jesuit Story in Huron to the Iroquois* (2004). With Bryan Cummins, he has co-authored *Full Circle: Canada's Native People* (2001) and *Aboriginal Policing: A Canadian Perspective* (2002). His areas of specialization are Aboriginal languages (primarily Huron) and Aboriginal history. John has a master's degree in anthropology from Memorial University of Newfoundland and a doctorate in postsecondary education from the Ontario Institute for Studies in Education (OISE) at the University of Toronto. He was adopted into the Wyandot tribe of Kansas in 1999 and given the name *Tehaondechoren* ("He splits the country in two").

INDEX

A

ability to pay, 68
Aboriginal peoples
   bands, 141–142
   the beginnings, 127–128
   and Canadian literature, 332–333, 334, 335
   contemporary picture, 131–136
   demographics, 130–136
   diversity of Native cultures, 129–130
   four medicines, 148–149
   future, 152–153
   historical picture, 130–131
   hunting rights, 135–136, 137–138
   and justice system, 111, 144–148
   land claims, 139–141
   language diversity, 129–130
   and the media, 316
   medicine dances, 150
   Native policing services, 145–147
   Nisga'a agreement, 110
   potlatch, 150–151
   potlatch ban, 150–151
   and racism, 110
   registered Indians, 131
   religion, 148–151
   reserves, 137, 142–144
   residential schools, 151–152
   sentencing circles, 147–148
   smallpox, 131
   "starlight tour," 144–145
   "status Indians," 131
   stereotypes, 143
   sweat lodge, 145, 149
   theoretical positions, 128–129
   treaties, 136–139
   urban reserves, 143–144
absence of voice, 332
abuse, 294–296
Accessibility for Ontarians with Disabilities Act, 212
advertising, 320, 322
African Canadian Online, 337
age and inequality, 78–79
age at first marriage, 41–42, 42f
age profile of population, 45
age-specific birth rates, 288t
age-specific marriage rate, 41
age-specific mortality rates, 43
aggregation of data, 53
AIDS, 259
Air Canada, 25
Air Force Headquarters, 21
Alberta, 352
alphabet, 307–309, 310t
American College of Pediatricians, 253–254
American media, power of, 318–322
American Psychological Association (APA), 252
Americans with Disabilities Act, 213
analysis of demographic information, 51–53
Angelini, Paul U., 4–37, 94–125, 304–329, 401
annuities, 137
anomie, 171
anticommunism, 320
antidiscriminatory legislation, 200
Arar, 114–115
Arron, Laurie, 248
ascribed statuses
   age, 78–79
   described, 74
   disability, 79
   ethnicity, 74–76
   and inequality, 74–79
   race, 74–76
   sex, 76–77
   social background, 76
Asper, Izzy, 319
assimilation, 116–117
Atlantic Canada
   bulk sale of fresh water, 31–32
   Canadian literature, 344–347
   and Confederation, 6, 30
   fishing dispute, 31–32
   flag removal, 31
   natural resources in, 10
   and St. Lawrence Seaway, 22–23
   view of, 14
Atwood, Margaret, 331, 333, 357–358

B

Bahdur, Kadir Khan, 181
band, 141–142
Bank of Canada, 21
baptisms, sex ratio of, 51
Barber, Benjamin, 322
Benedict XVI, Pope, 250
the Beothuk, 96
Berger, Brigitte, 105
Berger, Peter, 105, 168, 169
Beringia, 56
Bernardo, Paul, 317
Berton, Pierre, 99
bestiality, 251
Biblarz, Timothy, 254–255

Index

bibliography, 380–400
Bickenbach, Jerome, 201, 203
"big R" religion, 168
Bilingualism and Biculturalism Report on Immigration, 102
Bill 101 (Quebec), 107
Bill 178 (Quebec), 107
Bill C-31, 131, 132, 141
Bill C-33, 244–245
Bill C-38, 247, 296
Bill C-43, 31
Bill C-55, 244
Bill C-150, 243
Bill C-242, 244
Bill C-250, 245
Bill of Rights (1960), 187, 188
biological inferences, 53–55
biomedical model, 201
bisexual, 242
    see also LGBT community
Bissoondath, Neil, 120
Black people
    Canadian literature, 336–337
    and racist immigration policy, 100
    slavery in Canada, 47, 96, 97, 100, 101
    United Empire Loyalists, 96
Bloc Québécois, 17
Block, Susan, 241
bluebloods, 88
bondage, 251
Bourassa, Robert, 139
Brazil, 324–325
Bristol Aerospace, 21
British Columbia, 352–353
British North America Act, 13, 196
Broderick, Michelle, 38–63, 94–125, 401
Brown, Dee, 114
Buddhist tradition, 161, 167, 171
Bush, George W., 249
Bush, Laura, 250

butch, 258
Butler, Leslie, 216–239, 401

C
Cabinet positions, 14
Calder, Frank, 140
Calment, Jeanne Louise, 42
Canada
    age at first marriage, 42*f*
    crude birth rate, 40*f*
    crude death rate, 43*f*
    crude marriage rate, 41*f*
    electoral system, 15–17
    income. *See* income
    infant mortality rate, 44*f*
    mobility, 48*f*
    as multicultural, multiracial society, 95
    myth of "white" country, 36
    population in. *See* population
    sex-specific mortality rates, 44*f*
    slavery in, 47, 96, 97, 100, 101
    social class structure, 87–89
    social mobility, 89–90
    surviving the global village, 323–325
    in a world perspective, 28*t*
Canada: A People's History, 28–29
Canada Pension Plan Act (1965), 200
Canadair, 21
Canadian Broadcasting Corporation (CBC), 28
Canadian Citizenship Act, 186
Canadian Health and Social Transfer (CHST), 15
Canadian Human Rights Act (1985), 202, 244
Canadian literature
    Atlantic Canada, 344–347
    Canadian war stories, 343

    contemporary writing and diversity, 338–340
    and demographics, 340–343
    English voices, 349–350
    family, role of, 355–358
    feminist literature, 331
    gender, role of, 358–361
    Native people, perceptions of, 332–333
    Native voices, 334, 335
    nonfiction, 359
    Northern Canada, 354
    Ontario, 350–351
    personal voices, 334–335
    as postcolonial global trend, 332–340
    Quebec, 347–349
    and regionalism, 344–354
    role in recording and defining culture, 331–332
    romantic stereotypes, 333
    sexuality, role of, 358–361
    and social inequality, 344–354
    stories of race and ethnicity, 336–337
    universal themes, 332
    untold stories, 333–334
    voice, 331
    voices from the margins, 332
    voices of experience, 360
    Western Canada, 351–353
Canadian Pacific Railway (CPR), 25
Canadian religious experience
    challenge to Protestant domination, 179
    Chinese immigration, 179–180
    continuous journey regulation, 99, 185–186

intolerance, history of, 187–189
opposition to immigration, 184–186
overview, 177–179
religious trends, 182
South Asian immigration, 180–187
the struggle and the legislation, 186–187
CanWest Global, 319
capitalism, 11–12
capitalist class, 83
the Castro (San Francisco), 250–251
causes of regionalism, 9–12
celibacy, 41
census, 50
central Canada
contempt for, 32–34
demographic research focus on, 55
dominance of, 32–34
CF-18 fighter aircraft contract, 21
Charter of Rights and Freedoms, 118, 122
Cheal, David, 277
Cheney, Dick, 249
child custody, 225
childbirth, 286
childless and childfree couples, 286–289
Chinese people
exclusion law, 100
head tax, 98, 180
Hong Kong Chinese, 340–342
immigration to Western Canada, 49
opposition to immigration, 184–186
railway building, 179–180
Chomsky, Noam, 319
Choquette, Robert, 177
Chrétien, Jean, 20, 33, 246

Chrisler, Jennifer, 250
The Civil Marriage Act, 247
civil partnerships, 249
civil registries, 49–50
civil unions, 249
Clement, Wallace, 76
cohort, 53
Coleman, Maureen, 330–363, 401
come out (of the closet), 243
community, 256–257
comparative method, 51
Comuzzi, Joe, 247
Confederation, 6, 13, 30, 181, 196
conflict theory
Aboriginal peoples, 128–129
ethnicity and race, 104–105
the family, 280–283
gender spheres, 229–231
Karl Marx, 82–86
Max Weber, 86–87
social inequality, 82–87
conjugal family, 278
conservative ideology, 227–228
Constitution, 18, 188
Constitution Act of 1867, 13
consumer movement, 201
consumerism, 322
continuous journey regulation, 99, 185–186
Coser, Rose Laub, 274
Crompton, Susan, 77
cross-sectional analysis, 53
crosschecking data, 50–51
Crow's Index of Selection, 55
crude birth rate, 40, 40f
crude death rate, 43f
crude marriage rate, 41, 41f
cults, 171, 172–173
cultural dominance, 33–34
cultural identity, 105–109
cultural imperialism, 332–333

cultural symbols, 108
Cutler, Allan, 19

D
Daniels, Roger, 181
Dawson, Lorne, 171
de-assimilation, 258–259
decennial census, 50
deinstitutionalization, 205
demographic information
analysis of, 51–53
problems with sources, 50–51
sources of, 49–50
demographic variables
fertility, 39–40
meaning of, 39
migration, 46–49
morbidity, 45
mortality, 42–45
nuptiality, 40–42
demography
Aboriginal peoples, 130–136
biological inferences, 53–55
and Canadian literature, 340–343
demographic variables, 39–49
evolution by natural selection, 54–55
historical demographic analysis, 55–57
marriage partner choices, 53–54
meaning of, 39
Department of Indian Affairs (DIA), 141–142
dependency ratio, 46
developmental disabilities, 200
dialectical process, 168
dialects, 311
Diefenbaker, John, 186–187, 188
"direct passage" stipulation, 99

disability
- advocacy, 201
- biomedical model, 201
- characteristics of disabled Ontarians, 207f, 209
- comparisons with general population, 210t
- Confederation and *BNA Act*, 196
- developmental disabilities, 200
- early years, 195–196
- economic model, 202–203
- the future, 206–213
- historical context, 195–201
- Human Rights Codes, 200–201
- and inclusivity, 195
- and inequality, 79
- learning disability, 206
- Medicare, 199–200
- most difficult goal to achieve, 208t
- normalization theory, 205–206
- positive trends, 212–213
- priorities for improvement, 209t
- public pressure for services, 197
- rehabilitation planning and the labour market, 199
- sociopolitical model, 203–204
- 21st century trends, 207–213
- type and severity of most limiting disabilities, 211f
- the world wars, 198–199

discrimination, 110–113
discriminatory hiring practices, 112–113, 232
disposable income, 72

diversity
- *see also* specific topics
- *versus* conformity, 218
- and contemporary writing, 338–340
- and the global village, 313–325
- language diversity, 129–130
- of Native cultures, 129–130
- regionalism, 5
- television, 313–318
- through child's eyes, 355
- and tribalism, 312
- two solitudes, 331

Diversity Planning for Inclusive Employment (DPIE), 212
divorce, 291–293
documentary method, 51
domestic labour, 289
domestic partnerships, 248–249
domestic violence, 295–296
double ghetto, 77
the Doukhobors, 183, 188
Durkheim, Emile, 168

E

East African immigrants, 120
East Indian immigration, 99
the Eaton family, 88
ecclesiastical registries, 49–50
economic dimension of family, 293–294
economic dominance, 33
economic inequality, 222–224
economic model, 202–203
education level, 76–77, 90
Eichler, Margrit, 275, 280, 296
electoral system, 15–17
electronic stage, 311–312
emigration, 46
emigration rate, 47
emotional dimension of family, 294–296

The Empire Writes Back (Ashcroft, Griffiths and Tiffin), 332
empiricist, 162
employment, by industry and sex, 222t
employment equity, 118, 122, 234–235
Employment Equity Act, 122
empowerment, 203
empty nesters, 281
Engels, Friedrich, 280–281, 282
equality of opportunity, 81
Equality (Turner), 82
"Eskimo," as term, 132
essentialize, 160
ethnic groups, 105, 115
ethnicity
- and cultural identity, 105
- and inequality, 74–76
- and sociological theory, 104–105

ethnocentrism, 113–114, 248, 321
Europeans, 97–99
evolution, 54–55
executive federalism, 14
expectation method, 51
exploitation, 83–84
extended family, 274–275
extensions of humanity, 309, 310t
external migrants, 48
externalization, 168

F

failed states, 175
Fairclough, Ellen, 187
false consciousness, 83, 84f
familia, 281
familism, 284
the family
- balancing responsibilities, 289–290
- changing patterns, 286–296

conflict theory, 280–283
conjugal family, 278
definitions of, 274–275
divorce, impact of, 291–293
economic dimension, 293–294
emotional dimension, 294–296
extended family, 274–275
family life course perspective, 281
feminist theory, 283–285
five internal dimensions, 276
general systems theory, 278–280
inclusive approach, 275
liberation theories, 280
nuclear families, 274
poverty, 294
procreative dimension, 286–289
residential dimension, 290–293
role of, in Canadian literature, 355–358
same-sex marriage, 296–297
see also same-sex marriage
socialization dimension, 289–290
structural functionalism, 277–278
symbolic interaction theory, 281
theoretical perspectives, 276–285
unemployment, 294
unification theories, 277–280, 281
violence, 295–296
family life course perspective, 281
farm crisis, 21–22
Farrow, Jane, 256
fecundity, 39
federal election, 2004, 16
federal election, 2006, 16
federal government
   physical distances, reducing, 25
   promoting understanding, 27–29
   radio, TV and film involvement, 28–29
   regional differences, reducing, 25–34
   royal commissions, 27, 29
   transfer payments, 11, 12t, 25–27, 26t
federalism
   defined, 14
   executive federalism, 14
   and regionalism, 13–15
female circumcision, 120
The Feminine Mystique (Friedan), 220
feminist literature, 331
feminist theory, and the family, 283–285
feminization of poverty, 77
femme, 258
fertility, 39–40
fertility rate, 40
film making, 28–29
financial institutions, 21
First Ministers' Conference (2001), 15
First Nations Policing Policy, 145–147
"first past the post," 15
first people in North America, 56
fishing dispute, 31–32
flag removal, 31
flak, 320
Fraser, Sheila, 19
free market, 10–11
Friedan, Betty, 220
Frye, Northrop, 331
fundamentalism, 171–177, 189
Furnish, David, 249

G
gay, 219, 242
   see also LGBT community
gay marriage, 249
   see also same-sex marriage
Gay Pride Parade, 251–252
Geertz, Clifford, 160
gender
   and Canadian literature, 358–361
   challenges to traditional gender identity, 218–219
   diversity *versus* conformity, 218
   feminist theory, and the family, 283–285
   glass ceiling, 220
   meaning of, 217–221
   patterns, 219–221
   roles, 290
   versus sex, 217
gender identity, 217–218
gender reassignment surgery, 261
gender-role-bound, 257
gender spheres
   conflict theory, 229–231
   conservatism, 227–228
   construction of fair society, 234–236
   described, 223f
   discriminatory hiring practices, 232
   economic inequality, 222–224
   education, 234
   educational spheres, 221
   equity laws, 234–235
   and individual choice, 233–234
   interpersonal sphere, 224–225

gender spheres (*continued*)
    laws, 232
    leisure spheres, 221–222
    meaning of, 221
    merit *versus* social connections, 231–233
    myth of progress, 226–227
    nature and nurture, 227
    occupational spheres, 221
    progressivism, 227–228
    proportional representation, 235
    psychological barriers, 233
    sexual politics, 225–226
    structural-functionalist theory, 228–229
    why it matters, 233–236
    why they exist, 227–228
gender stereotypes, 218
genealogical analysis, 54
genealogies, 50
general systems theory, 278–280
geographical regions, 6
glass ceiling, 220
global village
    and diversity, 313–325
    emergence of, 312
    ethnic participation in media industry, 323
    language of, 322–323
    micro-global village, 340–343
    surviving, 323–325
    television, 313–318
glossary of terms, 364–379
Golden Horseshoe, 7–8, 21, 24
Gomery Report, 20
Gorlick, Carolyne, 292
Gosnell, Joseph Sr., 141
Grant, George, 324
Grattan, Eddie, 66–93, 401
Great Depression, 85, 100
Greater Toronto Area (GTA), 33

Green Party, 17
Grey Nuns, 196
Grimes, Roger, 15, 32
gross migration rate, 47
group distinctiveness, 257
Guetter, Brigitte, 240–271, 401
Gutenberg, Johannes, 309–311

H
Hamm, John, 30
"handicaps," 198
Harper, Elijah, 18
Harper, Stephen, 247
Harris, Mike, 11, 15, 30, 227
hate propaganda, 244
Havers, Grant, 304–329, 401
Herman, Edward, 319
heteroerotic, 261
heterosexuals, 242
Hewitt, Foster, 28
hieroglyphics, 307
Higgins, Paul, 206
Hill, Daniel, 111
Hinduism, 161
historical demographic analysis, 55–57
homoerotic, 255
Homolka, Karla, 317
homophobic, 245
homosexual, 257
homosexuality, 219
Hong Kong Chinese, 340–342
Hope, Adrian, 134
household, 50
Howe, C.D., 21
Hudson's Bay Company (HBC), 133–134, 137
human and sexual rights
    backlash effect, 245
    Canadian progress, 243–244
    described, 241–242
    sexual orientation as protected identifiable group, 244–245
    world progress, 242–243

Human Rights Codes, 200–201
Hunter, Alf, 72, 294

I
ideological legitimation, 284
ideology of the family, 284
immigrant and refugee stories, 338–340
immigration
    see also migration
    administrative fee, 99
    and Canada's religious experience. *See* Canadian religious experience
    education and occupation criteria, 103–104
    history of, to Canada, 95–104
    meaning of, 46
    point system, 187
    racist policies, 98–99, 100, 101–102, 180
    top ten countries of birth, 103t
Immigration Act (1906), 185
Immigration Act (1910), 185–186
Immigration Act (1952), 187
Immigration and Refugee Protection Act 2002, 104
immigration rate, 47
impairment, 201
inclusive societies, 195
income
    average total and after-tax incomes, 69t
    described, 68–69
    disposable income, 72
    distribution of total income, 71t
    government transfers and income tax, 70t
income inequality, 68–72
Indian Act, 127, 131, 151

Indigenous peoples. *See* Aboriginal peoples
individual discrimination, 111–112
inequality
    economic inequality, 222–224
        social inequality. *See* social inequality
infant mortality rate, 43–45, 44f
Innis, Harold, 309
"insidious" myth, 36
institutional completeness, 256–257
institutional discrimination, 112–113, 255
intensification of regionalism
    electoral system, 15–17
    federalism, 13–15
internal migrants, 48
internalization, 168
International Year of Disabled Persons, 200
Internet, 322, 325
interventionist approaches, 11
Inuit, 132–133
inventions, 312t
Islamic community, 117, 121, 317
Islamic tradition, 161
isonomy, 54

J

Jain tradition, 161, 165, 171
James, Carl E., 107, 109, 119
James Bay and Northern Quebec Agreement (1975), 139–140
Japanese-Canadians, 98, 99, 101–102, 184–186, 188
jingle dress dance, 150
jiva, 165
John, Elton, 249
John Paul II, Pope, 250
Judeo-Christian tradition, 161
Jung, Carl, 228
Just Desserts shooting, 317
justice system, and Aboriginal peoples, 144–148

K

karma, 161
"killing God," 171, 174
Kinsey's seven-point continuum of sexual orientation, 261–262
kinship coefficients, 54
Kirkbride, Fred, 251
Klein, Fred, 262
Klein, Ralph, 11
Klein Sexual Orientation Grid (KSOG), 262–263, 264f
Klippert, Everett, 243
Kurian, George, 181

L

Labrador, 31
land claims, 139–141
language, 105, 119, 129–130
    of the global village, 322–323
    Internet, 322
    lingua franca, 322
language families, 129
language isolates, 129
Laumann study, 263–265
Laurier, Wilfrid, 183, 184–185
Lawrence, Bonita, 132
learning disability, 206
leather-fag, 261
lesbian, 242
    see also LGBT community
LGBT community
    see also sexuality
    AIDS, 259
    de-assimilation, 258–259
    development of, 257–258
    emergence, 241
    families, 250–252
    institutional completeness, 256–257
    organization of, 257–258
    public morality, 259–261

Liberal Party, 174
liberation theories, 280
life expectancy, 42
life span, 42
lingua franca, 322
lipstick lesbians, 258
literature. *See* Canadian literature
"little r" religion, 168
live births, 39
lone mothers, 78
    see also single-parent families
longitudinal analysis, 53
Lundrigan, Paul, 246–247
Luther, Martin, 311
Luxton, Meg, 284
Lynch, Jessica, 318
Lynn, Marion, 295

M

Macdonald, John A., 20, 25
Manitoba, 351–352
Manitoba Act, 134
Manufacturing Consent (Herman and Chomsky), 319
market approaches, 10–11
maroons, 101
Marriage and Child Wellbeing, 253
marriage partner choices, 53–54
Martin, Paul, 20, 247
Marx, Karl, 82–86, 168, 280–281
Marxist approaches, 11–12
mass media
    Aboriginal peoples, 316
    American media, power of, 318–322
    court judgments, 316
    definitions of, 305–306
    effect of, 313
    facts, 326
    global village. *See* global village

mass media (*continued*)
   history of, 307–312
   important historical dates, 308t
   politics of, 314
   propaganda model, 319–320
   space-oriented media, 309
   television, 313–318
   time-oriented media, 309
mate selection, 53–54
Mathews, Ralph, 10
matriarchy, 281
McLachlin, Beverley, 247
McLuhan, Marshall, 305, 306, 309, 321, 324
McQuaig, Linda, 320
Meadowcroft Rockshelter, 127–128
means of production, 83
media. *See* mass media
media think, 318
median employment income, selected Canadian cities, 8t
Medicare, 199–200
medicine dances, 150
Meech Lake Accord (1987), 14, 18
Memorial University, 31
men, 76–77
"mentally deficient," 197
meritocracy, 231–233
Métis, 133–136, 333
Metropolitan Toronto, 32–34
micro-global village, 340–343
middle class, 89
migration, 46–49, 56–57
minority crimes, 317
the Mississauga, 138–139
mobility in Canada, 48f
mode of production, 281
modernism, 171
moksha, 165, 167
monolithic entity, 115
morbidity, 45
Morse, Samuel Finley Breese, 311–312
mortality, 42–45
Müller, Max, 160
Mulroney, Brian, 14, 18, 21
multicultural tyranny, 119–121
multiculturalism
   in Canada, 95
   and employment equity, 122
   meanings of, 115–116
   official policy of, 119, 120–121
   policy of, 187
   purpose of, 117–118
   and Quebec, 121–122
Multiculturalism Act, 117, 175–176, 187
Munro, Alice, 355–356
Murdock, George, 274
Murray, David, 255–256
Murray, Stephen O., 257, 259
Muskeg Lake Cree Nation, 143–144
Muslims. *See* Islamic community

N
Naistus, Rodney, 145
National Association for Research & Therapy of Homosexuality (NARTH), 260
National Film Board (NFB), 28
National Policy (1879), 20–21
Native peoples. *See* Aboriginal peoples
Native policing services, 145–147
natural resources approach, 9–10
natural selection, 54–55
nature/nurture, 227
NDP, 17
Nees, Barbara, 31
net migration rate, 47
net worth differences, 72–74, 73t
Nett, Emily, 276, 288
New Brunswick, 346–347
"new money" capitalists, 88
Newfoundland, 31, 344–345
Nicholls, Nancy, 194–215, 402
Nietzsche, Friedrich, 171, 174
Night, Darrell, 144
nirvana, 167
Nisga'a treaty (2000), 140–141
nominative records, 50
nomos, 169
nonfiction, 359
nonrandom mate selection, 53
normalization theory, 205–206
Northern Canada, 23–24, 354
Northwest Territories, 354
Nova Scotia, 47, 345–346
nuclear families, 274
Nunavut, 23, 133
nuptiality, 40–42

O
objectivation, 168–169
Obstacles, 200, 204
Oliver, Frank, 185
Omnibus Bill, 243
Ondercin-Bourne, Geoff, 272–301, 402
one-industry towns, 24
O'Neill, Eimear, 295
Ontario
   Canadian literature, 350–351
   dominance of, 33–34
   financial institutions in, 21
   Golden Horseshoe, 7–8, 21, 24
Ontario Court of Appeal, 246, 248
Ontario Human Rights Commission, 114
orality, 307
outed, 245

P

pack journalism, 318
paleodemography, 55
pantheistic, 162–163
parchment, 309
Parenti, Michael, 313, 318
Parsons, Talcott, 277–278, 282
Parti Québécois, 18
patriarchal, 281
"Peace and Friendship Treaties," 137
pedophilia, 244
Petros, C.I., 181
physical symbols, 107
pictograph, 307
pluralism, 116
political dominance, 33
political power, 86
polytheistic, 164
population
    defining populations, 51
    ethic origins, 52f
    historical demographic analysis, 55–57
    immigrants, proportion of, 102–104
    proportion over age 64, 45f
postcolonial writing, 332
potlatch ban, 150–151
poverty
    in Canadian families, 294
    feminization of poverty, 77
Powley test, 135–136
prejudice, 109–110
Prince Edward Island, 347
printing press, 309–311
procreative dimension of family, 286–289
products of the human cultural project, 169
Progressive Conservatives, 10–11, 25
progressive ideology, 227–228
propaganda model, 319–320
proportional representation, 235
prostitution, 80
provinces
    see also Atlantic Canada; central Canada; Western Canada; specific provinces
    as regions, 6–8, 9
    transfer payments, 11, 12t, 25–27, 26t
public morality, 259–261

Q

Quebec
    and the Constitution, 18
    demands, 18
    dominance of, 33–34
    language protection bills, 107
    literature in, 347–349
    and multiculturalism, 121–122
    Quiet Revolution, 174
    separatism, 17–18
Queer Subcultures, 260
quintile, 69

R

race
    and cultural identity, 105–109
    and immigration, 98–99
    and inequality, 74–76
    meanings of, 105–106
    and sociological theory, 104–105
    and "special treatment," 108–109
racial divide, 75
racial jokes, 112
racial profiling, 114–115
racism, 107–109, 110, 111
Radford, Mikal Austin, 156–193, 402
radio involvement, 28–29
Rae, Bob, 227
regional analysis
    geographical regions, 6
    provinces as regions, 6–8, 9
    regions as explanatory tool, 8–9
    regions of Canada, 7f
regional survey, 55–57
regionalism
    Atlantic Canada, 22–23
    and Canadian literature, 344–354
    causes of, 9–12
    and electoral system, 15–17
    and federal government actions, 25–34
    and federalism, 13–15
    as form of diversity, 5
    intensification of, 12–17
    meaning of, 5–6
    Northern Canada, 23–24
    Quebec, 17–20
    reality and outcomes, 17–24
    regional analysis, problems with, 6–9
    sociopsychological dimension, 30–34
    study of, 5–6
    subjective differences, 5
    Western Canada, 20–23
regionalism theories
    interventionist approaches, 11
    market approaches, 10–11
    Marxist approaches, 11–12
    natural resources approach, 9–10
regions of Canada, 7f
registered Indians, 131
rehabilitation medicine, 198
rehabilitation planning, 199
rehabilitation team, 198

religion
    Aboriginal peoples, 148–151
    "big R" and "little r," 167–168
    in Canada. *See* Canadian religious experience
    as cultic project, 171
    cults, 171, 172–173
    defining religion, 159–167
    as destroyer of worlds, 170
    and fundamentalism, 171–177, 189
    the infinite, 161–167
    as response to meaningless, 169–170
    and social change, 171
    and world-building, 168–169
    world religions, 162–165, 166–167
religious postmodernism, 164–165
religious symbols, 121
religious wars, 177–178
representation by population, 15
reserves, 137, 142–144
residential dimension of family, 290–293
residential schools, 151–152
reverse discrimination, 122
Richler, Mordecai, 357
Ricketts, Christine, 317
Riel, Louis, 134, 333
Robinson, Svend, 244, 245
Robinson Treaties of 1850, 138
romantic stereotypes, 333
Rousseau, Jacques, 135
Royal Commission on Bilingualism and Biculturalism, 95
Royal Commission on Dominion-Provincial Relations (1937), 29

Royal Commission Report on Aboriginal Peoples (1997), 29
royal commissions, 27, 29
Royal Proclamation of 1763, 137

S
sacred canopy, 170, 174, 175
The Sacred Canopy (Berger), 168
sadomasochism, 251
same-sex divorce, 249
same-sex marriage, 245–249
    and evolution of family, 296–297
same-sex parenting, 252–256
samsara, 165
Saskatchewan, 351
Saskatchewan Treaty Land Entitlement Framework Agreement, 143
Saskatoon Police Service (SPS), 144–145
Sauvageau, Florian, 32
Savoie, Donald, 30
Sayer, Pierre-Guillaume, 134
Schlesinger, Benjamin, 288
Schlesinger, Rachel, 288
Schneiderman, David, 32
scrip, 134
script, 309
Seevaratnam, Pushpa, 121
self-determination, 207
sentencing circles, 147–148
September 11, 71–72, 114–115, 157, 317
sex
    versus gender, 217
    and inequality, 76–77
Sex in America study, 265
sexology, 242
sexual inversion, 258
sexual orientation, 242, 244–245, 261–265, 262f
sexual politics, 225–226
sexual predators, 244

sexual rights. *See* human and sexual rights
sexuality
    and Canadian literature, 358–361
    ethnocentrism, 248
    factors involved, 265–266
    families of LGBT parents, 250–252
    human and sexual rights, 241–245
    LGBT community, 256–261
    LGBT emergence, 241
    public morality, 259–261
    same-sex marriage, 245–249
    same-sex parenting, 252–256
    sexual orientation, 261–265, 262f
Shadd, Adrienne, 36
Shariah law, 117, 121, 243
Shefferville (Quebec), 24
Sifton, Clifford, 183–184
Simeon, Richard, 9
Singh, Kesur, 181
single-member plurality system, 15
single-parent families, 77, 291–292
slavery, 47, 96, 97, 100, 101
Smart, Ninian, 157
smudging, 149
Snow, Judith, 205
social assistance, 70–71, 70t
social background, 76
social class structure, 87–89
social connections, 231–233
social controls, 218
social dominance, 33–34
social inequality
    and ascribed statuses. *See* ascribed statuses
    in Canada, 68
    and Canadian literature, 344–354

conflict theory, 82–87
ideological legitimation, 284
income inequality, 68–72
Karl Marx, 82–86
Max Weber, 86–87
meaning of, 67
structural functionalism perspective, 80–82
views of, 80–87
wealth inequality, 72–74
social interaction, 109–115
social mobility, 89–90
social movement, 199
socialization dimension of family, 289–290
socially constructed roles, 258
sociological theory, 104–105
sociopolitical model, 203–204
sociopsychological dimension to regionalism
contempt for central Canada, 32–34
fishing dispute, 31–32
flag removal in Newfoundland and Labrador, 31
meaning of, 30
sodomy, 242
South Asian immigration, 180–187
Southam newspapers, 319
space-oriented media, 309
Spanish fishing vessels, 31–32
special education, 197
Spencer, Larry, 247
spiritism, 164
Sponsorship Program and Advertising Activities, 18, 19–20
sports events, 67
St. Lawrence Seaway, 22–23
Stacey, Judith, 254–255
staples theory, 9–10, 11
Starowicz, Mark, 29
statistically representative sample, 263

"status Indians," 131
Steckley, John, 126–155, 402
stereotypes, 110, 143, 218
Stl'atl'imx Tribal Police, 146–147
straight trade, 258
stratification, 67
see also social class structure; social inequality
structural functionalism
Aboriginal peoples, 128
ethnicity and race, 105
the family, 277–278
gender spheres, 228–229
social inequality, 80–82
subculture, 260–261
subjective differences, 5
subworking class, 89
Supreme Court, 14, 32, 316
Survival: A Thematic Guide to Canadian Literature (Atwood), 331, 333
sweat lodge, 145, 149
symbolic interaction theory, 104, 281
systemic barriers, 232
systems theory, 278–280

T
Taoism, 161
Taras, David, 32
tariff, 20–21
Teaching Prejudice, 113
technology, 309, 310t
telegraph, 311
television, 313–318
television imperialism, 319
theistic, 163
time-oriented media, 309
Tipton, Steven, 171
Toronto firefighters, 118
Touchstone Conference (1951), 199, 200
Trans-Canada Airlines, 25
Trans-Canada Highway, 25
transcendent God, 161

transfer payments, 11, 12t, 25–27, 26t
transgender/transsexual, 242
see also LGBTcommunity
transsexuals, 219
transvestites, 218–219
treaty, 136–139
Treaty of Paris, 178
tribalism, 312
Trudeau, Pierre Elliott, 15, 18, 187, 188, 243
Turner, Bryan S., 82
TV involvement, 28–29

U
Ukrainian immigrants, 183
Ultimate Reality, 161
underground railroad, 47
unemployment, 294
unification theories, 277–280, 281
United Chiefs and Councils of Manitoulin Justice Project, 147
United States media, 318–322
Universal Declaration of Human Rights, 241–242
universal themes, 332
untold stories, 333–334
"up north," 23–24
upper class, 88
urban reserves, 143–144

V
variable, 39
see also demographic variables
vernaculars, 311
Via Rail, 25
violence in families, 295–296
vital events, 51
Vocational Rehabilitation of Disabled Persons Act (1961), 199
voice, 331

W

wage gap, 224t
wealth, 72–74
Weber, Max, 86–87, 157, 160, 168
Wegner, Lawrence Kim, 145
Western Canada
    Canadian literature, 351–353
    and CF-18 fighter aircraft contract, 21
    Chinese immigration to, 49
    the farm crisis, 21–22
    interest rates, and financial institutions, 21
    and National Policy (1879), 20–21
    perceptions of contribution to Canadian development, 22f
    view of, 14–15, 20–23
White House Easter Egg Roll, 250
Wiccans, 167–168
Wilks, James, 183
Williams, Danny, 31
Winter, James, 318
Wiseman, Adele, 356–357
Wolf, Alison, 232
women
    balance of family responsibilities, 289–290
    see also gender
    and divorce, 292–293
    lone mothers, 78
    violence against, 295–296
    in the workforce, 76–77, 293
word blindness, 206
working class, 83, 89
world religions, 162–165, 166–167
the world wars, 198–199
writing, 307–309

Y

Yukon, 354